THE CONCEPTS AND PRACTICE OF MATHEMATICAL FINANCE

Mathematics, Finance and Risk

Editorial Board

Mark Broadie, *Graduate School of Business, Columbia University*
Sam Howison, *Mathematical Institute, University of Oxford*
Neil Johnson, *Centre for Computational Finance, University of Oxford*
George Papanicolaou, *Department of Mathematics, Stanford University*

Modern finance in theory and practice relies absolutely on mathematical models and analysis. It draws on and extends classical applied mathematics, stochastic and probabilistic methods, and numerical techniques to enable models of financial systems to be constructed, analysed and interpreted. This methodology underpins applications to derivatives pricing for equities and fixed income products, asset-liability modelling, volatility, risk management, credit risk, insurance analysis and many more. This new series will consist of books that explain the processes and techniques of the new applied mathematics, and how to use them to model financial systems and to understand the underlying phenomena and forces that drive financial markets.

The audience for mathematical finance ranges from mathematics and probability through econophysics to financial economics, and the series will reflect this breadth of appeal, while maintaining a firm footing in the tradition of applied mathematics. Books will be pedagogical in style, enabling them to be used for teaching in universities, business schools and financial institutions, and sufficiently self-contained for stand-alone use. Mathematical techniques will be motivated by examples and their use illustrated through applications, and complemented by computation.

THE CONCEPTS AND PRACTICE OF MATHEMATICAL FINANCE

MARK S. JOSHI
Royal Bank of Scotland Group

PUBLISHED BY THE PRESS SYNDICATE OF THE UNIVERSITY OF CAMBRIDGE
The Pitt Building, Trumpington Street, Cambridge, United Kingdom

CAMBRIDGE UNIVERSITY PRESS
The Edinburgh Building, Cambridge CB2 2RU, UK
40 West 20th Street, New York, NY 10011–4211, USA
477 Williamstown Road, Port Melbourne, VIC 3207, Australia
Ruiz de Alarcón 13, 28014 Madrid, Spain
Dock House, The Waterfront, Cape Town 8001, South Africa

http://www.cambridge.org

© M. S. Joshi 2003

This book is in copyright. Subject to statutory exception
and to the provisions of relevant collective licensing agreements,
no reproduction of any part may take place without
the written permission of Cambridge University Press.

First published 2003
Fifth printing 2005

Printed in the United Kingdom at the University Press, Cambridge

Typeface Times 11/14 pt.   *System* LaTeX 2_ε [TB]

A catalogue record for this book is available from the British Library

Library of Congress Cataloguing in Publication data
Joshi, M. S. (Mark Suresh), 1969–
The Concepts and practice of mathematical finance / M. S. Joshi.
p. cm. – (Mathematics, finance and risk)
Includes bibliographical references and index.
ISBN 0 521 82355 2
1. Derivative securities – Prices – Mathematical models.   2. Options (Finance) – Prices – Mathematical models.   3. Interest rates – Mathematical models.   4. Finance – Mathematical models.   5. Investments – Mathematics.   6. Risk management – Mathematical models.   I. Title.   II. Series.
HG6024.A3J67 2003
332′.01′51 – dc22   2003055594

ISBN 0 521 82355 2 hardback

To my parents

Contents

Preface		*page* xiii
Acknowledgements		xvii
1	Risk	1
1.1	What is risk?	1
1.2	Market efficiency	2
1.3	The most important assets	4
1.4	Diversifiable risk	8
1.5	The use of options	9
1.6	Classifying market participants	12
1.7	Key points	13
1.8	Further reading	13
1.9	Exercises	14
2	Pricing methodologies and arbitrage	15
2.1	Some possible methodologies	15
2.2	Delta hedging	17
2.3	What is arbitrage?	18
2.4	The assumptions of mathematical finance	19
2.5	An example of arbitrage-free pricing	21
2.6	The time value of money	23
2.7	Mathematically defining arbitrage	26
2.8	Using arbitrage to bound option prices	28
2.9	Conclusion	38
2.10	Key points	38
2.11	Further reading	38
2.12	Exercises	39
3	Trees and option pricing	41
3.1	A two-world universe	41
3.2	A three-state model	46

3.3	Multiple time steps	47
3.4	Many time steps	50
3.5	A normal model	52
3.6	Putting interest rates in	55
3.7	A log-normal model	57
3.8	Consequences	64
3.9	Summary	66
3.10	Key points	67
3.11	Further reading	67
3.12	Exercises	68

4 Practicalities 70
4.1	Introduction	70
4.2	Trading volatility	70
4.3	Smiles	71
4.4	The Greeks	74
4.5	Alternate models	81
4.6	Transaction costs	86
4.7	Key points	87
4.8	Further reading	87
4.9	Exercises	88

5 The Ito calculus 89
5.1	Introduction	89
5.2	Brownian motion	89
5.3	Stochastic processes	92
5.4	Ito's lemma	96
5.5	Applying Ito's lemma	101
5.6	An informal derivation of the Black–Scholes equation	104
5.7	Justifying the derivation	105
5.8	Solving the Black–Scholes equation	109
5.9	Dividend-paying assets	111
5.10	Key points	113
5.11	Further reading	114
5.12	Exercises	114

6 Risk neutrality and martingale measures 117
6.1	Plan	117
6.2	Introduction	118
6.3	The existence of risk-neutral measures	119
6.4	The concept of information	130
6.5	Discrete martingale pricing	135
6.6	Continuous martingales and filtrations	144
6.7	Identifying continuous martingales	146

	6.8	Continuous martingale pricing	147
	6.9	Equivalence to the PDE method	151
	6.10	Hedging	152
	6.11	Time-dependent parameters	154
	6.12	Completeness and uniqueness	156
	6.13	Changing numeraire	157
	6.14	Dividend-paying assets	159
	6.15	Working with the forward	159
	6.16	Key points	163
	6.17	Further reading	163
	6.18	Exercises	164
7		The practical pricing of a European option	166
	7.1	Introduction	166
	7.2	Analytic formulae	167
	7.3	Trees	168
	7.4	Numerical integration	173
	7.5	Monte Carlo	176
	7.6	PDE methods	181
	7.7	Replication	181
	7.8	Key points	183
	7.9	Further reading	184
	7.10	Exercises	184
8		Continuous barrier options	186
	8.1	Introduction	186
	8.2	The PDE pricing of continuous barrier options	189
	8.3	Expectation pricing of continuous barrier options	191
	8.4	The reflection principle	192
	8.5	Girsanov's theorem revisited	194
	8.6	Joint distribution	197
	8.7	Pricing continuous barriers by expectation	200
	8.8	American digital options	203
	8.9	Key points	204
	8.10	Further reading	204
	8.11	Exercises	205
9		Multi-look exotic options	206
	9.1	Introduction	206
	9.2	Risk-neutral pricing for path-dependent options	207
	9.3	Weak path dependence	209
	9.4	Path generation and dimensionality reduction	210
	9.5	Moment matching	215
	9.6	Trees, PDEs and Asian options	217

9.7	Practical issues in pricing multi-look options	218
9.8	Greeks of multi-look options	220
9.9	Key points	223
9.10	Further reading	223
9.11	Exercises	224
10	Static replication	225
10.1	Introduction	225
10.2	Continuous barrier options	226
10.3	Discrete barriers	229
10.4	Path-dependent exotic options	231
10.5	The up-and-in put with barrier at strike	233
10.6	Put-call symmetry	234
10.7	Conclusion and further reading	238
10.8	Key points	240
10.9	Exercises	241
11	Multiple sources of risk	242
11.1	Introduction	242
11.2	Higher-dimensional Brownian motions	243
11.3	The higher-dimensional Ito calculus	245
11.4	The higher-dimensional Girsanov theorem	248
11.5	Practical pricing	253
11.6	The Margrabe option	254
11.7	Quanto options	256
11.8	Higher-dimensional trees	258
11.9	Key points	261
11.10	Further reading	262
11.11	Exercises	262
12	Options with early exercise features	263
12.1	Introduction	263
12.2	The tree approach	266
12.3	The PDE approach to American options	267
12.4	American options by replication	270
12.5	American options by Monte Carlo	272
12.6	Upper bounds by Monte Carlo	275
12.7	Key points	276
12.8	Further reading	277
12.9	Exercises	277
13	Interest rate derivatives	279
13.1	Introduction	279
13.2	The simplest instruments	281

13.3	Caplets and swaptions	288
13.4	Curves and more curves	293
13.5	Key points	295
13.6	Further reading	296
13.7	Exercises	296

14 The pricing of exotic interest rate derivatives — 298

14.1	Introduction	298
14.2	Decomposing an instrument into forward rates	302
14.3	Computing the drift of a forward rate	309
14.4	The instantaneous volatility curves	312
14.5	The instantaneous correlations between forward rates	315
14.6	Doing the simulation	316
14.7	Rapid pricing of swaptions in a BGM model	320
14.8	Automatic calibration to co-terminal swaptions	321
14.9	Lower bounds for Bermudan swaptions	324
14.10	Upper bounds for Bermudan swaptions	328
14.11	Factor reduction and Bermudan swaptions	331
14.12	Interest-rate smiles	334
14.13	Key points	337
14.14	Further reading	337
14.15	Exercises	338

15 Incomplete markets and jump-diffusion processes — 340

15.1	Introduction	340
15.2	Modelling jumps with a tree	341
15.3	Modelling jumps in a continuous framework	343
15.4	Market incompleteness	346
15.5	Super- and sub-replication	348
15.6	Choosing the measure and hedging exotic options	354
15.7	Matching the market	357
15.8	Pricing exotic options using jump-diffusion models	358
15.9	Does the model matter?	360
15.10	Log-type models	362
15.11	Key points	364
15.12	Further reading	365
15.13	Exercises	366

16 Stochastic volatility — 368

16.1	Introduction	368
16.2	Risk-neutral pricing with stochastic volatility models	369
16.3	Monte Carlo and stochastic volatility	370
16.4	Hedging issues	372

16.5	PDE pricing and transform methods	373
16.6	Stochastic volatility smiles	377
16.7	Pricing exotic options	377
16.8	Key points	378
16.9	Further reading	378
16.10	Exercises	379

17 Variance Gamma models 380
    17.1 The Variance Gamma process 380
    17.2 Pricing options with Variance Gamma models 383
    17.3 Pricing exotic options with Variance Gamma models .. 386
    17.4 Deriving the properties 387
    17.5 Key points 389
    17.6 Further reading 389
    17.7 Exercises 390

18 Smile dynamics and the pricing of exotic options 391
    18.1 Introduction 391
    18.2 Smile dynamics in the market 392
    18.3 Dynamics implied by models 394
    18.4 Matching the smile to the model 400
    18.5 Hedging 403
    18.6 Matching the model to the product 404
    18.7 Key points 407
    18.8 Further reading 407

Appendix A Financial and mathematical jargon 409
Appendix B Computer projects 414
Appendix C Elements of probability theory 438
Appendix D Hints and answers to exercises 449
References 462
Index 468

Preface

There are many different emphases and approaches to presenting the basics of mathematical finance. My objective in this book is to do two things, the first is to impart to the reader a conceptual understanding of the basic ideas in mathematical finance. The second is to show the reader how these ideas are translated into practicalities.

There is an aphorism that goes "Don't think of the problem, think of the solution." I believe that this aphorism is often taken too much to heart when presenting mathematical material: the solution is often presented without stating the problem. We therefore spend a couple of chapters going over the basic ideas of finance. In particular, we first introduce the concept of risk in order to give the reader an understanding of why risk is important before proving the surprising and fundamental result that ignoring risk is the key to pricing many products, which comes later in the book.

There are at least three approaches to mathematical finance, trees, PDEs and martingales. Rather than plump for one of these, we try to examine each problem from the viewpoint of each one and attempt to use the multiple approaches to emphasize the underlying ideas.

Mathematical finance is a burgeoning field and no book can cover everything, nor should it try to do so. My guiding principle has been to include what I think a good quant ought to know. Inevitably many topics are not covered in depth or at all. Where possible, I have tried to indicate other textbooks which cover the topics and where not possible the original papers. Let me stress at this point, that this is a text book not a research letter so the absence of a reference does not mean that I believe a result is new. However, on the more cutting-edge topics I have tried to indicate the original papers. If any reader is offended by the lack of a reference my apologies and please let me know for the second edition. Three books which are very strong on references are [42], [79] and [96].

After introducing risk, we move on in Chapter 2 to the concept of arbitrage which is the fundamental idea of modern derivatives pricing theory. The principle of no arbitrage is then used to develop model-free bounds on option prices, and to show that there exist certain relationships between option prices.

To pass beyond bounds to definite prices requires the introduction of a model of how asset prices change. Although a fundamental assumption is the random character of asset price movements, one must model the nature of this randomness in order to develop pricing models. In Chapter 3, we introduce the simplest of models: the binomial tree. The binomial tree is an essentially discrete model which posits that in each time period the asset moves up or down by a fixed amount. We analyze pricing on binomial trees from various points of view including replication, risk-neutral pricing and hedging. We examine the surprising result that the probabilities underlying the asset's movements have little effect on the price of options. We then see how this discrete model can be used as an approximation to a continuous model, and we deduce the Black–Scholes formula for the price of a call option via a limiting argument.

Having developed the Black–Scholes formula, we then discuss in Chapter 4 its flaws and how these flaws affect its use in practice. This chapter is very much a foretaste for chapters near the end of the book where we study alternative models of price evolution which try to compensate for the shortcomings of the Black–Scholes model.

In Chapter 5, we step up a mathematical gear and introduce the Ito calculus. With this calculus we introduce the geometric Brownian motion model of stock price evolution and deduce the Black–Scholes equation. We then show how the Black–Scholes equation can be reduced to the heat equation. This yields a derivation of the Black–Scholes formula.

In Chapter 6, we step up another mathematical gear and this is the most mathematically demanding chapter. We introduce the concept of a martingale in both continuous and discrete time, and use martingales to examine the concept of risk-neutral pricing. We commence by showing that option prices determine synthetic probabilities in the context of a single time horizon model. We then move on to study discrete pricing in martingale terms. Having motivated the definitions using the discrete case, we move on to the continuous case, and show how martingales can be used to develop arbitrage-free prices in the continuous framework. We show that the Black–Scholes PDE can be found as a consequence of the martingale method. We then move on to studying changes of numeraire and market completeness.

After the rigours of Chapter 6, we shift back to the practical in Chapter 7. In this chapter, we examine how the price of European option can be developed using the various possible pricing approaches. In particular, we discuss analytic formulas, trees, Monte Carlo, numeric integration, PDEs and replication.

In Chapter 8, we study the pricing of the simplest of exotic options, the continuous barrier option, and develop analytic formulas for its price in the Black–Scholes world using both PDE and martingale techniques. As part of the study, we examine the concept of change of measure and the reflection principle.

In Chapter 9, we commence the study of non-vanilla options by analyzing the pricing of path-dependent exotic options depending on the value of the underlying at a finite number of times. We concentrate on Asian options and discrete barrier options for concreteness. We discuss pricing using Monte Carlo and PDE methods. We also look at the computation of Greeks by Monte Carlo.

In Chapter 10, we study the use of static replication as a tool for pricing and hedging. Under a variety of assumptions, we examine the replication of continuous barrier options, discrete barrier options, and general path-dependent exotic options.

In Chapter 11, we extend the theory to cope with several sources of uncertainty and develop pricing models which can cope with derivatives whose price depends on the price behaviour of several assets. As applications of the theory, we study the pricing of Margrabe options and quanto options.

We look at how to introduce early optionality in Chapter 12. We discuss the use of tree and PDE methods before looking at the difficulties involved in pricing using Monte Carlo. We develop methods for both lower and upper bounds using Monte Carlo.

We shift our emphasis in Chapter 13 to look at the pricing of simple interest rate derivatives. We introduce forward-rate agreements and swaps, and their optional analogues the caplet and the swaption. We develop pricing formulas under simple assumptions.

In Chapter 14, we study the pricing of exotic interest rate derivatives using the LIBOR market model. Our study includes both calibration and implementation. This chapter draws on a lot of what has gone before, and we finish up with an examination of the pricing of Bermudan swaptions by Monte Carlo.

We commence our study of alternative pricing models in Chapter 15. Here we analyze the Merton jump-diffusion model and develop a pricing formula. We also discuss the additional issues raised by pricing in a model that does not allow perfect hedging.

We continue our study of alternative models in Chapter 16 where we introduce stochastic volatility. We develop pricing approaches using PDE and Monte Carlo techniques for vanilla and exotic options.

In Chapter 17, we introduce the Variance Gamma model and use it to study the pricing of vanilla and exotic options.

To round off the main part of the book, we finish with a chapter on the philosophical and practical issues inherent in using sophisticated models to price exotic options. We look at the relationship between models and smile dynamics, and compare these dynamics to those found in the market. We also see that for certain products there are features which are crucial to capture.

The website for this book is www.markjoshi.com/concepts. Updates, corrections and news of future editions will be posted there.

Acknowledgements

Writing this book has been a project of vastly greater magnitude than I contemplated when I started out with the objective of writing a book that stressed the ideas of mathematical finance more than the mathematics. I am grateful to the Royal Bank of Scotland for providing a stimulating environment in which to learn, study and do mathematical finance. My views on and understanding of the topic have come from daily discussions with current and former members of the Group Risk Management Quantitative Research Centre including Chris Hunter, Peter Jäckel, Dherminder Kainth, Jan Kwiatkowski and Jochen Theis, and particularly Riccardo Rebonato. I am also grateful to numerous people for their many comments on the manuscript, particularly to Alex Barnard, Dherminder Kainth, Alan Lewis, Sukhdeep Mahal, Riccardo Rebonato and Jochen Theis. David Tranah, my editor at Cambridge University Press, has done a careful job of editing and has succeeded in removing the worst quirks in my style. My wife has been very supportive during a project that at times seemed neverending.

1

Risk

1.1 What is risk?

It is arguable that risk is the key concept in modern finance. Every transaction can be viewed as the buying or selling of risk. The success of an organization is determined by how much return it can achieve for a given level of risk. Before we can justify these statements, we need to achieve some understanding of what risk is.

In a typical pure mathematical ploy, let us start by trying to understand the absence of risk. A riskless asset is an asset which has a precisely determined future value. Do such assets exist? The fundamental example is that of a government bond. We can buy a government bond for say a £100 today and know that we will receive say £5 a year, (called a coupon payment), until a pre-determined date, when we receive our £100 back. Is this asset truly riskless? There is of course a possibility that the government will renege on its promise to pay. (This is known as defaulting.) But if we pick the right government this possibility is sufficiently remote that we can for practical purposes neglect it. If this seems unreasonable, consider that if the British, American or German government reached such straits, the world's financial system would be in such a mess that there would be precious few banks left to employ financial mathematicians. In fact, the existence of such riskless assets is so fundamental both to financial mathematics and to the modern finance industry, that the fiscal policy of the American and British governments of running budget surpluses, and therefore reducing the number of bonds they have issued, caused great consternation. The reader who is tempted to chuckle at the predicament of the finance industry should consider that financial institutions fund pensions by buying long-maturity government bonds and using the interest coupons to pay the pension. The shortage of long-maturity bonds therefore makes pensions harder to fund and ultimately results in smaller pensions.

We can now define a risky asset to be an asset which is not riskless. That is it is an asset of uncertain future value; risk can be regarded as a synonym for uncertainty. The most basic example of such an asset is a share of a public limited company and we shall return to this example again and again. However, it is important to realize that almost anything except a riskless government bond is such an asset. For example, we could hold foreign currency and be exposed to the risk that the exchange rate will change against us, or we could buy a flat in London and be exposed to the possibility that there is a property crash, as occurred in the early 1990s.

The sharp reader will have noted that the definition in the paragraph is not quite right, in that an investor would not actually care about the riskiness of an asset if the worst possible future value of the asset was greater than today's value.

However, we have to be slightly careful about what we mean by value here. Unless there is no inflation, £1 a year from now will buy less than £1 today. This means that £1 a year from now is effectively worth less than £1 today. In addition, even in a non-inflationary world, most people prefer jam today to jam tomorrow and so would not be happy to receive the same amount of money back in a year with no compensation. A better view of riskiness is that the asset can return less than the same amount invested in a riskless government bond for the same period. A good example of such an asset is the premium bond. In the United Kingdom, one can buy a government bond, called a premium bond, redeemable at any time which pays no coupon but instead the holder gets a free entry in a prize draw paying up to a million pounds a month. This seems too good to be true at first, but the issue is, of course, that the bond is not very different from investing some money and using the interest to buy lottery tickets. That said, the expected winnings for the amount of interest foregone is much better for premium bonds than for lottery tickets. The investor is effectively buying risk.

1.2 Market efficiency

Before we can understand why risk is so important we first have to understand the concept of market efficiency, which underlies most of financial mathematics and modern economics. This concept roughly states that in a free market, all available information about an asset is already included in its price. Therefore there is no such thing as a good buy – the only value an asset has is its market value and it is meaningless to attempt to think otherwise.

Is this hypothesis correct? To see that it cannot be wholly so, consider the apocryphal story of the two economists who see a ten dollar bill lying in the gutter. The first one goes to pick it up but the second one tells him not to be silly as if it were real, someone else would have already picked it up. However, the true moral of this story is that market efficiency only works if someone does not believe in it – the first person who does not believe in market efficiency picks up the ten dollar bill and thereafter it is no longer lying in the gutter. Warren Buffett is the most famous example of a non-believer who has very effectively made a lot of money through his disbelief. He has largely done so by buying shares in companies he believes are undervalued by the market. Indeed, until Bill Gates overtook him he was the richest man in the world, and he made his money by beating the market.

So although market efficiency is not wholly correct, there are enough people attempting to be the next Buffett, for it to be sufficiently correct that we can work under the assumption that it is true. What does this mean for us? Well, the first thing it means is that it is pointless to for us to try to predict the future price of a share by looking at graphs of its past prices. All this information is already encoded in the share price. This is sometimes called the Markov property, and is also called the *weak efficiency of markets* as it's a consequence of the strong form mentioned above.

It is interesting to note that the modern white-collar crime of insider trading is really based on the principle of market efficiency. Insider trading is trading based on knowledge which is not publicly available and therefore not included in the share price. For example an employee of a company might know that the company was about to announce unexpectedly large losses or profits which would move the share price in an obvious direction, and take advantage in advance. The perception of this as a crime rather than a natural action is fairly recent, and is based on the ubiquity of the concept of market efficiency.

Given that all assets are correctly priced by the market, how can we distinguish one from another? Part of the information the market has about an asset is its riskiness. Thus the riskiness is already included in the price, and since it will reduce the price, the value of the asset without taking into account its riskiness must be higher than that of a less risky asset. This means that in a year from now we can expect the risky asset to be worth more than the less risky one. So increased riskiness means greater returns, but only on average – it also means a greater chance of losing money. From this point of view, an asset's price reflects the value it is likely to have in the future reduced by a factor depending upon its riskiness.

To illustrate these ideas, let us consider a simple game. Suppose we toss a coin, if it comes up heads I give you £3, if it comes up tails you give me £1. Unless beset by moral qualms, you would consider this game a very good deal and play it – your expected winnings would be

$$\frac{1}{2}3 - \frac{1}{2}1 = 1,$$

and your maximum losses would only be 1. Suppose we play a slightly different game, I pay you £13 on heads, you pay me £11 on tails. Your expected winnings are still £1 but are you still so keen to play? If not why not? If you are still keen, let's take the payment on heads to be £103, and on tails to be £101. At some point, when the stakes become high enough you will stop regarding the game as a good deal. The point where you stop depends upon personal risk preferences; the stopping point is where the expected gains stop outweighing your aversion to the possibility of losing money.

Now suppose the game is changed a little again. The sum you lose is paid to me today and we toss a coin a year from now. If the coin comes up heads, I return your money to you and pay you my losses, otherwise I keep your money. What has changed? During the year in between, I have put the money on deposit with a bank and earned some interest. If you were not playing the game you could have done so also. The amount of return you would want from the risky game would increase to express the interest foregone. And since you could have made money from the interest payment without taking any risks, you will demand that the expected winnings be greater than the amount of interest you could have earned.

The moral is that there is no such thing as a guaranteed high return. The reader would be well-advised to remember this the next time he sees a guaranteed high return in a newspaper or Internet advertisement.

Let's return to the concept of weak market efficiency. This says that all the past movements of an asset's market price is already expressed by today's price. At this point, the prospects of a financial mathematician could be regarded as being pretty bleak. Why? This tells us that trying to predict the future price from past data is a waste of time – there is no periodicity nor trends to be read. The only mathematical information is today's price which tells us very little. In that case, why is financial mathematics a burgeoning field? The job of a financial mathematician is not to predict prices but instead to relate the movements of price in one asset to that of another. These price movements are viewed as being driven by information arriving in the market and since that information is by definition unknown until it arrives, we can view it as being random.

The key point in mathematical finance is to use market instruments which are affected by the same information in such a way as to cancel out randomness. This process is called hedging. The objective of mathematical finance is to understand how to do this and to understand the consequences.

1.3 The most important assets

We have been discussing an asset rather vaguely so let's look at the basic assets in finance from the point of view of risk.

1.3.1 Bonds

The simplest asset, already mentioned, is a government bond issued by a reliable government. Typically, the government issues a bond of, say, thirty years in length, which pays every year a sum called the *coupon* and gives the investor his original investment back at the end of the thirty years. The original investment is called the *principal*. The day the investor gets the principal back is called the *date of maturity*, when the bond is said to *mature*. In the meantime, each year the investor receives interest payments to compensate him for the fact the government has his money. From a mathematical point of view however the coupons just confuse things, so mathematicians typically study zero-coupon bonds although they are in fact rather rare. A *zero-coupon bond* is a bond which pays no interest but instead just returns the investor his investment on the date of maturity. Why would anyone buy such a bond? Suppose the principal is one dollar. The point is that the investor does not pay a dollar for the bond, but instead pays a smaller amount which if it was invested in interest-paying bonds would give a dollar in total (including the compound interest) at maturity.

The interesting thing about the riskless bond is that there is some risk in it. Not in the possibility of default, but instead in the possibility that interest rates might change in the meantime. At the time of writing, interest rates are quite low by historical standards so an investor might be wary of buying a bond with a long time to maturity: he has locked in today's interest rate and if rates go up he loses out. Of course, if rates go down he gains. There are well-established markets in the major government bonds so our investor need not hold his bond until maturity instead he can just sell it in the market. But what price will he get? If interest rates have gone up, the market price will be less than he paid for it as other investors will want the fixed sum on maturity to reflect how much money they could have got it by investing the money in a newly issued riskless bond. Thus as well as the interest rate reflected by the coupon payment, the price of a bond reflects today's interest rates and indeed reflects today's expectations of future interest rates. The effective interest rate implied by the market price is called the *yield* of the bond and this can be very different from the coupon. Since the yield of a bond reflects expectations of future rates, bonds of different maturities can have different yields implied by the market.

As the date of maturity of a bond approaches, there is less and less uncertainty left. The principal of the bond is known and will be paid on the date of maturity, and the interest rate is unlikely to move much in a short period of time so there is less and less uncertainty as the maturity date gets closer and closer. A bond of longer maturity will be exposed to more uncertainty, that is risk, than one of short maturity. We can therefore expect long-dated bonds to have higher yields to compensate investors for this additional risk. This is generally, but not always, true. Indeed at the time of writing, gilts (gilts is the financial jargon term for UK government bonds) display a hump: the yields first rise and then decrease. See Figure 1.1. Recall that the yield also reflects expectations of future rates so one explanation is that the market expects UK interest rates to rise in the short term, but to decrease in the long term. An alternative and probably more correct explanation is that there is

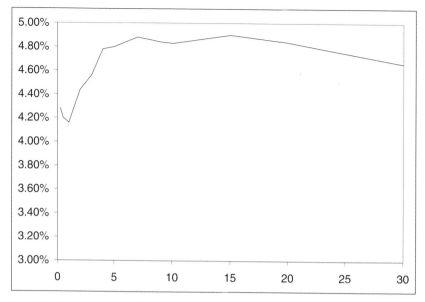

Fig. 1.1. The yields on UK government bonds (gilts) in October 2001 as a function of years.

currently a shortage of long-dated bonds which drives their prices up and hence their yields down.

1.3.2 Stocks and shares

Probably the most ubiquitous sort of traded asset is the share or stock in a company, (share and stock are equivalent terms.) While the reader is almost certainly familiar to some extent with shares, it is worth examining precisely what that term means. The holder of a share of a company owns a fraction of that company. Companies traded on the stock exchange typically have plc after their name reflecting the fact that they are public limited companies. Public just means that anyone can buy shares in them. Limited means that they are of limited liability; the owners of such a company have no liability for its debts if it goes bankrupt. The importance of this fact should not be underestimated. In the author's opinion the existence of limited liability companies is the foundation of modern capitalism. Why? Because it reduces the riskiness of investing in a company by capping the total losses to be the amount invested. If the company is sued for a billion dollars and goes bankrupt, you the shareholder are not liable. If you could be liable would you still buy shares?

To emphasize this point, consider the 'Names' at Lloyds. The Lloyds insurance market worked in the following way. A person would agree to set aside at least

£100, 000 against claims by people insured in the Lloyds' market, in return for which they would receive a lucrative stream of insurance premiums. The £100, 000 could also be happily invested in other assets allowing the Name to do particularly well. The Name got a very high rate of return but high returns equal high risk. Indeed, the particular feature of the Lloyds market which made investing very different from buying shares in an insurance company was that a Name had unlimited liability. Not only could he lose the £100, 000, he could lose everything including his house and his shirt. And in a particularly bad year that is precisely what happened. Perhaps the most interesting aspect of this story is how surprised many Names were; somehow they had failed to appreciate the connection between risk and return.

We have seen that the holder of a share owns part of a company and the sole risk he bears is that the value may drop to zero. The share may bring the investor money in two ways. The first is simply that the value of the share may go up, the other is that the company will generally pay dividends that is payments, often annual, to shareholders dispensing the profits of the company. As one might expect, the total return to shareholders on average is much greater than the rate received on depositing the money in a riskless bond or a bank account to compensate for the danger that the company will go under or just not do very well.

1.3.3 The corporate bond

Another asset commonly traded in the markets is the corporate bond. This lies somewhere between a share and a government bond. A company wishing a loan issues bonds in the market paying some coupon in interest. The coupon is generally higher than that of a riskless bond. The investor's risk is that the company may default on its payments as it has gone bankrupt in the meantime. However, bondholders have more claim on the company's assets than shareholders so the riskiness is reduced. The main disadvantage of a bond is that if the company share price soars the bondholder does not gain at all. Thus both the returns and the riskiness of bonds are lower than those of shares. In order to entice investors to buy bonds, companies sometimes issue *convertible bonds*, that is, bonds that can be converted into shares if the investor so chooses. This allows the investor the upsides of both bonds and shares – of course, typically the coupon or yield on such a bond would generally be less than that of an ordinary bond.

1.3.4 Positivity

All the assets discussed so far have one similarity, they all carry rights without liabilities. This has an important consequence for the mathematician: their values are always positive. In an extreme case, the value could become zero on bankruptcy but the important point is that they will never be negative. There are plenty of market instruments that do not share this property, and we shall encounter many of them in this book.

1.3.5 The risk paradigm outside the markets

It is important to realize that the concept of risk is inherent in all investment decisions not just those of what to purchase in the financial markets. For example, suppose a company wishes to invest in a new plant to produce a new product. It will estimate the amount of return it will receive on the invested capital. What level of return should it demand? Well to answer that question it needs to assess the riskiness of the project and it should demand the same return as a market instrument of comparable riskiness. Otherwise, it would do better just to buy the relevant market instrument and forget about the plant.

1.4 Diversifiable risk

We have treated all risks as being equal, however some risks are better than others. Consider a contract that pays £100 if a coin flip is heads and zero otherwise. From what we have said so far, we would expect this to trade for less than £50 depending on the risk aversion of investors. We can also consider the complementary contract that pays £100 if the *same* coin flip is tails. We also expect this to be worth less than £50. However, if both these contracts are trading in the market, we can buy both and be guaranteed £100 whatever happens. The risk has disappeared and since we do not expect to be able to make riskless profits, we conclude that the original contracts were worth £50 after all. We have a paradox, after arguing at great length that such a contract would have to be worth less than £50, we conclude that we were wrong and that it is worth £50. What has changed? The addition of a second contract has removed the risk. This process is called *diversification*. As long as only one of the contracts is tradable, its value is less than £50 but as soon as both are, the risk stops being undiversifiable and the risk premium disappears. The lesson of this example is that the market will only compensate investors for taking risks which are not diversifiable. Undiversifiable risk is known as *systemic* risk. The job of an investor is therefore to achieve the maximum amount of systemic risk for a given level of riskiness by diversifying his portfolio.

One paradoxical side-effect of investors' need to diversify is the increasing demand for products which express *purity of risk*. An investor who feels he is overexposed to one sort of risk will want to buy other risks to offset it. To achieve precisely the balance of risks desired, he will therefore look for products which express only one sort of risk. Options are one source of such products.

1.5 The use of options

Given that risk is inherent to all financial decision making, a bank or company will want to manage its risk carefully. In particular, it may want to buy certain sorts of financial instruments to increase or decrease a certain type of risk. This is where options and related products enter the picture. These products can be used to reduce risk or to increase it. Whether they are a good or a bad thing depends purely on the way they are used.

To give an example before we start making definitions, consider an American company which exports to Japan. The Japanese importer pays the company in yen but the company prefers dollars. The company estimates that it will receive between one and two billion yen next year, which it will need to exchange for dollars. The variability of the exchange rate between dollar and yen means that the company is exposed to some extra risk which it would prefer to avoid. One solution is to enter a forward contract. This is a contract to exchange a fixed amount of yen for dollars at a fixed future date at today's exchange rate (modified slightly to take account of interest rates.) The company's problem is that it does not know precisely how many yen it will wish to sell so it cannot remove (or hedge) all the risk. One solution would be to enter a forward contract for a billion yen, since it is sure it will need to sell at least that much and treat the rest separately. To deal with the rest of the yen which is a variable amount between zero and one billion, the company could buy an option.

What is an option? An option is typically an instrument which gives the holder the right to buy or sell a quantity of some fixed asset during a specified period of time at a price fixed today. The important point is that unlike a forward contract there is no obligation to buy or sell. The option carries rights but not obligations, and therefore will have always have a positive value before the time of expiry.

In this case, the company could therefore buy an option to sell a billion yen at today's price a year from now. Buying the option would, of course, cost the company a fee but it would cap the amount of losses it might make if the exchange rate moved in the wrong direction, thus reducing the company's risk. The important thing to realize is that the company will not decide whether to use the option on the basis of how many yen it needs to convert but instead on the basis of whether the market price is higher or lower than the price which the option guarantees. Any excess yen can always be sold in the market.

The financial derivatives market is full of jargon. An option to buy is called a *call* option. An option to sell is called a *put* option. (The easiest way to remember which is which is that C is close to B for buy in the alphabet.) Using an option is called *exercising* it. The price which is guaranteed by the option is called the *strike* price and the option is said to have been *struck* at that price. There are many different sorts of rules for when the option can be exercised. The simplest sort is the *European* option which can be exercised on one specified date in the future. An *American* option can be exercised on any day before a specified date in the future. Note that since an American option carries all the same rights as a European option and more on top, it will always be worth at least at much as a European option and generally more. One thing we will demonstrate later in the book is the surprising result that under certain quite natural circumstances American and European call options have the same value. The options we have mentioned so far are very much the beginning of the list, and the list of possible options goes on and on, growing every day. The options we have mentioned above are generally called *vanilla* options to express the fact that they are standardized and less interesting than *exotic* options. Banks sometimes have different teams of mathematicians for the pricing of vanilla options and of exotics. Many exotic options are not really options in the sense that the holder does not get a choice but instead receives a payoff, which is possibly negative, dependent on the behaviour of some asset. This asset is generally called the *underlying*. The generic term for all instruments whose value is defined in terms of the behaviour of some other asset is *derivatives*. This name expresses the idea that their value is derivative of the behaviour of the price of the underlying.

Ultimately, an option is a powerful instrument to change one's exposure to risk. The big difference between buying an option on a stock and buying the stock is that if the stock price moves the wrong way the option will be valueless whereas the stock will not, but on the other hand if the stock moves the right way, the option holder will have made much more money for the amount of money spent than the stock holder. One attraction of speculating using options is that the maximum downside is the loss of the initial premium, whereas the up-side is unlimited. Of course, it is important to appreciate that for the option seller the position is reversed.

Another way an investor could use an option to reduce risk is as follows. Suppose he holds a large number of stocks which he knows he might need to sell a year from now, but he is worried about a crash in the meantime. Whilst he knows that he can always wait for the market to come back up in the long term he may need to sell straight after the crash which would lose him a lot of money. This investor could buy a put option on his stock guaranteeing him a price, thus capping his risk for the cost of the premium today. This approach is particularly employed by fund managers who are worried about performance targets in the short term.

Currently, one of the greatest growth areas is in credit derivatives. The simplest example is a contract where company 'A' pays company 'B' a regular premium until a company 'C' defaults on some of its debts. Upon default company 'B' pays 'A' a payment fixed in advance. More generally the size of the payment on default could be based on how much company 'C' defaults on its debts. Why is this useful? The bank 'A' may have made a large loan to company 'C' which it feels represents too high level of risk and in particular, undiversified risk. It can use the credit derivative to reduce this risk without affecting its business relationship with 'C' either by refusing the loan or trying to sell on the loan which would be difficult in any case. In order to fund the derivative, 'A' could then write a credit derivative on another company 'D' to which it has no exposure. The bank has thus reduced its overall risk at no cost to itself, by diversifying risks using credit derivatives. This then allows the bank to charge company 'C' a lower rate of interest, thus reducing the cost of capital for 'C' and allowing it to function more profitably which of course then allows the company to undercut its rivals by charging less. Thus the development in banking feeds into better value for consumers.

Another example is an airline company which is heavily exposed to the price of aviation fuel and thus effectively the price of oil. A couple of years ago, crude oil was very cheap by historical standards and trading around eight dollars a barrel. A general shortage drove the price up to around thirty dollars a barrel, and airlines really felt the pain of the increased costs. However, a smart airline could have used derivative contracts to lock in the price of oil, and become immune to price changes. Indeed, Ryanair have just announced a very successful year and attribute part of their success to the hedging of oil prices.

As the popularity of derivatives grows, they are traded on wider and wider products. One current growth area is weather derivatives. The holder of a weather derivative receives a pay-off based on the temperature on certain days or the amount of rainfall. For example, a City of London winebar, Corney and Barrow, noticed that its profitability was largely dependent how many sunny Thursdays and Fridays there were in July and August. On such days, traders after a hard day's work would come and sit in the sun and guzzle beer.[1] Corney and Barrow therefore bought a weather derivative which would pay them a sum of money each Thursday or Friday which was not hot. Their profits are therefore no longer dependent on the whims of the weather. Their average profits are, of course, the same as before but there is less variation i.e. less risk, which allows better financial planning and makes the business more valuable.

At this point, the reader may feel the line between insurance and derivatives is becoming a little blurred and indeed it is. The principal difference between derivative products and insurance is that derivative products are hedgeable. That is the seller can reduce his risk by holding the underlying asset or a similar asset, or selling another derivative product to another client which cancels out much of the risk. For example, a farmer or a wine-bar might want sunny weather, but a company paying a fortune for air-conditioning will want cool weather. The derivative products they would purchase to reduce risk would cancel each other. A second difference is that derivatives are always specified in terms of the occurrence of events rather than in terms of loss. If a farmer bought a derivative against too little rain or too

---

[1] Why they drink beer in a wine bar has never been fully explained.

much, he would receive the pay-off according to how much rain fell rather than according to how much damage his crops suffered.

From the point of view of risk, we can regard an option as an attempt to encapsulate a specific piece of risk. As the option is purer in its risk, its value is more sensitive to market changes, and therefore the amounts to be gained and lost on options are much larger. However, it would be a mistake to view an option as a risky asset which only the foolhardy would buy. The purpose of an option is to allow the buyer to guard against certain events and thus reduce his risk. The best metaphor for an option is to regard it as concentrated acid – handled carefully a very important tool, but used carelessly very dangerous.

1.6 Classifying market participants

There are many different reasons to participate in the markets, but we can make a broad classification according to their attitudes towards risk.

The *hedger* uses market instruments to reduce his risks.

The *speculator* uses market instruments to increase his risks – remember more risk equals more return.

The *arbitrageur* tries to spot discrepancies in the pricing of risks. By selling one risk and buying it elsewhere at a different price, he tries to make profits without risks.

The role of a bank is a mixture of speculator and arbitrageur. Every time it makes a loan, it is speculating that the return on the loan will outweigh the credit risks taken on. Its better access to the markets also allows it to sell products to companies at a margin above what it can buy them for. This is essentially a form of arbitrage.

Private investors are, from this point of view, speculators. They buy risky products such as shares in order to increase their returns.

Companies, in general, are hedgers. They are exposed to the market because they need to buy and sell commodities, or exchange currency. They wish to reduce these risks, and they can do so by the use of derivatives. Indeed, it has now become expected that they do so, and stock analysts will include in their reports critiques of companies' hedging strategies. Some companies even use options because of analysts' criticisms of their failure to do so.

The reader has no doubt heard of various cases where the use of options has led to vast sums of money being lost. These cases tend to occur in two ways, the first is that a trader has explicitly chosen to break the rules set him, and taken on much riskier positions than he has been set by his bank. Leeson's behaviour at Barings is the classic example of this. He made bigger and bigger bets to try and recoup his losses. The risk managers were not doing their jobs properly and failed to notice what he was doing. Eventually, everything went horribly wrong and the bank crashed.

The second sort of case is where a company which should be using derivatives to hedge its positions starts to use them to speculate instead. In particular, companies have a tendency to *overhedge*, that is they buy so many derivative contracts that instead of hedging their risks they cancel out all their risk, and in addition create an exposure in the opposite direction. Sometimes they even start selling options instead of buying them. This is really just speculation. For example, Ashanti, the gold mining company, 'hedged' its exposure to falls in the gold price in such a way that they lost a huge amount of money when the price of gold increased. This was seen as a sign by many gold mining companies that hedging is bad, but it was not. It was simply a sign that hedging should be for hedging not speculation.

This book is mainly about the bank's role as arbitrageur. The ability to spot market mispricings of derivative products depends upon some complicated mathematics which is the topic of this book.

1.7 Key points

- Risk is key to investment decisions as the only way to make money is by taking risky positions.
- Market efficiency means that all information is already encoded in the price of an asset so we cannot foretell stock prices.
- The risk premium is the amount of money we receive for taking on a risk.
- Hedging is the process of taking positions in different assets which reduce the total risk.
- Diversifiable risk does not receive a risk premium as it can be hedged away.
- A bond is an asset that pays a regular *coupon* and returns the *principal* at its maturity.
- A stock or share is a fraction of the ownership of a company. It is of limited liability and so carries rights without obligations.
- A forward contract is the right and obligation to buy an asset at an agreed day in the future at a price agreed today.
- A call option is the right but not the obligation to buy an asset at an agreed day in the future at a price agreed today.
- A put option is the right but not the obligation to sell an asset at an agreed day in the future at a price agreed today.

1.8 Further reading

A lot of learning about finance is about getting a good feel for how the market and its participants behave. I list a few books which are good background material and enjoyable reads.

- *Against the Gods: the Remarkable Story of Risk*, by P. L. Bernstein, [16], a history of risk management from ancient times
- *The Great Crash*, by J. K. Galbraith, [49], an account of the causes and aftermath of the stock-market crash of 1929.
- *Wriston: Walter Wriston, Citibank, and the Rise and Fall of American Financial Supremacy*, by P. L. Zweig, [122], a biography of the former head of Citibank. It's really a history of the evolution of modern banking.
- *Buffett: the Making of an American Capitalist*, by Roger Lowenstein, [87], a biography of Warren Buffett, it gives a fair amount of insight into how he made so much money simply by investing.
- *F.I.A.S.C.O.* by F. Partnoy, [101], described by the Head of the Financial Services Authority as a "nasty little book," it should not be taken too seriously but does give some insight into what goes on in the markets.
- *The New Financial Capitalists* by G. Baker and G. D. Smith, [9], an account of the modern revolution in corporate finance and how it changed the way companies are run.

1.9 Exercises

Exercise 1.1 Suppose an asset pays £1 if a roll of a fair die is 6 and zero otherwise. How much would you expect it to trade for?

Exercise 1.2 Suppose we have six assets, $E_1, \ldots, E_6$, which pay off according to the roll of a fair die. If the die roll is equal to the asset's index it pays one and zero otherwise. How much would you expect each asset to trade for? How much will the sum of all the assets trade for?

Exercise 1.3 If bond yields increase, what happens to bond prices?

Exercise 1.4 A gilt and a corporate bond have the same principal and the same coupons and coupon dates. How will their prices compare?

Exercise 1.5 A bond can be converted into a share of the issuer one year from now. How will its price compare to the price of a bond with the same principal and coupons which is not convertible?

2

Pricing methodologies and arbitrage

2.1 Some possible methodologies

In the previous chapter, we introduced the concept of risk and examined its relationship to various products including options. In this chapter, we want to examine how to price options and more general 'derivative' products. Recall that a derivative product is a product whose value is determined by the behaviour of another asset, generally called the underlying, and thus any option is a derivative. The interesting thing about pricing derivatives is that their price is closely related to that of the underlying asset, and we can expect to find relationships between their prices. The objective of mathematical finance is to find these relationships. It is not immediately obvious where to start. Let's work our way through various possible approaches in one special case.

We want to price an option to buy a particular stock for £1 five years from now. The stock's price today is £0.95.

A first approach might be to study the stock and estimate its growth potential (affected by its riskiness of course.) Suppose we think the stock will be worth £3 at the option's expiry date. We could say, well since we expect the option holder to make £$(3 - 1) = $ £2, we should charge him £2. Of course, we would charge less than £2 in order to take account of the fact that we could invest the premium for five years, and therefore would charge the amount of money it would take to have £2 in five years by investing in a five-year riskless bond. A more sophisticated version of this pricing model would be estimate the future distribution of the stock price and take the average value of the option under this distribution. For example, we might estimate that there's a 10% chance of the stock being worthless, and hence the option being worthless. Another 10% chance of the stock being worth less than £1 and the option being valueless. Say a 10% chance of the stock being around £2 and the option worth £1 and so on. We can then average the pay-offs and obtain an expected value for the option. We then discount this, as above, to compute the amount of money we need to invest today to match the expected pay-off.

To see the flaws in this approach, suppose we have sold such an option for £1.5. Instead of investing the money in a riskless five year bond, we just buy the stock today for £0.95. If the stock is below £1 in five years, then the option will not be exercised and we have made $(150 - 95) = 55$ pence plus the stock price on the date of expiry. If the stock is above 100 pence we sell it to the option buyer for £1 and we have made $(55 + 100) = 155$ pence. Thus, whatever happens we have made at least £0.55. We therefore have a riskless profit. This example shows that the cost of a call option should never be more than today's value of the stock, since the seller can then use the option's premium to buy the stock today, and cover himself in all possible outcomes. This is totally independent of any opinions about future movements.

The issue here is that the seller has adopted a hedging strategy. His strategy is to buy the underlying stock immediately. An alternative pricing methodology might be to estimate the expected loss to the seller under every possible hedging strategy. And having done so, adopt the strategy that minimizes the expected loss. Alternatively, we could adopt a strategy that minimizes the maximum possible loss. We should emphasize at this point that the hedging strategy can be dynamic: that is, the option seller can buy and sell the underlying according to its price movements during the life of the option.

Let's consider a new hedging strategy. The option will only be exercised at expiry if the value of the stock is above £1. The stock price is initially below this level. When the stock price crosses £1 we buy the stock, and if the stock price crosses £1 again we sell it. If it then crosses £1 again, we buy again and so on. If the stock price is below £1 at the end, we hold no stock and have no liability. If we assume interest rates are zero for simplicity, any money spent buying the stock on upcrosses will have been regained on selling on downcrosses, so we have lost nothing. If the stock price is above £1 on expiry, we sell the stock to the option holder for £1, and this repays us the £1 we used to buy the stock. Hence we can hedge the option for free, but the option holder will have paid us a positive amount for buying it. We pocket his premium and have done rather than well. Where is the flaw?

The flaw is that when the stock price is at £1, it is equally likely to go up as down; the fact that the stock price has just come from say £0.99 does not mean it will be £1.01 next: it could just as well be £0.99, so how do we know whether to buy or sell at £1?

Let's modify the strategy to be buy at £1.01 and sell at £0.99. Our strategy is now well-defined and appears almost as good. However there's a difference if we always buy at £1.01 and sell at £0.99 we lose £0.02 every time the stock crosses the interval. This means that the cost of hedging the option under this hedging strategy would depend upon the number of times the option crosses the intervals. Our expected loss would therefore depend on the expected number of crossings.

Interestingly, if our stock has a high growth component it is less likely to cross the interval many times since it will quickly have a price far away from £1. Stocks which have high growth should have cheaper options. This is a little paradoxical as we expect an investor to be keener to buy a call option if the stock price is likely to be much higher than the strike price. One answer lies in the connection between risk and return. A stock which is expected to have very high growth will also be very risky. We can therefore expect the stock price of a risky stock to bob up and down a lot, so we can expect it to cross the interval more often and thus the expected cost of hedging will be higher. Another answer is that there are better ways of pricing and we shall soon encounter them.

While this hedging strategy appears quite good, one downside of it is that there is no upper limit to the hedging cost. If the option seller was unlucky the stock could cross the interval thousands of times which would be much costly than the strategy of just buying the stock today. This raises another issue, what does the seller wish to achieve from his hedging strategy? Some possibilities are:

(i) Make a good return on average.
(ii) Cap the total amount that can be lost.
(iii) Minimize the variance (i.e. the riskiness) of the outcome.
(iv) Invest an amount today that will always precisely cover the cost of the option's pay-off at expiry.
(v) Avoid mispricing any risk.

The first objective is really that of a speculator. The others are those of hedgers and arbitrageurs. The purpose of financial mathematics is to achieve objectives (ii) through (v). This 'stop loss' strategy is better at (i) than at (ii) through (v).

2.2 Delta hedging

Under certain assumptions, which are arguably dubious, on the behaviour of the stock, it can be shown that there is a mathematically correct price, and an optimal hedging strategy which guarantees that the option's value at expiry will always be covered precisely by the option's premium at purchase. To see how this works, suppose that the value of the option is known and depends on the time to expiry of the option and the current value of the stock. The assumption of knowing the value may seem a little circular but this approach can be made rigorous. The value of the option will depend on other things such as interest rates but assume that these are fixed throughout. If we know the value of the option for any stock price, then we also know the rate of change of the option price with respect to the stock price. In mathematical terms, we know the derivative with respect to the stock price. We buy an amount of stock equal to this rate of change. Then the rate of change of our total portfolio, which is the amount of stock minus one option, will be zero. (This may involve buying a fractional amount of stock but that is no big deal from a mathematical point of view.) We have effectively removed all risk from the portfolio. Of course, as soon the stock price changes, the rate of change changes too, and the amount of stock needed to be held changes also. This hedging strategy is called *Delta hedging*. If one assumes that the rehedging can be carried out continuously, then it can be shown that it leads to a riskless portfolio, that is a portfolio of totally predictable value, which allows the option always to be hedged and this yields a correct price. This price is called the *Black–Scholes price* and the argument which led to it is the starting point of all modern mathematical finance.

The curious thing about the Black–Scholes price is that there is an alternative way of arriving at the same price which goes as follows. We estimate the future distribution of the stock price in terms of how risky it is, but we take the average expected price to be the same as could be achieved by holding a riskless bond. That is we assume that the stock buyers are not risk-averse but instead are *risk neutral*, that is they do not demand a discounting of the price to take account of risk. This is a little paradoxical after everything we have said so far about the importance of risk. Note that the model does not actually require the stock holders to be risk-neutral, it simply says that it is valid to price as if they were. The point is that the option seller having hedged his risk precisely, holds a riskless asset which should therefore grow at the same rate as a riskless bond. Since all risk has been removed we no longer have to worry about the effect of risk on pricing, and we can simplify things by assuming that investors are risk-neutral. Surprisingly, of all the pricing methodologies we have encountered, it is risk-neutral pricing that is the most pervasive in the markets. The reason is essentially that in certain circumstances it gives a price which can be shown to be necessarily correct in a truly real and practical sense. However, the paradigm has now become so standard that it is often used without any real justification.

2.3 What is arbitrage?

We shall return to the Black–Scholes price and risk-neutrality in later chapters and give a fuller mathematical treatment of them. Before doing so we need to understand the concept of arbitrage which is another fundamental concept in mathematical finance, and in particular, is a way of guaranteeing a correct price in certain circumstances. Arbitrage basically expresses the concept that one cannot make money for nothing. It is sometimes called the 'no free lunch' principle.

Arbitrage is probably simplest to explain in the context of foreign exchange. Suppose a £1 is worth $1.5 and a $1 is worth 100 yen, which is approximately true at the time of writing. How many yen is a pound worth? It has to be worth exactly

150 yen. If it is worth more than 150 yen, we sell pounds for yen, sell yen for dollars and sell dollars for pounds. We end up with more pounds than we started with. We keep on doing this for as long as we can. If £1 is worth less than 150 yen, we do the same thing but go round the triangle in the opposite direction, and make money in the same way. This process is called taking advantage of an arbitrage opportunity. The important point is that this process will cause the opportunity so disappear. This is closely related to the concept of market efficiency.

In particular, in the case where £1 is worth more than 150 yen, the action of buying yen will drive the pound/yen rate down and buying dollars will drive the yen/dollar rate down and so on. Thus the arbitrage opportunity will be short-lived. In the real financial markets, arbitrage opportunities can exist but they will generally be very small and disappear quickly, as someone will always be ready to pounce when they appear. In the mathematical theory of markets, it is therefore customary to assume that there is no arbitrage. Another way of looking at this, is that the mathematician's job is to find the possible prices in an arbitrage-free market. If the observed prices are not in agreement then there is an arbitrage opportunity to be exploited. Whilst the foreign exchange arbitrage was easy to spot, more complicated instruments may imply the existence of arbitrage opportunities which are anything but obvious, and that is why the banks employ their mathematicians.

Mathematical bankers generally search for the arbitrage opportunities under some assumptions. Whilst all of these assumptions can be criticised, they provide a good starting point for modelling. The objective is more to come up with a good model than a perfect description. There is a certain similarity to physics here. Newtonian physics makes certain assumptions about the nature of space and time which are demonstrably wrong. However, bridges are built with Newtonian physics and they do not fall down (or at least not very often.) The reason is that Newtonian physics provides a good approximation in the everyday world which only breaks down in the small subatomic world and the huge astronomical scale. Similarly, the models of mathematical finance provide good approximations under what one might call 'normal' conditions, but they may perform less well in extremities. However, just as in physics, the fact that models are not universally valid actually keeps people in work.

2.4 The assumptions of mathematical finance

What assumptions do mathematicians generally make?

2.4.1 Not moving the market

The first is the assumption that our actions do not affect the market price. That is we can buy or sell any amount without affecting the price. The whole point of free markets is that this is not true – if demand increases then prices increase, encouraging more production; and if demand decreases then prices decrease, discouraging production. Thus our actions can affect the market price but as long as we are trading in small quantities the effects will be negligible.

2.4.2 Liquidity

The second is the assumption of liquidity. This says that we can at any time buy or sell as much as we wish at the market price whenever we want. This is more valid in some markets than others. For example, in the major foreign currency markets and in the large company markets this is basically true. However, in the bond markets and the small companies market this is not the case. Note that the speculators and traders in the banks are actually providing a public service – their frenetic buying and selling increases the liquidity of markets thus ensuring that the ordinary investor can buy or sell at anytime he wishes, rather than being forced to wait until a counterparty can be found.

2.4.3 Shorting

The third is the assumption that one can go 'short' at will. That is, one can have negative amounts of an asset by selling assets one does not hold. Whilst there are some restrictions on short-selling assets in the market, it is allowed. The opposite of going 'short,' holding an asset, is sometimes called being 'long' in it. Similarly, buying an asset is called 'going long.'

2.4.4 Fractional quantities

The fourth assumption is the ability to purchase fractional quantities of assets. Whilst one can clearly not do this in the markets, when one is dealing in quantities of millions, which trading banks generally do, this is not so unreasonable – the smallest unit one can hold is a millionth of the typical amount held, so any error is pretty small in comparison.

2.4.5 No transaction costs

The fifth assumption is that there are no transaction costs. That is one can buy and sell assets without any costs. In the market, there are two typical ways to incur transaction costs. The first is just that doing something costs money. The second is that typically buy and sell prices differ slightly (or in the case of high street foreign exchange differ greatly.) This is called the *bid–offer spread*. The size of the bid–offer spread is closely related to liquidity, in a very liquid market it will be tiny but in less liquid markets it can be a substantial proportion of the asset's value. Taking transaction costs into account is currently an active area of research; we will work in a world without transaction costs. Note that the bid–offer spread is how banks make money. They will buy or sell from you as you wish, when you wish. The difference between the price you trade at, and the 'true' price, half-way between the bid and offer prices, is their fee for this service.

2.4.6 Models

There are other assumptions which are made in certain models. These are generally related to how the asset prices changes over time and we will introduce them as necessary. At this point, after such a long list of disputable assumptions the reader may start to feel that our models are a long way from reality and wonder what use they are. However, we must emphasize that our purpose is to build a model which will have the functions of providing a reasonable, but not perfect, price and of helping the bank understand its exposure to risks. One can view mathematical finance as an arms race, each bank is continually attempting to build more accurate models than its competitors in order to make money from trading, and to achieve the maximum return for a given level of risk.

2.5 An example of arbitrage-free pricing

Within the context of our assumptions, we now look at a few examples of arbitrage-free pricing. Suppose a company wishes to enter into a contract to exchange a fixed number of dollars, say one million, for yen one year from now. This is called a forward contract. Note that the company has no choice once it has entered the contract – this is not an option. How do we set the exchange rate? One could attempt to estimate the exchange rate in a year, and use this to price this contract. Alternatively, we could try to decompose the trade into instruments we already know the price of. We can do this by selling a zero coupon riskless bond in dollars today, which matures in a year with value one million dollars which corresponds precisely to the payment we get in a year. The money we get in exchange for the bond today is one million divided by $1+r$ where r is the yield of the bond. We exchange that money at today's exchange rate and use the proceeds to buy a zero coupon bond maturing in a year in yen, which will of course grow according to the yield, d. The value of this bond upon maturity will therefore be one million divided by $1+r$, multiplied by the exchange rate and then multiplied by $1+d$ and it is precisely this amount which we give the company in exchange for one million dollars in a year from now. In conclusion, if today's exchange rate is K, then the forward exchange rate for a year in advance is

$$K' = K \frac{1+d}{1+r}.$$

The important point to realize here is that any other rate leads to an arbitrage opportunity, because we can synthesize this product precisely at this exchange rate. In particular, if there is an alternative rate L available in the market we can make unlimited amounts of money by entering forward contracts at the rate L, and then hedging them using bonds as we have indicated above. Of course, whether we enter forward contracts buying or selling yen will depend on whether L is bigger or less than K. In conclusion, L must be equal to K' not because we believe the exchange rate in a year will be L, but because any other value leads to an arbitrage.

This example really contained two different and very important concepts as well as illustrating arbitrage. The first is that of *replication*. If we can decompose an instrument into other instruments then we can price it simply by stringing those instruments together, any other price will lead to an arbitrage. A first approach to any pricing problem should therefore always be to attempt to decompose an instrument. A curious side effect of this idea has been the growth of primitive building block instruments which strip out just one aspect of an instrument. For example, a corporate bond pays a premium above riskless bonds because of the risk of default. Thus one could regard a corporate bond as a riskless bond plus a risky asset which pays out an annual coupon and demands a payment in the event that the issuer defaults. This risky asset is now traded in its own right and is called a *credit default swap*. Note that it can be synthesized by going long a corporate bond and short a riskless bond.

We can use similar replication and arbitrage arguments to price a forward contract on any asset. Indeed to price a forward contract on a stock, provided it does not pay a dividend, we can use the same argument just by setting the foreign interest rate to zero. If we want to include the stock's dividend in the model, we pick the foreign interest rate to reflect the rate at which the stock pays dividends. This approximation is, of course, not quite right, as generally, though not always, a stock will pay its dividend in cash rather than stock, which is what the model suggests. Nevertheless this approximation is commonly used in option pricing, the big advantage being that it allows us to use the same arguments simultaneously for both stocks and foreign exchange. It's important to realize that the forward price of the stock is not the same as the expected future value. Indeed the argument shows that the forward price of the stock just grows at the same rate as a riskless bond and so it is a risk-neutral price, that is the future price if investors did not expect compensation for additional risk. We will see this phenomenon in more general contexts – arbitrage implies that a perfectly-hedged contract will be valued as if investors were risk-neutral.

Another simple contract we can price this way is a *forward-rate agreement*, generally called a FRA. A forward-rate agreement is simply an agreement to take some money on deposit, or to borrow some money, at an interest rate fixed today for a fixed period of time starting at a specified future time. For example, a company may be paid for some goods on a known future date, and will buy some other goods on a fixed date after that. The company wishes to make plans on the basis of this and so enters a FRA, thus ensuring that there is no interest-rate risk. How would the bank decide at what rate to offer the FRA? The FRA is easily synthesized by going long and short the appropriate bonds. If the FRA starts at time t_0, we go short a bond expiring at time t_0 to bring the money back to the present – thus multiplying the sum deposited by $(1 + r_0)^{-t_0}$ where r_0 is the yield of the t_0 bond. To take the money forward to time t_1, the end of the deposit period, we go long bonds maturing at time t_1 with yield r_1 and thus multiply the sum by $(1 + r_1)^{t_1}$. In conclusion, in return for the £1 deposit at the start of the FRA, the company receives

$$X = (1 + r_0)^{-t_0}(1 + r_1)^{t_1}$$

at the end. One can then convert this into an equivalent compounding annual interest rate, r_2, by solving

$$(1 + r_2)^{t_1 - t_0} = X.$$

2.6 The time value of money

The second important aspect of pricing the forward contract is the concept of time value of money. Jam today is better than jam tomorrow – an investor will prefer a pound in his pocket today to a pound in his pocket one year from now. In effect, a pound a year from now is therefore worth less than a pound today. The interest paid on a riskless loan expresses this. We can quantify precisely how much less by using risk-free bonds. A zero-coupon bond with principal £1 maturing in a year is precisely the same as receiving the sum of £1 in a year. We can therefore change the timing of a cashflow through the use of zero-coupon bonds. (A cashflow is a flow of money that occurs at some time.) If we are to receive a definite cashflow of £X at time T, then that is the same as being given X zero-coupon bonds today, and we can convert it into a cashflow today by simply selling X zero-coupon bonds of maturity T. The two cashflows at time T will then cancel each other. If the market value of a T-maturity bond is $P(T)$, then £X at time T is equivalent to £$XP(T)$ today. Similarly a cashflow of £Y today is worth $Y/P(T)$ pounds at time T.

The conversion of sums through time is therefore dependent on the market value of zero-coupon bonds. It is generally easier, though sometimes misleading, to think in terms of interest rates. The cost of bonds is generally quoted in terms of yield, that is, the effective annually compounded interest rate which would give the same value on maturity. If the yield is r which could be a number written as 0.05, or more often as 5%, and the bond runs for T years then £1 invested in it today will be worth £$(1+r)$ after a year and because of compounding £$(1+r)^2$ after two years and so. In particular, after T years, it will be worth £$(1+r)^T$. Similarly, £1 in T years from now will be worth £$(1+r)^{-T}$ today. To see this, consider that we can take out a loan of £$(1+r)^{-T}$ today with the knowledge that we can pay off the loan with the £1 when it arrives. In these formulas, it is important to realize that r and T are not as independent as they look: r is the yield of a bond which matures at time T, and bonds of different maturities may have different yields. Clearly, if we know that the price of a bond maturing at time T is $P(T)$, then there will be a unique r such that

$$P(T)(1+r)^T = 1, \qquad (2.1)$$

and this number will be the yield of the bond. This definition of yield is often called the *annualized* yield. Note that one consequence of (2.1) is that yields go down when prices go up and vice-versa.

Whilst annualized yield is convenient for quoting in the markets, it is cumbersome to work with mathematically since a year is a rather arbitrary time scale, and mathematicians prefer to be able to break up time into smaller and smaller time pieces. Suppose we divide a year into n pieces of equal length and over each piece we receive r times the length of the piece as an interest payment, and the interest is compounded. If we start with £1 after T years we have the sum of £$(1+\frac{r}{n})^{Tn}$. We can make the compounding period shorter and shorter by letting n go to infinity. It is an elementary theorem of mathematical analysis that the limit is

$$e^{rT}.$$

In mathematical finance, it is this form of interest-rate that is used when pricing equity and foreign-exchange (FX) options. The quantity r is then called the *short rate* or *continuously compounding* rate as it's the interest rate for investing over very short periods of time. When working with foreign exchange or with dividend-paying stocks there will be a corresponding interest rate on the asset, or dividends on the stock. This rate is generally called the *dividend rate* and denoted d. We typically model the dividend on the stock as a *scrip dividend*. This means that rather than receiving a cash sum as a dividend payment, we receive extra units of stock. (Since the dividend is not precisely divisible by the stock price, the amount left over is paid in cash, but we ignore these issues as the effects are tiny for large holdings.) Thus if we take a dividend rate of d and we start with one stock or unit of currency, we will have e^{dt} units at time t.

Each zero-coupon bond will therefore imply a different rate r over the period of its life given by the value r such that

$$e^{rT} P(T) = 1.$$

2.6 The time value of money

Much of the time, we will assume that there is a single unique r reflecting constant interest rates. This means that a sum of £1 invested today in riskless bonds will always be worth £e^{rT} at time T. This investment is often called the *money-market account* or the *cash bond*. It reflects the notion that one is continuously buying a very short-dated bond, letting it mature and then reinvesting in another short-dated bond. For clarity, we reproduce our forward price argument in a more mathematical fashion using continuously compounding rates.

Theorem 2.1 *If a liquid asset trades today at S_0 with dividend rate d and the continuously compounding interest rate is r then a forward contract to buy the asset for K with expiry T is worth*

$$e^{-rT}\left(e^{(r-d)T}S_0 - K\right).$$

In particular, the contract will have zero value if and only if

$$K = e^{(r-d)T}S_0.$$

Proof We first show that if

$$K = e^{(r-d)T}S_0, \tag{2.2}$$

then the forward contract has zero value.

Suppose we have sold the forward contract. At time zero, we set up a portfolio consisting of $-e^{-dT}S_0$ pounds and e^{-dT} assets. We can do this at zero cost. At time T, we will hold one asset, since the asset grows by e^{dT}, and be short £$e^{(r-d)T}S_0$ because of interest charges.

The forward contract turns our single asset into £$e^{(r-d)T}S_0$ by construction. This cancels the negative pounds in our portfolio and we hold nothing.

Thus whatever the price of the asset at time T, we have no holdings. This means that we have precisely hedged the forward contract at zero cost, so the contract must be worth zero or there would be an arbitrage.

If we have a forward contract struck at K', we can decompose it as a forward contract struck at K, with K as above, and the right to receive £$(K - K')$ at time T. The right to receive £$(K - K')$ is the same as holding $K - K'$ zero-coupon bonds. Note that if $K < K'$, we are really borrowing $K' - K$ zero-coupon bonds.

The forward contract struck at K has zero value so the value of the contract must be the value of the zero-coupon bonds, that is

$$e^{-rT}(K - K') = e^{-rT}\left(e^{(r-d)T}S_0 - K'\right), \tag{2.3}$$

and we are done. □

The second part of the theorem motivates a definition. The *forward price* of a stock for a contract at time T is $e^{(r-d)T}S_0$.

2.7 Mathematically defining arbitrage

We have seen that arbitrage can price various simple contracts precisely in a way that allows for no doubt in the price, and the price is independent of our views on how asset prices will evolve. The great revolution in modern finance is based on the observation that such arguments can be extended to cover the pricing of vanilla options, and that is the topic of the rest of this book.

Before proceeding to the valuation of options, we discuss the concept of arbitrage a little more. How can we make a more rigorous definition capturing our notion of no money for nothing?

Definition 2.1 A portfolio is said to be an *arbitrage portfolio*, if today it is of non-positive value, and in the future it has zero probability of being of negative value, and a non-zero probability of being of positive value.

For now, we will ignore the probabilities in the statement of arbitrage and simply regard an arbitrage portfolio as one that is of zero cost to set up, has non-negative value in the future, and may be of positive value in the future. By creating such a portfolio, an investor would receive at no cost the possibility of receiving money in the future. An important consequence of the no-arbitrage principle is the following monotonicity theorem,

Theorem 2.2 *If portfolios A and B are such that in every possible state of the market at time T, portfolio A is worth at least as much as portfolio B, then at any time $t < T$ portfolio A is worth at least as much as portfolio B. If in addition, portfolio A is worth more than portfolio B in some states of the world, then at any time $t < T$, portfolio A is worth more than portfolio B.*

Proof The proof of this theorem follows simply by applying the no-arbitrage principle to a portfolio C constituted by being long portfolio A and short portfolio B. Portfolio C then has non-negative value in all world states at time T as its value is just the value of portfolio A minus that of portfolio B, and so must be of non-negative value at time t. If A can be worth more than B at time T then C can have positive value at time T, and C must have positive value at time t, or there would be the possibility of making money from a portfolio of zero cost with no risk. But C having positive value is the same as saying that A is worth more than B and our theorem follows. □

Whilst the monotonicity theorem is easy to state and prove, it is at the heart of most arguments in mathematical finance. An easy corollary is

Theorem 2.3 *A vanilla call or put option always has positive value before expiry.*

Proof To prove the theorem, simply let portfolio A be the option and let portfolio B be empty. Then at the expiry time, A is either worth a positive quantity if it is advantageous to exercise the option, or zero otherwise. We therefore have that A is worth more than portfolio B in some world states, and at least as much in all world states. The monotonicity theorem then says that at all previous times A is worth more than B, that is the option has positive value. □

Another consequence of the monotonicity theorem is that there can only be one riskless asset of a given maturity.

Theorem 2.4 *If P and Q are riskless zero-coupon bonds with the same maturity time, T, then they are of equal value at all previous times.*

Proof Suppose both bonds P and Q are guaranteed to be worth exactly £1 at time T. Then Q is worth as much as P in all possible worlds at time T, so Q is worth at least as much as P in all possible worlds at all previous times. By symmetry, we conclude that P is also worth as much as Q, and thus that P and Q have the same price in all possible worlds at all times. □

If instead of paying £1, Q paid £A, then considering holding A units of the bond P, and applying the monotonicity theorem, we conclude that Q is worth AP in all possible worlds at all previous times.

A third simple consequence of the monotonicity theorem is

Theorem 2.5 *If two portfolios, P and Q, are of equal value today and if at some future time, T, P is worth more than Q in some world states, then Q is worth less than P in some world states.*

Proof If two portfolios, P and Q, are worth the same today, and if at some future time in some possible world state, P would be worth more than Q then there must be a world state at that time in which Q is worth less than P. Otherwise, the monotonicity theorem would imply that P was worth more than Q today. □

Whilst these examples of the monotonicity theorem have been very simple and are easily understood, the theorem becomes more subtle and more useful when applied to dynamically changing self-financing portfolios. A *dynamic self-financing portfolio* is a portfolio in which a certain sum of money is initially invested, and no money is either extracted or added thereafter, but the buying and selling of stocks and bonds at the prevailing market price, according to a strategy depending on the world state, is allowed. The money raised from such buying and selling is kept within the portfolio via conversion into riskless bonds. We shall see that the pricing of options is based on such arguments.

For a very simple example of this, consider portfolio A to contain one American call option struck at K and expiring at time T, and portfolio B to contain one European call option also struck at K. For portfolio A our 'dynamic' rule is "do not exercise before expiry." At expiry, portfolios A and B are then of equal value in all possible worlds. We conclude that portfolio A with this rule is worth the same as portfolio B in all worlds at all previous times. Since we are not constrained to follow this rule, we conclude that an American option is always worth at least as much as a European one with the same expiry and strike. There was of course nothing special about a call option in this argument, and it holds equally for put options. This example demonstrates the simple point that adding extra rights to an option can only increase its value, as the holder can always ignore them if he sees fit. We shall see below that extra rights can be worthless – in a strict arbitrage sense, they add no value to certain options.

The concept of arbitrage has become so ubiquitous that it is often used in senses that are not strictly correct. For example, at the time of writing, NatWest made an offer for Legal and General shares at an offer price of 210p a share. The share price for Legal and General reacted to this information by immediately jumping to about 200p. So there was an 'arbitrage opportunity' to purchase Legal and General shares for 200p and sell them for 210p to NatWest. Many 'arbitrage houses' therefore bought lots of Legal and General shares and financed the purchase by short-selling NatWest. However there was a good reason for the market's pricing the shares at 200p – there was a still a possibility that the deal would fall through. Indeed, Bank of Scotland launched a bid for NatWest and urged the shareholders to reject the Legal and General merger. The NatWest share price jumped up, the Legal and General one fell and the arbitrage houses had their fingers badly burnt. (The final outcome was that the Royal Bank of Scotland launched a second bid, and took over NatWest.) The real point here is that there was not an arbitrage opportunity in the strict sense, only in the weaker sense of a likely profit.

2.8 Using arbitrage to bound option prices

In this section, we study bounds on option prices which do not involve any assumptions on the way the assets move. Instead we prove upper and lower bounds on prices, and prove relationships between the prices of differing options. Recall that a European call option on an asset is the right to buy the asset for a fixed price K at some fixed time, T, in the future. A rational investor who owns the option will exercise the option if and only if the market price of the asset, S, is more than K at time T, in which case he makes $S - K$. Otherwise the option expires worthless. Hence at expiry the option is worth precisely

$$(S - K)_+ = \max(S - K, 0).$$

2.8 Using arbitrage to bound option prices

This means that instead of thinking of the call option as the right to buy the stock for K, we can think of it as an asset that pays the sum $(S - K)_+$ at time T. The function $(S - K)_+$ is then called the option's *pay-off*. Similarly, a put option will have the pay-off

$$(K - S)_+ = \max(K - S, 0).$$

Whilst these two pay-offs are the most common ones, there is nothing to stop us considering an option that has any pay-off; that is, the pay-off could be an arbitrary function of S and K.

If we hold a call option and sell a put option both struck at K, what is our pay-off? If $S \geq K$ it is $S - K$, and if $S \leq K$ it is $-(K - S)$ which is of course $S - K$. This means that if we hold the portfolio of plus one call and minus one put, we always receive $S - K$. A forward contract struck at K will also have pay-off $S - K$. This means that the price of a call option minus the price of a put option must equal the price of a forward contract, if all are set at the same strike. Otherwise, one can make money by selling the more expensive of the two contracts and buying the cheaper, thereby making an instant profit and having no exposure. We conclude

Theorem 2.6 (Put-call parity) *If a call option, of price C, a put option, of price P, and a forward contract, of price F, have the same strike and expiry then*

$$C - P = F.$$

Put-call parity is very useful as it means that if we can price a forward contract and one of the call or the put, we can immediately deduce the price of the other. This has both mathematical and practical advantages. For example, the value of a put is never more than K at exercise time, whereas the value of a call can be arbitrarily large, which makes some mathematical convergence arguments involving calls tricky, but these can be simply avoided by invoking put-call parity.

An option is said to be *in-the-money* if it would be worth something at expiry provided the underlying's price did not change, and *out-of-the money* otherwise. An option at the cross-over point is said to be *at-the-money*. From the practical point of view, it is generally much quicker to compute the value of out-of-the money options than in-the-money ones as the value, being much smaller, means that sums will converge faster. One therefore simply values whichever of the call or put is out-of-the money and then deduces the value of the other. Indeed, many markets will only quote the value of out-of-the money options leaving the cost of in-the-money options to be deduced. Note that our argument has not depended in any way on the nature of the asset, or any assumptions about how its price will evolve. There is some inconsistency in the markets as to the definition of at-the-money. We have defined at-the-money in terms of the *spot price*; that is, the price of the asset today, but it is often defined in terms of the forward price. Sometimes an option is said to be struck *at-the-forward*. There is a certain mathematical appeal in working with the forward, in that the at-the-forward price is where the forward contract's value changes sign, and therefore it is the point where call and put options have equal value.

This result has an interesting consequence. It means that our views on where the future value of the asset price is likely to be at expiry, cannot affect the price of a call or put option struck at the forward. For if we believe the spot is more likely to finish in-the-money, and therefore conclude that the call ought to be worth more than the put, we have a contradiction, since we know that the call and put must have the same value. This observation is a central point of mathematical finance:

> Our views on the mean of the stock price at expiry of an option do not affect the price of an option.

As well as relating the prices of calls and puts to each other, we can prove bounds on the prices of options. Recall that we argued above that a call option could never be worth more than the current spot value of the stock, since the option can be hedged by buying the stock today. This means that we have an upper bound on the price. In fact, we can prove

Theorem 2.7 *At time t, let C_t be the price of a call option on a non-dividend-paying stock, S_t, with expiry T and strike K. Let Z_t be the price of a zero-coupon bond with maturity T. We have*

$$S_0 > C_0 > S_0 - KZ_0. \tag{2.4}$$

Proof We reprove the upper bound for completeness. At expiry, the stock is worth S_T and the option is worth $S_T - K$, so at that time the call option is always worth less than the stock. It therefore follows from the monotonicity theorem that the option is worth less than the stock at time 0. That is

$$S_0 > C_0.$$

To prove the lower bound, consider the portfolio consisting of one option, and K zero-coupon bonds which mature at time T. At the time of the option's maturity, we have the option and K pounds. If the stock price, S_T, is greater than K, we can exercise the option and spend the K pounds to buy the stock. Our portfolio is then worth S_T. Otherwise, we do not exercise the option and our portfolio is worth K.

At time T, our portfolio is therefore worth

$$\max(S_T, K) = (S_T - K)_+ + K;$$

it is always worth as much as the stock and in some circumstances is worth more.

This means that at all previous times the portfolio must be worth more than the stock, as it carries all the same benefits (remember the stock is not dividend-paying) and at time T is worth at least as much. We deduce, using the monotonicity theorem, that

$$C_0 + K Z_0 > S_0, \tag{2.5}$$

which is equivalent to

$$C_0 > S_0 - K Z_0 \tag{2.6}$$

$\square$

If we make the assumption that interest rates are non-negative then we have $Z_0 < 1$, and

$$C_0 > S_0 - K.$$

This relation is very important as it implies that before expiry a European call option on a non-dividend-paying stock is always worth more than its intrinsic value, i.e. the value that would be obtained by exercising it today, if that were possible.

An important consequence of this is that if we consider an American call option on a non-dividend-paying stock, we can deduce that it has the same value as a European call option. To see this, observe that the American option carries all the rights that a European option does and more, so must be worth at least as much, which means that it is always worth more than its intrinsic value, before expiry. One should therefore never exercise an American call option before expiry. But if one is never going to use the additional rights, they are worthless and an American call option on a non-dividend-paying stock is worth the same amount as a European call option. We therefore have

Theorem 2.8 *If interest-rates are non-negative, then a European call option and an American call option, on the same non-dividend-paying stock with the same strike and maturity, are of equal value.*

Note that if one had purchased the American call option to hedge a particular risk, and therefore needed to exercise it at some earlier time, the thing to do would be to sell the option in the market rather than to exercise it, as this is guaranteed to raise more money. Arguably, it is preferable to sell American call options rather than European ones as there is always the possibility that the buyer will exercise it early, thereby costing the seller less than the European one. However, this is more in the realms of psychology than mathematics, as we have shown that the rational investor would not exercise early.

An immediate corollary is

Theorem 2.9 *Let S be a non-dividend-paying stock. Let C_1 and C_2 be European call options on S struck at K with expiries T_1 and T_2, with*

$$T_1 < T_2.$$

If interest rates are non-negative then C_2 is worth at least as much as C_1.

Proof Consider an American call option, A, which expires at time T_2; it has the same value as C_2 by our argument above. However, it can also be exercised at time T_1 so it carries all the rights of C_1. Thus A must be worth at least as much as C_1. Consequently, we have that C_2 is worth at least as much C_1. Since the times T_1 and T_2 were arbitrary, this shows that for options with the same strikes, the value is an increasing function of expiry date. □

If we make a mild assumption that option prices are time-homogeneous, we can say more. By time-homogeneous, we mean that if the stock price does not change then, for a given strike, the cost of buying an option only depends on the difference of the current time t and the expiry time, T and not on t and T individually. An option expiring in 18 months from now will cost the same a year hence as an otherwise identical 6-month option does today. This implies that if the stock price does not change then an option we buy today expiring at time T, will at time $t < T$, be worth the same as an option bought today expiring at time $T - t$. Since $T - t < T$, we conclude that the option will be worth less at time t than it is today. Thus time-homogeneity implies that the value of a European call option on a non-dividend-paying stock is a decreasing function of time for a fixed stock price. We illustrate this time-dependence in Figure 2.1.

What does the assumption of time-homogeneity mean financially? It essentially says that the future is not qualitatively different from the present. A model which is not time-homogeneous should generally be regarded with suspicion, as it implies that the future will be different from the present. Of course, we are not saying that in general the future will be precisely the same as the present but that the model should not imply a specific different form of behaviour without good reason. A good reason would be, for example, the formation of a currency union. An option to buy francs for marks would have had a rather lower price if the exercise time was after the formation of the euro than if it was before.

We prove by a simple no-arbitrage argument that the difference in price of two call options of different strikes but the same expiry is less than their difference in strikes. In particular, suppose the two options, C_1 and C_2, expire at time T and have strikes K_1 and K_2, respectively. We take $K_1 < K_2$. At expiry, we necessarily have that C_1 is worth more than C_2 as it allows us to buy a stock for less money. This means that at all previous times, we know that C_1 is worth more than C_2 to avoid the possibility of arbitrage.

2.8 Using arbitrage to bound option prices

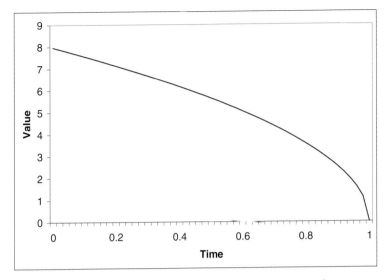

Fig. 2.1. The Black–Scholes value of a call option struck at-the-money as a function of time with spot fixed.

However, a portfolio consisting of C_2 and $K_2 - K_1$ zero-coupon bonds will be worth, at time T,

$$(K_2 - K_1) + \max(S - K_2, 0),$$

which is greater than the pay-off of C_1: $\max(S - K_1, 0)$. We therefore have, if $Z(t, T)$ is the value of the bond, that at all previous times

$$C_2 < C_1 < C_2 + (K_2 - K_1)Z(t, T). \tag{2.7}$$

If we let K_2 tend to K_1, then the price of C_2 must converge to that of C_1, which means that the price of an option is a continuous function of strike. (In fact, we have that it is a Lipshitz-continuous function with constant $Z(t, T)$. See Figure 2.2.)

An alternate approach to proving this result is to show that put option prices are an increasing function of strike and to invoke put-call parity. However, this would only apply if we were considering options on a tradable asset, whereas our results hold regardless of the tradability of the underlying.

We have shown that call option prices of fixed maturity are a strictly decreasing function of strike, and that the prices cannot decrease too rapidly. What other properties can we prove? A much less obvious property is convexity. Recall that a function is *convex* if the line between any points on the graph lies on or above the graph. This is equivalent to saying that if $C(K)$ denotes a call option struck at K,

34   *Pricing methodologies and arbitrage*

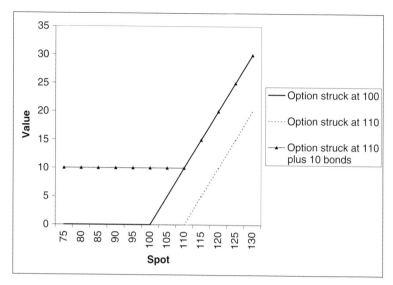

Fig. 2.2. The portfolios required to prove that call option prices are Lipshitz-continuous and a decreasing function of strike.

and $K_1 < K_2$, we have

$$\theta C(K_1) + (1-\theta)C(K_2) \geq C(\theta K_1 + (1-\theta)K_2), \quad \text{for } 0 \leq \theta \leq 1. \quad (2.8)$$

Fixing K_1, K_2 and θ, consider the portfolio consisting of θ call options struck at K_1, $1 - \theta$ call options struck at K_2 and -1 call option struck at $\theta K_1 + (1-\theta)K_2$. As the final pay-off of the call option is convex for any fixed value of the underlying (see Figure 2.3), we have from (2.8) that our portfolio is always of non-negative value at expiry. This means that at all previous times the portfolio must be of non-negative value, or there would be an arbitrage opportunity. The portfolio being of non-negative value is equivalent to (2.8) and we are done. We illustrate the portfolio constructed in this argument in Figure 2.5, and the result in Figure 2.4.

Note that in this argument the only crucial point was that the final pay-off of the call option was a convex function of strike for any fixed value of spot. The argument therefore carries over immediately to any instrument with a convex pay-off, and in particular we have that put option prices are also a convex function of strike. (This could also be proven using put-call parity provided the instrument was tradable.)

To summarize, we have proven

Theorem 2.10 *Let S be an indeterminate quantity. Let $C(K, T)$ denote the price of a call option on S of strike K and expiry T. We then have*

2.8 *Using arbitrage to bound option prices* 35

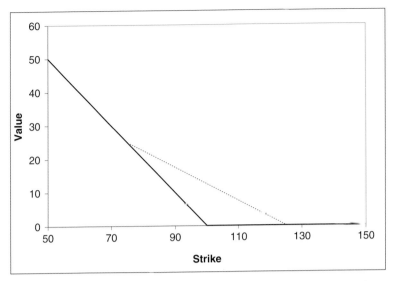

Fig. 2.3. The chord to the final pay-off as a function of strike lies above the graph.

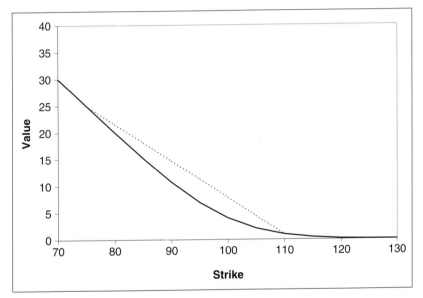

Fig. 2.4. The chord to the value of a call option as a function of strike lies above the graph.

(i) $C(K, T)$ is a decreasing function of K,
(ii) $C(K, T)$ is a (Lipshitz-) continuous function of K with Lipshitz constant $Z(T)$,
(iii) $C(K, T)$ is a convex function of K.

If S is a non-dividend-paying stock, we also have that $C(K, T)$ is an increasing function of T.

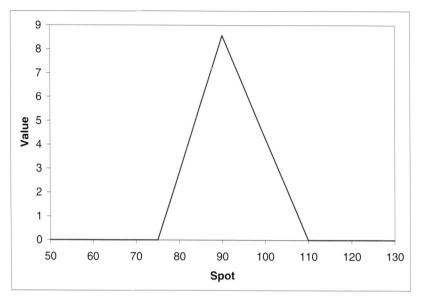

Fig. 2.5. The final pay-off of the portfolio used to prove convexity. This pay-off is generally called the butterfly.

Note that for the first three properties, we made no assumptions on the nature of S. It could be a non-dividend-paying stock but it could be a lot of other things too, for example the temperature on a given day or an exchange rate. Note that the first and second properties are conditions on the first derivative of C when it exists. The third condition is equivalent to C having a non-negative second derivative when it is differentiable.

Theorem 2.7 was a bound on the price of a call option; we can prove similar bounds for any option with a single pay-off time by using holdings of the underlying instrument and a riskless bond. The fundamental idea is that we want to create a portfolio which dominates the pay-off in the sense of being worth more at expiry than the option whatever the value of spot is. We therefore search for the cheapest portfolio of stocks and bonds which dominates the option pay-off at expiry, and for the most expensive portfolio which is dominated by it at expiry.

As a portfolio consisting of α stocks and β zero-coupon bonds is a linear function (or to be more precise an affine function), we are really trying to find the closest straight lines above and below the pay-off. Thus Theorem 2.7 is the observation that

$$f(S) = S > (S - K)_+, \qquad (2.9)$$

and that

$$g(S) = S - K < (S - K)_+. \qquad (2.10)$$

Of course, when $S < KZ_0$, we have the better lower bound of the zero function.

2.8 Using arbitrage to bound option prices

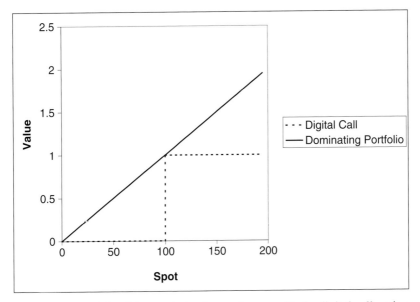

Fig. 2.6. A multiple of the stock dominates the pay-off of a digital call option.

For example, suppose we have a *digital call* option struck at K on a non-dividend-paying stock which will pay 1 if the stock finishes above K, and zero otherwise. Our pay-off is therefore

$$H(S - K),$$

where $H(t)$ is the Heaviside function which is 1 for $t \geq 0$ and 0 otherwise.

We want to find the set of dominating portfolios. It is clear geometrically that the two critical points are 0 and K; if the Heaviside function is dominated at those two points by an upwards sloping line, it will be dominated for all $S > 0$.

If the portfolio is α stocks and β zero-coupon bonds then domination at zero is achieved if $\beta \geq 0$ for any α. Domination at K is achieved if

$$\alpha K + \beta \geq 1. \tag{2.11}$$

The initial value of our portfolio is

$$\alpha S + \beta P,$$

where P is the cost of a zero-coupon bond expiring at time T. If we take the solution which passes through the two crucial points, then we get $\alpha = K^{-1}$ and $\beta = 0$. (See Figure 2.6.) This portfolio will have set-up cost SK^{-1}. If the stock price is greater than K we have not achieved much as the set-up cost will be greater than 1 which is the maximum pay-off of the option.

An alternative upper bound is attained by taking $\beta = 1$ and $\alpha = 0$. This gives an upper bound for the digital option equal to P. We conclude that the digital option

must be worth less than
$$\min(P, SK^{-1}).$$

We would also like a lower bound on the option price. Since the pay-off is always zero or more, clearly zero is a lower bound. In fact, this lower bound is optimal – try to prove this.

2.9 Conclusion

We have seen that, by making some quite mild assumptions on the market, no-arbitrage arguments lead to bounds on the value of options and forward contracts which are not immediately obvious, The bounds we have proven involve *static* portfolios; that is, we set up a portfolio at time zero and do not change it until the maturity of the option. These bounds are sometimes called *rational bounds* as they hold without taking any view on the behaviour of the underlying. To prove sharper bounds, we will need to carry out *dynamic* trading strategies which will involve trading en route. To justify these strategies we will need to make assumptions about the underlying's behaviour – we will have to quantify not the asset's trends but instead the nature of its randomness.

2.10 Key points

- An *arbitrage* is an opportunity to make money for nothing. An arbitrage portfolio is a portfolio of zero value which may be of positive value in the future, and will never be of negative value.
- A *hedging strategy* is a method of reducing the uncertainty in the value of the pay-off of an option by trading in the underlying.
- *Rational bounds* on the price of an option are arbitrage bounds which can be proven without making assumptions on the future behaviour of the asset.
- The *short rate* is the interest rate we obtain if we continuously reinvest cash. This is sometimes called the cash bond or the money-market account.
- A portfolio *replicates* an option if whatever happens, it has the same value as the pay-off of the option at the expiry of the option. If a portfolio replicates the option then the option's value is the price of setting up the portfolio.

2.11 Further reading

The basic model-free inequalities, known as rational bounds, were proved by Merton. His book *Continuous Time Finance*, [94], is a collection of his papers including the original proofs, and is well worth acquiring.

Cochrane's *Asset Pricing*, [35], is an account of pricing methodologies for assets including alternatives to no-arbitrage when no-arbitrage pricing is too weak to provide effective bounds.

2.12 Exercises

Exercise 2.1 If a dollar is 120 yen and £1 is $1.4, what can we say about the pound/yen exchange rate?

Exercise 2.2 Each of the following products pays a function of the spot price, S, of a non-dividend-paying stock one year from now. If there are no interest rates and spot is 100, give optimal upper and lower bounds on their prices today.

(i) The pay-off is 1 between 110 and 130 and zero otherwise;
(ii) The pay-off is $S - 80$;
(iii) The pay-off is zero below 80, increases linearly from zero at 80 to 20 at 120 and then is constant at 20 above 120;
(iv) The pay-off is $(S - 100)^2$.

Exercise 2.3 Let P be a digital put struck at K_1 and C be a digital call struck at K_2. (A digital put pays 1 if spot is below the strike at expiry, and a digital call pays 1 if spot is above the strike.) What can we say about the prices of C and P in each of the following cases?

(i) $K_1 = K_2$;
(ii) $K_1 < K_2$;
(iii) $K_1 > K_2$.

Exercise 2.4 If interest rates increase how will the forward price of an asset change? How will the value of a forward contract change?

Exercise 2.5 Suppose no-arbitrage bounds for an option price show that the price lies between L_1 and L_2 in a world without transaction costs. What can we say about the bounds if we take transaction costs into account?

Exercise 2.6 Show that if interest rates are zero and call option prices are a differentiable function of strike then the derivative of the prices with respect to strike must lie between -1 and 0. What if interest rates are non-zero?

Exercise 2.7 Let $D(K)$ pay $(S - K)^2$ if $S > K$, and zero otherwise. Show that if $D(K)$ is a differentiable function of K then the third derivative of D with respect to K is non-negative.

Exercise 2.8 Show that if the current spot price is S_0, and the continuous compounding rate is r then a call and a put both struck at $S_0 e^{rT}$ and expiring at time T are of equal value.

Exercise 2.9 Prove that zero is the optimal lower bound for a digital call option.

Exercise 2.10 Interest rates are non-negative. An asset is worth 100 today and in the future the value is constant, except at random times when the asset's value drops by a random amount. Construct an arbitrage.

Exercise 2.11 Let S_t be the price of a non-dividend-paying stock. Suppose derivatives A and B pay functions f and g of the stock price at expiry. Suppose that we have

$$f(x) \leq \alpha + \beta x + \gamma g(x).$$

What can we say about the relative prices of A and B today?

Exercise 2.12 Asset A pays 1 if the stock price over the next year is at some point above 100. Asset B pays 1 if the stock price is above 100 a year from now. What can we say about the relative prices of A and B?

Exercise 2.13 Formulate analogues to Theorem 2.7 and Theorem 2.10 for put options.

Exercise 2.14 Suppose it costs αX to buy or sell α shares. What forward prices for a stock are non-arbitrageable?

Exercise 2.15 Let S be a non-dividend-paying stock. The riskless bond has value e^{rt}. A contract pays $S_{t_2} - S_{t_1}$ at time t_2. Show that this contract can be replicated by trading in the stock and riskless bond, and gives its price today.

3

Trees and option pricing

3.1 A two-world universe

In this chapter, we start the pricing of vanilla options using the concept of a tree. We commence with a highly stylized situation and gradually extend our model to make it more accurate. We start by considering an option on an asset which can only take two values in the future. For concreteness, suppose we have an asset which is worth 100 today, and will be worth either 110 or 90 tomorrow. Suppose we are a bank and someone wishes to purchase a call option today. Suppose the option is struck at 100. We assume that interest rates are zero for simplicity.

The option buyer will exercise the option if and only if the stock price is greater than the strike price, that is he will exercise the option only if the stock price is 110 and then will make 10. If the stock price is 90, he does not exercise and thus he makes nothing. This means that in the first state of the world, the bank is down 10 and the other it is down nothing. We conclude that the value of the option must be between zero and 10p.

In one state of the world, which we henceforth call 'A', the option is worth 10 and the stock 110, whereas in the other, 'B', the option is worth zero and the stock 90.

We, the bank, wish to hedge our risk. The simplest hedging strategy would be to buy the stock if the stock was going to be worth 110 tomorrow and do nothing otherwise. However, this requires foretelling the future, and if we could do that there would be plenty of arbitrage opportunities! We require a hedging strategy independent of the future – another way of saying this is that we must make a decision on the basis of the information available today. To make this concept precise we will require a notion of information: a topic to which we will return in Chapter 6.

3.1.1 Pricing in a one-step tree by hedging

Without foreknowledge, we must buy a fixed number, δ, of stocks today. Denoting the stock price by S and the option's value by Opt, our portfolio can be written

$$\delta S - \text{Opt}.$$

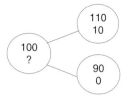

Fig. 3.1. A one-step, two-state model. Time goes from left to right.

In state of the world A, the portfolio will be worth

$$110\delta - 10$$

and in state of the world B, it will be worth

$$90\delta.$$

To hedge our risk means to remove all uncertainty; that is, our portfolio should be worth the same in either state of the world. To achieve this, we need to choose δ so that

$$110\delta - 10 = 90\delta.$$

That is

$$\delta = 1/2.$$

Assuming it makes sense to hold half a share, we buy half a share today to hedge our risk, and whatever happens our portfolio is worth $90/2 = 45$ tomorrow. Since we are assuming no interest rates and our portfolio is riskless, this means that the portfolio must be worth precisely 45 today also. The crucial point here is that the no-arbitrage condition enforces the prices. The portfolio agrees with 45 riskless bonds in every state of the world tomorrow, so it must have the same value as 45 riskless bonds today. That is it must be worth 45 today.

Today's share price is 100 so our portfolio is worth, $(100/2) -$ Opt, which must be equal to 45. This implies that our option is worth 5 and that 5 is the only arbitrage-free price.

3.1.2 Risk-neutral valuation

A notable point about the above argument is that probabilities appeared nowhere. The argument remains valid regardless of what the probability of an up-jump is. If the reader thinks 'Ah, but what if the probability of an up-jump is 1?' we observe that a probability of 1 would lead to a simple arbitrage opportunity – the value of the stock tomorrow cannot grow faster than the risk-free interest rate. Otherwise, one could borrow at the risk-free rate and use the money to buy a stock and

3.1 A two-world universe

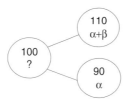

Fig. 3.2. A more general one-step, two-state model. Time goes from left to right.

achieve a certain profit. Thus in an arbitrage-free world a probability of 1 is not possible.

The only role of probabilities is thus to ensure that both world states are possible.

Suppose we did try a probabilistic approach. In particular, suppose that world A occurred with probability p, and world B occurred with probability $1-p$. The expected value of our stock tomorrow would then be

$$110p + 90(1-p)$$

and the expected value of our option would be

$$10p + 0(1-p) = 10p.$$

The only probability which gives the same answer as our arbitrage-free argument is

$$p = 1/2.$$

This gives an expected value to the stock of 100, which is the same as today's price and since we have taken interest rates to be zero, the same as a risk-free bond of value 100 today. Hence, if we wish to use a probabilistic approach, we must assume that investors are risk-neutral; that is, they do not require a premium for the riskiness of the stock over a risk-free bond. Of course, we do not actually believe the investors are risk-neutral – this is a mathematical sleight of hand to make the probabilistic approach give the correct answer.

Note the important point that in fact p will be bigger than $1/2$ because of risk-aversion, and the expected value of the option will therefore be greater than the guaranteed arbitrage-free price. The ability to hedge has removed the risk premium.

Lest we think the above argument was an artefact caused by the particular numbers chosen, or by the particular sort of option, let's consider an option (really a derivative) that pays $\alpha + \beta$ in world A and α in world B. If we sell one option today and buy δ stocks today then our portfolio is worth

$$110\delta - \alpha - \beta$$

in world state A tomorrow and

$$90\delta - \alpha$$

in world state B. In order to hedge all risk, we therefore need

$$110\delta - \beta = 90\delta.$$

This is, of course, solved by taking $\delta = \beta/20$ and our portfolio is worth $4.5\beta - \alpha$ tomorrow whichever world state occurs.

Our portfolio today is worth 5β − Opt which must equal the portfolio's value tomorrow. Solving for Opt we get

$$\text{Opt} = \alpha + 0.5\beta.$$

Let's compare this price with the risk-neutral price. As before, risk-neutral pricing means that the world states both occur with probability $1/2$. The risk-neutral expected value of the option is therefore

$$0.5(\alpha + \beta) + 0.5\alpha = \alpha + 0.5\beta,$$

which agrees with the arbitrage-free price. Note how much easier the risk-neutral argument is.

In this two-state model, we can price any option by picking α and β appropriately. We have deduced this price by showing that it is both the unique arbitrage-free price and the price implied by risk-neutral pricing.

Why does risk-neutral pricing work? Once we have picked p, and set the value of Opt to its risk-neutral value, we have, denoting the riskless bond by B, that

$$\mathbb{E}(B) = B_0, \tag{3.1}$$

$$\mathbb{E}(S) = S_0, \tag{3.2}$$

$$\mathbb{E}(\text{Opt}) = \text{Opt}_0. \tag{3.3}$$

In other words, every market instrument's value today is equal to its risk-neutral expected value tomorrow. As expectation is linear, this will be true of any combination of market instruments. i.e. if a portfolio is of zero value today, it must have expected value zero tomorrow. This means that either the portfolio will always be worth zero tomorrow, or, if it can be worth a positive amount, it is also possible for it to be worth a negative amount. If it were the case that the portfolio could be positive and never negative, the expectation would not be zero and so we know that this is not case for our portfolio, which means that it is not an arbitrage portfolio. This argument shows that if the probability of an up-move is the risk-neutral probability, and the price of the option is its expectation using that probability, then there can be no arbitrage.

3.1 A two-world universe

What if the probability of an up-move is not the risk-neutral probability? Arbitrage is still impossible with the risk-neutral price as the definition of arbitrage only uses zero and non-zero probabilities. This means that if an arbitrage exists for some value of p between 0 and 1, it exists for all values of p between 0 and 1. Conversely, if there are no arbitrages for some value of p then there are no arbitrages for all values of p.

We have therefore proved that the risk-neutral price for an option in this two-state interest-rate-free world is an arbitrage-free price. It is important to realize that the risk-neutral argument and the hedging argument are actually arguments in opposite directions. The risk-neutral argument showed that a certain price was arbitrage-free it did not prove that no other prices were arbitrage-free. The hedging argument showed that all prices except one were arbitrageable. It did not prove that the remaining price could not be arbitraged.

Thus the risk-neutral price gives a lower bound on the set of arbitrage-free prices, and the hedging arguments gives an upper bound. Importantly, in this two-state model, the upper and lower bound agree, and we are left with a unique price which is arbitrage-free.

3.1.3 Pricing by replication

There is an interesting third alternative interpretation of this price. It is the price guaranteed to cover the cost of replicating the option in all possible worlds. In the last chapter, we showed that if a portfolio dominated the option, in the sense of being worth at least as much as the option in all possible worlds at the time of expiry, then the option can be worth no more than the portfolio today. Similarly, if the portfolio is dominated by the option in all possible worlds, the option must be worth at least as much today. Thus if a portfolio is worth the same as the option in both states of the world tomorrow, it must have the same value as the option today.

In our two-state model, there will be a unique portfolio which matches the option's pay-off in the two worlds, and the price must therefore agree with that portfolio's value. To see this, observe that any combination of stock and bond, has a pay-off which is a straight line as a function of the stock price. This line will have slope equal to the number of stocks held, and its value when the stock price is zero will be the number of bonds held. There is a unique straight line through any two points, and so there will be a unique portfolio which agrees with the option's payoff in both of the possible world states.

We illustrate this point by returning to the option which paid $\alpha + \beta$ in state A and β in state B. We have sold the option, and receive a fee for doing so. Our argument above said that we must buy $\beta/20$ stocks to hedge our risk, and this will cost

$$100 \times \frac{\beta}{20} = 5\beta.$$

46   *Trees and option pricing*

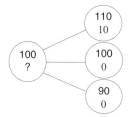

Fig. 3.3. A one-step, three-state model. Time goes from left to right.

We therefore borrow $4.5\beta - \alpha$, and combine this with the fee to purchase $\beta/20$ stocks. In state A, our portfolio consisting of the stocks and the loan, which is a liability and therefore of negative value, will be worth

$$110 \times \beta/20 - 4.5\beta + \alpha = \beta + \alpha,$$

and in state B it will be worth

$$90 \times \beta/20 - 4.5\beta + \alpha = \alpha.$$

In both states, our portfolio's final value is equal to the option's payoff. We have thus created a portfolio which precisely replicates the option's payoff and perfectly hedged our risk. The fact that the portfolio and the option agree in all world states tomorrow, thus guarantees that their prices must agree today. The price of an option is therefore the sum of money, which, by being invested appropriately today, is guaranteed to match the value of the option's payoff in all states tomorrow.

Note that one interpretation of the hedging argument is that we are using the option and stock to replicate the bond. In the replication argument, we use the stock and bond to replicate the option. Note also that one could equally well use the option and bond to replicate the stock.

3.2 A three-state model

We have found a price for an asset that takes one of two possible values in the future, and have given three different interpretations of the price thus found. However, an asset that takes precisely one of two prices tomorrow is hard to find, so our model is too primitive to be useful. How can we improve it? One naive approach might be simply to attempt the same argument with more possible world states. To illustrate the problems with this, suppose there is an additional world state 'C' where the stock takes value 100, and as before we wish to price a call option with strike 100.

If we buy γ stocks today and sell one option, then our portfolio will be worth

$$110\gamma - 10, \ 100\gamma, \ 90\gamma,$$

in world states A, C and B, respectively. As we have only one variable to play with, we cannot make these three quantities equal. In particular, if the last two are equal, then γ is zero and the first is -10, so we have not hedged at all. If we ignore C and hedge for world states A and B, then we obtain the hedge implied by the two-state model that is $\gamma = 1/2$. The portfolio is then worth 45, 50, 45, in states A, C, and B respectively. Whilst the portfolio is no longer risk-free, we do know it must be worth at least 45 tomorrow and could be worth more so it must be worth more than 45 today. This implies that

$$50 - \text{Opt} > 45.$$

The price of our option is therefore less than 5. However, this is only an upper bound not a necessary price.

The difference here is that if we set up the replicating portfolio from the two-world case, it dominates the pay-off rather than reproduces it.

Suppose we try risk-neutral valuation. In order for the expected value of the stock tomorrow to equal the price today, we must have that the probability of state A equals that of state B. Unlike before however, this can be achieved by any probability, p, between 0 and $1/2$ as the excess probability is mopped up by state C which thus has probability $1 - 2p$, which must of course be non-negative. The expected value of the option is now just $10p$ which ranges between zero and five, yielding the same bounds on the option price as the hedging argument.

Our hedging argument has shown that only prices between zero and five can be non-arbitrageable, whilst the risk-neutral argument shows that prices between zero and five are not arbitrageable. We therefore conclude that the set of arbitrage-free prices for the option is the set of prices between zero and five.

The three-world universe is an example of an *incomplete market*, that is, a market where portfolios cannot be arranged to give precisely the desired pay-off, and it is characteristic of incomplete markets that the price of an option can only be shown to lie in an interval rather than being forced to take a precise value. The market price of such an option would then be determined within the range of possible prices by the risk-preferences of traders in the market rather than mathematics.

3.3 Multiple time steps

3.3.1 More realism

At this point, option pricing is not looking very successful – clearly we will want to price options on assets that can have more than two values in the future. The solution to this problem is to catch the stock moving between the points. If we assume that stock prices move continuously, which is a reasonable though disputable assumption, we can do this.

48 *Trees and option pricing*

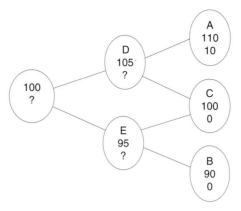

Fig. 3.4. A two-step tree. Time goes from left to right.

We previously considered an asset which went up or down 10 after one day. We can make this model finer by letting the asset go up or down 5 after each half day. The asset is therefore of price 100 today and can take prices 105 or 95 after half a day. If it takes price 105 after half a day, then it can take 110 or 100 after a day, and if it takes price 95, then it takes price 90 or 100.

We denote the states 110, 100, 90 as A, C and B respectively. The half-day states 105 and 95, we denote by D and E respectively. These are illustrated in Figure 3.4.

3.3.2 Pricing in a two-step model by hedging

To compute the value of our call option, we now compute backwards. We know the value of our call option in each of the final states as before. We price it after half a day.

If the price after half a day is 95, we are in state E and the pricing is easy, since whatever happens in the second half day the option will have zero value. Thus we conclude the price in this world state is zero.

At D, we apply the same scheme as in the one-step two-state case. We can use either risk-neutral valuation or a hedging argument. We do the hedging argument first. If we hold δ assets then the value of our assets minus the option in states A and C will be

$$110\delta - 10 \quad \text{and} \quad 100\delta$$

respectively. These two numbers are equal if and only if $\delta = 1$. The portfolio is then worth 100 in both world states at the final time, and therefore must be worth precisely 100 after half a day in state D. The portfolio is one stock minus one option; this says that

$$105 - \text{Opt}(D) = 100, \tag{3.4}$$

and hence that $\text{Opt}(D) = 5$.

3.3 Multiple time steps

On day zero, we have to hedge the possibilities D and E. In state D, the asset is worth 105 and the option is worth 5. In state E, the asset is worth 95 and the option nothing. To be hedged, we hold ϵ assets, and our portfolio will be worth

$$105\epsilon - 5 \quad \text{and} \quad 95\epsilon$$

in states D and E respectively. For these to be equal, that is for us to be hedged, we must have $\epsilon = 1/2$. The portfolio will then be worth 47.5 in both states D and E, and so must be worth 47.5 today. This means that

$$0.5 \times 100 - \text{Opt}(0) = 47.5,$$

that is $\text{Opt}(0) = 2.5$.

3.3.3 A two-step model and risk-neutral valuation

Whilst the hedging argument guarantees us a mathematically correct price, it is rather cumbersome to carry out. The risk-neutral price, whilst harder to justify, is much easier to actually use and will always agree with the hedging price. We compute as follows.

At D a risk-neutral asset will move up or down with probability $1/2$, as this is the only probability that makes the expected value 105. The option is therefore worth $(10+0)/2 = 5$ in state D. In state E, the expectation is zero whatever the probability is, and the value of the option is therefore zero.

Initially, the risk-neutral probabilities must be $1/2$ again to get the expectation to be 100. We can therefore deduce, using risk-neutral valuation again, that the initial value of the call option must be

$$\frac{1}{2}(0+5) = 2.5,$$

which, of course, agrees with the hedging price.

Note that an alternative way to compute this price is first to work out the risk-neutral probability of the stock arriving at each final node, and then compute the expectation of the final pay-off using these probabilities. In particular, the probability of attaining state A is $1/4$, as it requires two up-moves. The expectation of the final pay-off is therefore

$$10/4 = 2.5,$$

which of course agrees with the arbitrage-free price above.

We have demonstrated that it is possible to hedge an option precisely in a two time-step model. This hedging guarantees a unique arbitrage-free price for the option, and this price agrees with the price obtained by assuming that investors are risk-neutral. The prices are illustrated in Figure 3.5.

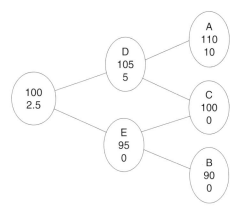

Fig. 3.5. A two-step tree with the option values filled in. Time goes from left to right.

The only important qualitative feature of the above argument was that at each time-step the asset could only move to two possible new values. This meant that by using only the asset, we could hedge the option totally. The number of time-steps was not important: one could do as many time-steps as one liked, provided the feature of having only two immediately succeeding states was retained.

Note that the third interpretation of the price is still valid in the two time-step model. The price of the option is the amount of money we need to invest today in order to match the pay-off of the option whatever happens. At each point, we hold the amount of assets suggested by the hedging and keep the rest (which is possibly negative) in riskless bonds, that is, cash in our interest-rate-free world. As the amount of assets held to hedge varies with the step, the investment is dynamic – the number of assets held is a function of time and asset price.

3.4 Many time steps

There is nothing magical about two steps. As long as each node has precisely two daughter nodes at the next step, the same arguments will work, no matter how many steps there are. The general structure of the set of nodes is called a *tree*. As in the two-step case, one always starts at the final time, where the value of the option is just its payoff, and work backwards so that when computing at a node, one always knows the value of the option at both the daughter nodes. Thus all our arguments work and there is a unique arbitrage-free price for the option. As before, we can choose how to compute. In particular, we can iterate back through the tree computing the value of the option in each of the nodes in the second last layer, and then the third last and so on. Alternatively, we can string the risk-neutral probabilities together to get the probability that the spot lands at each node in the

3.4 Many time steps 51

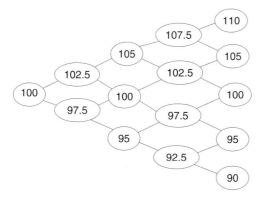

Fig. 3.6. A four-step tree. Time goes from left to right.

final layer and then take an expectation against them. In Figure 3.6, we give an example of a tree with four steps.

If we are in a zero-interest-rate world, then at each node in the tree the risk-neutral probabilities will, as in the one-step world, be 0.5 in order to ensure that the expectation at the next time-step is equal to the current value. This means that we can easily compute the probability of spot landing at each of the final nodes.

To land at 110, we must have 4 up-moves so the probability is

$$\frac{1}{2^4} = \frac{1}{16}.$$

To land at 105, we must have exactly 3 up-moves and one down-move (in any order) so the probability is

$$\binom{4}{1}\frac{1}{2^4} = \frac{1}{4}.$$

To land at 100, we must have exactly 2 up-moves and 2 down-moves (in any order) so the probability is

$$\binom{4}{2}\frac{1}{2^4} = \frac{3}{8}.$$

To land at 95, we must have exactly 1 up-move and 3 down-moves (in any order) so the probability is

$$\binom{4}{1}\frac{1}{2^4} = \frac{1}{4}.$$

To land at 90, we must have exactly 0 up-moves and 4 down-moves so the

probability is
$$\frac{1}{2^4} = \frac{1}{16}.$$

The value of a call option is then its expectation with these probabilities. Thus a call option struck at 100 is worth

$$(110 - 100)\frac{1}{16} + (105 - 100)\frac{1}{4} = \frac{10}{16} + \frac{5}{4} = \frac{15}{8}. \tag{3.5}$$

Whilst a call option struck at 95 is worth

$$(110 - 95)\frac{1}{16} + (105 - 95)\frac{1}{4} + (100 - 95)\frac{3}{8} = \frac{85}{16}. \tag{3.6}$$

More generally, suppose spot is S, we have N steps and at each step the spot goes up or down by x. Let S_j be the value of spot after j up-moves and $N - j$ down-moves. Thus

$$S_j = S + jx - (N - j)x = S + (2j - N)x.$$

The probability in the final layer that spot has value S_j is equal to

$$\binom{N}{j} \frac{1}{2^N}$$

as there are $\binom{N}{j}$ ways to reach that node. The value of a derivative that pays $f(S)$ after N steps will be

$$\sum_{j=0}^{N} \binom{N}{j} \frac{1}{2^N} f(S + (2j - N)x),$$

as it is the sum of the value of the derivative in each node times the probability that that node is achieved. We want to understand what happens as N goes to infinity. To do so, we will have to let the trees change with N in such a way that some properties are preserved.

3.5 A normal model

By creating a tree with more and more steps, that is by taking smaller and smaller time-steps, we can get finer and finer gradations at the final stage and thus hopefully a more accurate price. However, we have to be a little careful about how we do this in order to get the prices to converge to a meaningful value. Which limiting price we obtain will depend on how we make the trees finer – this essentially comes down to assumptions we make about the random process the asset price follows.

3.5 A normal model

As a warm-up, we consider how, for a very simple model, a limiting price can be obtained. In particular, we take a model in which the expected price at expiry is equal to today's price, interest rates are zero, and in every time-step the asset moves up or down by the same amount with equal probabilities. Let's suppose that the asset's value today is S_0, the exercise time of the option is T, the mean value of the asset at time T is S_0, and the variance of the asset price at time T is $\sigma^2 T$.

If we divide the time interval from time 0 to time T into k equal steps, then at each step we will want a mean change of 0 and a variance of $\sigma^2 T/k$, since for independent random variables the mean and variance just add.

At each step, we therefore have that the asset moves up or down by

$$\sigma_k = \sigma \sqrt{\frac{T}{k}},$$

with probability 0.5. Note that this gives the correct variance for each step.

As we have chosen a model in which the asset's average growth rate is zero and interest rates are zero, the risk-neutral probabilities are equal to the real-world probabilities, that is, 0.5. Let Z_j denote a sequence of independent random variables which take the values 1 and -1 each with probability 0.5. After k steps the asset will be distributed as

$$S_0 + \sum_{l=1}^{k} \sigma_k Z_l.$$

For any fixed value of k, we can apply risk-neutral valuation (or equivalently the hedging argument) to obtain the unique arbitrage-free price which is implied by this k-step process. We have done this by backward propagation through the tree; we could equally well string the probabilities together in a forward direction to obtain the probability of ending at a given final value at maturity, and then take the expectation of the option value against this resultant risk-neutral probability density. For k steps, we would then obtain for the price of an option that pays $f(S_T)$ at time T the expression

$$\mathbb{E}\left(f\left(S_0 + \sigma_k \sum_{l=1}^{k} Z_l \right) \right).$$

We want to understand what happens to this expression as we let k tend to infinity.

As k increases, our tree becomes finer and finer but the variance of the expression $\sigma_k \sum_{l=1}^{k} Z_l$ remains equal to $\sigma^2 T$ and it retains the mean 0.

Recall the Central Limit theorem (see for example [53] Section 5.10 or Appendix C)

Theorem 3.1 *Let $X_1, X_2, \ldots$, be a sequence of independent identically-distributed random variables with finite means μ and finite non-zero variances σ^2, and let*

$$S_n = X_1 + X_2 + \cdots + X_n.$$

Then the distribution of

$$\frac{S_n - n\mu}{\sqrt{n\sigma^2}}$$

converges to that of a Gaussian random variable of mean 0 and variance 1 as n tends to infinity.

In our case this means that

$$\frac{1}{\sqrt{k}} \sum_{l=1}^{k} Z_l$$

converges to a Gaussian distribution of mean 0 and variance 1. We denote such a Gaussian distribution by $N(0, 1)$.

Thus the distribution of the asset price converges to that of

$$S_0 + \sigma \sqrt{T} N(0, 1),$$

and the price of the option converges to

$$\mathbb{E}(f(S_0 + \sigma \sqrt{T} N(0, 1))) = \frac{1}{\sqrt{2\pi}} \int f(S_0 + \sigma \sqrt{T} x) e^{-\frac{x^2}{2}} dx. \quad (3.7)$$

For certain pay-offs, such as that of a call option, this integral can be evaluated explicitly.

Unfortunately, our simple model is a little too simple. If the modelled asset is a stock then we know that the price can never be negative. However, having the final price distributed as a normal distribution implies a positive probability of negative value which we know is impossible. Also, the absolute movements in price of a stock depend on its value. For example, for a stock with price 1000, a movement of 30 is minor whereas for a stock of value 100 it is large, and for a stock of value 30, it is huge. Furthermore, if a company's shares are valued at 1000 and the company decides to do a ten-for-one split, that is it replaces each share by ten new ones, then each new share will be worth 100, and we expect the new shares to move a tenth as much as the old ones.

It is therefore better to think in terms of percentage movements; so rather than our stock moving up or down 30, we instead let it move up or down 3%. One easy way to do this is to work with the log of the share price instead of with the price itself. As the log of a product is the sum of the logs, we can model percentage changes by adding terms to the logs. As the exponential function is inverse to the log function, and the exponential of any number, positive or negative, is positive, modelling changes to the log function guarantees that the share price remains positive. We can apply the same tree methodology with some slight modifications to obtain a process for the log of the stock price and a risk-neutral price for the option. The limiting distribution for the stock price in both the unadjusted and risk-neutral processes will then be log-normal; that is, the log follows a normal process. We will also have to take into account the fact that the real-world behaviour will encompass a risk premium and so the real-world behaviour will be different from the risk-neutral behaviour. In particular, the mean real-world value will not be today's value even when there are no interest rates.

3.6 Putting interest rates in

Before proceeding to the study of the limiting case for an asset whose log follows a normal process, we look at how to put interest rates back in. Suppose the continuous compounding interest rate is r. At time zero, we then have that the riskless bond is worth 1 and at time Δt it is worth $e^{r\Delta t}$.

Our stock is worth S_0 today and either S_+ or S_-, with $S_- < S_+$, at time Δt. We must have that

$$S_- < S_0 e^{r\Delta t} < S_+. \tag{3.8}$$

Otherwise, we can construct an arbitrage just by considering the portfolio consisting of the difference of the two assets.

As in the interest-rate free world, given a derivative that pays $f(S)$ at time Δt, we can construct a portfolio which precisely replicates it by considering a multiple of the stock and a multiple of the bond. However, rather than repeating that argument we look at the risk-neutral valuation approach.

We need to find the probability that makes the stock grow on average at the risk-free rate. In other words, we must find p such that

$$\mathbb{E}(S_{\Delta t}) = pS_+ + (1-p)S_- = S_0 e^{r\Delta t}, \tag{3.9}$$

or

$$p(S_+ - S_-) = S_0 e^{r\Delta t} - S_-. \tag{3.10}$$

We thus deduce that

$$p = \frac{S_0 e^{r\Delta t} - S_-}{S_+ - S_-}. \tag{3.11}$$

It follows from (3.8) that p lies strictly between zero and one.

We previously justified risk-neutral valuation by saying that the expectation value of every possible portfolio was equal to today's value, so no portfolio could be an arbitrage portfolio; a portfolio of zero value today with possible positive value tomorrow and no possibility of negative value would not have zero expectation.

Let $\mathbb{E}_{\text{RN}}$ denote expectation with the risk-neutral probability p. The bond no longer satisfies

$$\mathbb{E}_{\text{RN}}(B_{\Delta t}) = B_0$$

as the left side equals $e^{r\Delta t}$ and right side is equal to 1. It is therefore not possible to make the same argument.

We can rescue the argument by thinking in terms of discounted prices. Instead of requiring that every portfolio should have expectation equal to today's value, we require that its expectation should be equal to the asset's value invested at the risk-free growth rate, or equivalently that its discounted expectation is equal to today's value. We thus want

$$\mathbb{E}_{\text{RN}}\left(\frac{A_{\Delta t}}{B_{\Delta t}}\right) = \frac{A_0}{B_0}, \tag{3.12}$$

for every asset. This equation is trivially satisfied for the bond and we have chosen the risk-neutral probability so that it is satisfied by construction for the stock. This leaves us with the option we wish to price. We define Opt_0 to satisfy (3.12):

$$\text{Opt}_0 = \mathbb{E}_{\text{RN}}\left(\frac{A_{\Delta t}}{B_{\Delta t}}\right) = e^{-r\Delta t}\mathbb{E}_{\text{RN}}(f(S)), \tag{3.13}$$

where f is the option's pay-off.

If we now set up a portfolio which contains multiples of the stock, option and bond with initial value zero, then the expected value of its ratio with the bond, using the risk-neutral probability, at time Δt will be zero too. As the bond's value does not depend on the value of the asset, this means that the portfolio's risk-neutral expectation must be zero also. This again implies that it cannot be an arbitrage portfolio by the same argument. If it could be positive but never negative then its risk-neutral expectation could not be zero and so it would not be of zero value today.

In conclusion, we can find an arbitrage-free price in a one-step two-state world with interest rates by setting

$$\text{Opt}_0 = e^{-r\Delta t}(pf(S_+) + (1-p)f(S_-)),$$

with p given by (3.11).

Whilst we have arrived at this price via risk-neutral valuation, we could equally proceed via hedging or replication arguments. In a many-step tree, hedging and replication arguments require rebalancing the portfolio at every time-step so as we let the number of steps go to infinity, we will have to rebalance the portfolio

3.7 A log-normal model

3.7.1 The real world behaviour

Taking the results of the last two sections into account we now study a model in which the log of the asset price moves up and down by fixed increments instead of one in which the asset price does. We do so partially as an overture for the more technical results of Chapters 5 and 6 which deduce the same final price for a call option by using quite different techniques.

Thus suppose that we want to know the price of an option that pays off at time T and we divide time into N steps. We want to keep the mean and variance of the log of the stock price at time T fixed as we vary N. Thus suppose we take the mean change of the log at time T to be μT and the variance of the log to be $\sigma^2 T$. In each small time-step of length $\Delta t = T/N$, this means we want mean $\mu \Delta t$ and variance $\sigma^2 \Delta t$ since mean and variance add for independent variables. We shall call σ the *volatility* of the asset as it reflects how much the asset wobbles up and down. We call μ the *drift* as it expresses how much the asset drifts upwards. We will expect μ to be bigger than r, since the stock should grow more quickly than a riskless bond in compensation for its riskiness. We can say that the stock carries a *risk premium*.

We therefore take

$$\log S_{j\Delta t} = \log S_{(j-1)\Delta t} + \mu \Delta t + \sigma \sqrt{\Delta t} Z_j, \qquad (3.14)$$

where Z_j takes the values 1 and -1 each with probability $1/2$. It is easy to check that the mean and variance of Z_j are 0 and 1 respectively. This immediately implies that the mean and variance of the change in $\log S$ across the time step are $\mu \Delta t$ and $\sigma^2 \Delta t$ as desired. Adding the terms for each j together, we have for a given value of N that

$$\log S_T = \log S_0 + \mu T + \sigma \sqrt{\Delta t} \sum_{j=0}^{N-1} Z_j. \qquad (3.15)$$

We can rewrite the final term as

$$\sigma \sqrt{T} \frac{1}{\sqrt{N}} \sum_{j=0}^{N-1} Z_j.$$

It follows from the Central Limit theorem, just as in our normal model, that as N goes to infinity the random variable

$$\frac{1}{\sqrt{N}} \sum_{j=0}^{N-1} Z_j,$$

becomes distributed like a normal random variable of mean 0 and variance 1. Thus as N tends to infinity, $\log S_T$ becomes distributed like

$$\log S_0 + \mu T + \sigma \sqrt{T} N(0, 1),$$

and so the distribution of S_T is that of

$$S_0 e^{\mu T + \sigma \sqrt{T} N(0,1)}.$$

We shall then say that S_T is *log-normally* distributed, as its log is normally distributed.

3.7.2 The risk-neutral world behaviour

For pricing options, the real-world distribution of the asset is not so important. We already saw that the probability of up- and down-moves does not affect the price, and instead it is the distribution of the asset under risk-neutral probabilities that matters. Thus to price options we need to find out what the final distribution of the asset price is if we use the risk-neutral probabilities at each step. This means that we need to understand what the probability of up and down moves at each step are, and understand how they behave as N tends to infinity.

We can use (3.11) to compute the probabilities across each step. We know that $S_{j\Delta t}$ is given by

$$S_{j\Delta t} = S_{(j-1)\Delta t} e^{\mu \Delta t \pm \sigma \sqrt{\Delta t}}.$$

Using (3.11), we have that p_j, the risk-neutral probability at step j, is given by

$$\frac{S_{j-1} e^{r\Delta t} - S_{j-1} e^{\mu \Delta t - \sigma \sqrt{\Delta t}}}{S_{j-1} e^{\mu \Delta t + \sigma \sqrt{\Delta t}} - S_{j-1} e^{\mu \Delta t - \sigma \sqrt{\Delta t}}},$$

where we have denoted $S_{(j-1)\Delta t}$ by S_{j-1}. We cancel through to remove S_{j-1}. We deduce that the probability p_j is independent of j and is equal to

$$p = \frac{e^{r\Delta t} - e^{\mu \Delta t - \sigma \sqrt{\Delta t}}}{e^{\mu \Delta t + \sigma \sqrt{\Delta t}} - e^{\mu \Delta t - \sigma \sqrt{\Delta t}}}. \tag{3.16}$$

We are ultimately interested in what happens as the number of steps gets large, or

3.7 A log-normal model

equivalently what happens as the step-size goes to zero. If we cancel through by $e^{\mu \Delta t}$ we obtain

$$p = \frac{e^{(r-\mu)\Delta t} - e^{-\sigma\sqrt{\Delta t}}}{e^{\sigma\sqrt{\Delta t}} - e^{-\sigma\sqrt{\Delta t}}}. \qquad (3.17)$$

For the denominator, expanding e^x in powers of x and observing that the odd powers cancel, we have

$$e^{\sigma\sqrt{\Delta t}} - e^{-\sigma\sqrt{\Delta t}} = 2\sigma\sqrt{\Delta t} + \mathcal{O}(\Delta t^{3/2}). \qquad (3.18)$$

Where a quantity is $\mathcal{O}(t^\alpha)$, if it is less than or equal Ct^α. This implies that

$$\frac{1}{e^{\sigma\sqrt{\Delta t}} - e^{-\sigma\sqrt{\Delta t}}} = \frac{1}{2\sigma}\Delta t^{-1/2} + \mathcal{O}(\Delta t^{1/2}). \qquad (3.19)$$

Expanding the numerator in (3.17) we get

$$e^{(r-\mu)\Delta t} - e^{-\sigma\sqrt{\Delta t}} = 1 + (r-\mu)\Delta t + \mathcal{O}(\Delta t^2) - 1 + \sigma\sqrt{\Delta t} - \frac{\sigma^2}{2}\Delta t + \mathcal{O}(\Delta t^{3/2}). \qquad (3.20)$$

We deduce that

$$p = \frac{1}{2}\left(1 + \left(\frac{r - \mu - \frac{1}{2}\sigma^2}{\sigma}\right)\Delta t^{1/2}\right) + \mathcal{O}(\Delta t). \qquad (3.21)$$

The probabilities of up and down moves are no longer $\frac{1}{2}$ but instead are adjusted to take account of the difference between the mean rate of growth of the stock and the growth rate of the riskless bond. However, even if they are both zero we still get extra terms. These arise from the fact that

$$\frac{1}{2}(e^{\sigma\sqrt{\Delta t}} + e^{-\sigma\sqrt{\Delta t}}) \neq 1;$$

and we therefore need a probability adjustment to make the expectation of the stock value equal to the bond's future value.

We want to study the behaviour as the number of steps, N, goes to infinity; just as we did with the real-world probabilities, we can write

$$\log S_T = \log S_0 + \mu T + \sigma \sqrt{\frac{T}{N}} \sum_{j=1}^{N} \tilde{Z}_j, \qquad (3.22)$$

where now $\tilde{Z}_j$ denotes a random variable taking the value 1 with probability p and the value -1 otherwise. Life is a little more complicated now as the definition of the random variable $\tilde{Z}_j$ depends on the probability p, which depends on the step-size and hence on N.

What properties does $\tilde{Z}_j$ have? Its mean is no longer zero but is equal to

$$\nu = \left(\frac{r - \mu - \frac{1}{2}\sigma^2}{\sigma}\right)\sqrt{\Delta t} + \mathcal{O}(\Delta t)$$

instead. It immediately follows the mean value of $\log S_T$ is

$$\log S_0 + \left(r - \frac{1}{2}\sigma^2\right)T + \mathcal{O}(N^{-1/2}),$$

using the fact that $\Delta t = T/N$. We therefore have that the mean value of $\log S_T$ will converge to

$$\log S_0 + \left(r - \frac{1}{2}\sigma^2\right)T,$$

as N goes to infinity.

The variance of $\tilde{Z}_j$ is on the other hand equal to

$$1 - \nu^2 = 1 - \left(\frac{r - \mu - \frac{1}{2}\sigma^2}{\sigma}\right)^2 \Delta t + \mathcal{O}(\Delta t^{3/2}). \tag{3.23}$$

Recalling that $\Delta t = T/N$, we find that $\log S_T$ has variance equal to $\sigma^2 T + \mathcal{O}(N^{-1/2})$.

Thus as N goes to infinity, $\log S_T$ converges to a distribution with the same mean and variance as

$$\log S_0 + \left(r - \frac{1}{2}\sigma^2\right)T + \sigma\sqrt{T}N(0,1),$$

with $N(0, 1)$ a normal distribution with mean 0 and variance 1. In fact, by applying a suitably modified version of the Central Limit theorem one can prove that this actually is the limiting distribution and hence not only is the real-world distribution of S_T log-normal but the risk-neutral distribution too. However, the risk-neutral distribution has a shifted mean to take account of the absence of risk premia. We do not carry out the technical details as they are not particularly illuminating and we will undertake other more rigorous arguments in Chapters 5 and 6.

In fact, it is easy to show (exercise!) that

$$\mathbb{E}\left(e^{\sigma\sqrt{T}N(0,1)}\right) = e^{\frac{1}{2}\sigma^2 T}, \tag{3.24}$$

3.7 A log-normal model

which implies that

$$\mathbb{E}(S_T) = \mathbb{E}\left(S_0 e^{(r-\frac{1}{2}\sigma^2)T + \sigma\sqrt{T}N(0,1)}\right) = S_0 e^{rT}.$$

Note that the real-world drift of the stock μ has disappeared. The drift plays no part in derivatives pricing. Our pay-off is the expected value of the option's pay-off in a risk-neutral world in which risk premia play no part.

Since we know that the option price is just the risk-neutral expectation of the option pay-off suitably discounted, we can now value any option by integrating its pay-off function against the log-normal distribution. Thus for a call option with expiry T and strike K, we obtain

$$e^{-rT}\mathbb{E}\left(\left(S_0 e^{(r-\frac{1}{2}\sigma^2)T + \sigma\sqrt{T}N(0,1)} - K\right)_+\right).$$

With a little effort (see Section 6.8 for the details) this leads to the famous Black–Scholes formula,

$$C(S, K, \sigma, r, T) = SN(d_1) - Ke^{-rT}N(d_2) \qquad (3.25)$$

where

$$d_j = \frac{\log\left(\frac{S}{K}\right) + \left(r + (-1)^{j-1}\frac{1}{2}\sigma^2\right)T}{\sigma\sqrt{T}}, \qquad (3.26)$$

and N denotes the cumulative normal function, that is,

$$N(x) = \frac{1}{\sqrt{2\pi}} \int_{-\infty}^{x} e^{-\frac{1}{2}s^2} ds. \qquad (3.27)$$

There is a similar formula for put options which can be deduced immediately from the call option formula by using put-call parity:

$$P(S, K, \sigma, r, T) = -SN(-d_1) + Ke^{-rT}N(-d_2). \qquad (3.28)$$

One could also derive this formula directly by using a similar argument.

As the Black–Scholes formula is a little opaque we plot the value of a call option as a function of volatility in Figures 3.7, 3.8 and 3.9. Since the call option price must obey the rational bounds of the last chapter, no matter how low the volatility goes, the option is worth more than the intrinsic value, which explains the shape of the graph in Figure 3.9.

The remarkable fact about the at-the-money call option is that it is an almost linear function of volatility, and very neatly expresses the market's view on volatility. In fact, we can derive a simple approximation for at-the-money options. (Here we

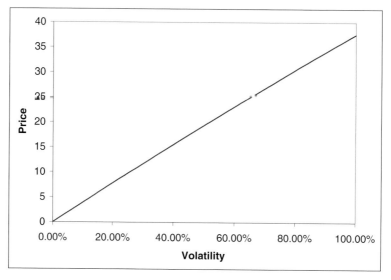

Fig. 3.7. The value of a one-year at-the-money call option as a function of volatility. Note how linear this graph is.

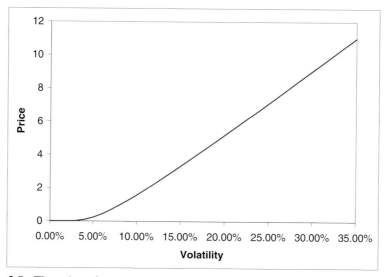

Fig. 3.8. The value of a one-year out-of-the-money call option as a function of volatility.

use at-the-money to mean that the strike of the contract is the forward price of the stock.) Thus if $K = Se^{rT}$, the call option is worth

$$S\left(N\left(\frac{1}{2}\sigma\sqrt{T}\right) - N\left(-\frac{1}{2}\sigma\sqrt{T}\right)\right).$$

3.7 A log-normal model

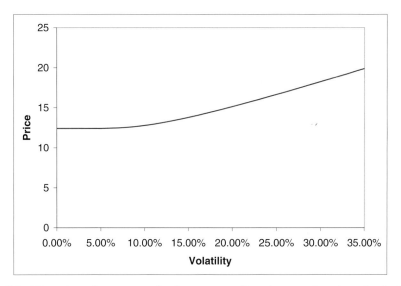

Fig. 3.9. The value of a one-year in-the-money call option as a function of volatility.

Now by Taylor's theorem, we have

$$N(x) = N(0) + N'(0)x + \frac{x^2}{2}N''(0) + \mathcal{O}(x^3). \tag{3.29}$$

Substituting and observing that even terms cancel, we see that the call option is worth

$$S\left(N'(0)\sigma\sqrt{T} + \mathcal{O}\left(\sigma^3 T^{\frac{3}{2}}\right)\right).$$

As $N'(0) = \frac{1}{\sqrt{2\pi}}$, this means that if $\sigma\sqrt{T}$ is small, we have the approximation

$$\frac{S\sigma\sqrt{T}}{\sqrt{2\pi}}.$$

For mental calculation, we use

$$0.4 S\sigma\sqrt{T}.$$

Note that as we are at-the-money, the formula applies equally to puts as well as calls, by put-call parity.

Example 3.1 If spot is 100, volatility is 10%, strike is 100, expiry is three months and there are no interest rates, price a call option which expires in three months. The approximation gives

$$0.4 \times 100 \times 0.1 \times 0.5 = 2.$$

Using instead the Black–Scholes formula, we obtain

$$1.995.$$

The error is less than 1%.

We have deduced an arbitrage-free price for a call option via risk-neutral valuation. We will study in detail a different approach using a hedging argument in Chapter 5. We know that for trees, hedging and replication arguments both yield the same price as risk-neutral evaluation so we can expect the same thing to happen in the limit. What hedge will we need to hold? One method of finding the hedge is to compute it for a given value of Δt and then let Δt go to zero. However, it is easier if we just think about what the hedge is supposed to achieve.

When we use a hedging argument, we replicate the riskless bond by holding a mixture of stock and option. We therefore wish our portfolio value to be immune to small changes in the stock value in the limit. If we are short one call option and long Δ stocks then our portfolio is worth

$$-C(S, t) + \Delta S.$$

For a fixed value of Δ the rate of change of this with respect to S is

$$-\frac{\partial C}{\partial S}(S, t) + \Delta.$$

The only value of Δ which will make the rate of change zero is therefore

$$\Delta = \frac{\partial C}{\partial S}(S, t).$$

Thus we should always hedge by holding Δ stocks. This process is called *Delta hedging*. As Δ depends on S and t, this certainly means that the hedge will need to change continuously.

3.8 Consequences

Whilst we have used the tree methodology to deduce the Black–Scholes formula, we should note that other pricing methods are a natural consequence of our arguments. We sketch these briefly here as a foretaste of later chapters. The first is simply that rather than trying to pass to the limit in our trees, we can simply pick a sufficiently fine tree and then apply the argument above to compute the option's value. Whilst there is little point for a vanilla European call in doing this, this argument will work for any pay-off function including ones for which we cannot solve the integral analytically – though we could always compute it numerically. More generally, there are various sorts of exotic options which can be tackled by tree methods. Recall that an American option is an option that can be exercised at any point up to its time of maturity, i.e. it can be exercised early. Since it carries all the same rights as a European option, it must clearly be worth as much. In general, not surprisingly it is worth more, though not if it is a call on a non-dividend-paying stock. To value such an option using a tree, we can work backwards as before, the only difference being that at each node we have two different methods of valuation. The first is the arbitrage-free method outlined above, which corresponds to not exercising the option. The other is the intrinsic value obtained by exercising at that time – that is, just the difference between the strike and share price. As we can assume that our investor will maximize his assets, we take the maximum of the two. Working backwards, we can compute the price all the way back to the start as before. Note that the arbitrage-free price computed at each node takes into account not just the intrinsic value of exercising at that time but also the possible intrinsic value obtainable by exercising at any future time.

Another sort of option we can value with a tree is a *knock-out* or *barrier* option. This is an option that has a pay-off at some fixed time in the future, unless the share price drops below a given price, the barrier, at any time. We can then adapt our tree by setting the value at all nodes corresponding to prices below the knock-out barrier to zero and then compute in the usual arbitrage-free manner. In practice, we might want to adapt our trees slightly to ensure numerical stability at the barrier.

In applying these tree methods to price an option, we want to have some idea of how accurate our price actually is. Typically some idea can be achieved by evaluating the price for various numbers of time steps and seeing how the price changes. When increasing the number of steps no longer affects the price significantly, we can regard the tree as converged and take that price. Alternatively, we can use another pricing method and compare prices. If they agree, then the prices are probably correct.

A second method we can use is *Monte Carlo simulation*. The price of a European option is just the expectation of the option's pay-off under the risk-neutral log-normal distribution. We can therefore value the option simply by repeatedly drawing a share price from the risk-neutral log-normal distribution and averaging the resultant option pay-offs. The law of large numbers, see Appendix C, tells us that this will eventually converge to the correct price. The worth of this technique is not so much to value ordinary European options, but to price exotic options where alternative methods are not obviously applicable. We would then have to simulate the entire path, approximating it in little time-steps and then computing the final value of the option along the path. Note however that if the option involves some choice on the part of the holder, there is still an issue of deciding what decision the holder would make – for example, in the case of an American option, when should the holder exercise.

A third method is to observe that the Black–Scholes price satisfies a certain partial differential equation known as the Black–Scholes equation. Let $C(S, t)$ denote the price at time t of an option struck at K with expiry T when the stock price is S; then straightforward differentiation shows that C satisfies

$$\frac{\partial C}{\partial t} + rS\frac{\partial C}{\partial S} + \frac{1}{2}\sigma^2 S^2 \frac{\partial^2 C}{\partial S^2} = rC \tag{3.30}$$

with the boundary condition that C_T is equal to the pay-off of the call option. As this equation is linear and holds for any call option, it must hold for any linear combination of call options. Any option pay-off can be approximated arbitrarily well (in a certain sense) by call option pay-offs so it follows that the price of any option with pay-off depending on spot at time T can be priced by solving the Black–Scholes equation with appropriate boundary conditions.

3.9 Summary

In this chapter, we started with a simple model of stock price evolution in which an asset moves up or down by a fixed amount only; we showed in this simple case that the principle of no arbitrage guaranteed a single unique price for an option which did not in any way depend on the probability that the stock price took a particular value. We saw that there were multiple ways to arrive at the arbitrage-free price including hedging, replication and risk-neutral evaluation.

This model was extended to be more accurate by stringing together copies of the one-step model, and these extended models still resulted in single prices which were guaranteed by no-arbitrage arguments.

We then went on to construct a continuous-time model by letting the size of a time-step converge to zero; this resulted in a final risk-neutral distribution for the stock price which did not depend on the original mean of the stock price. The price of a call option was then obtained as a discounted expectation of the call option's pay-off using this distribution. This option price did not depend in any way on the mean value of the stock price but instead the mean of the risk-neutral distribution was precisely the future value of a risk-free bond with the same initial value.

The price on a tree of any number of steps was enforced by the principle of no-arbitrage which applied because at each step and point in the tree, the option could be hedged in such a way to ensure that any other option value would lead to an arbitrage. The price therefore depended on a rebalancing of the portfolio at every time-step. Thus in the limit, the option price was enforced by continuously trading in the underlying asset.

3.10 Key points

- For a one-step tree with two branches, every option has a unique price.
- The probability of an up-move does not affect the price of an option in a one-step tree.
- A one-step three-branch tree does not lead to a unique price for an option
- A no-arbitrage price for an option can be found by hedging, replication and risk-neutral evaluation.
- Replication and hedging arguments show that certain prices are not arbitrage-free, but do not guarantee that the remaining prices are arbitrage-free.
- Risk-neutral valuation arguments show that certain prices are arbitrage-free but do not show that no other prices are arbitrage-free.
- Any tree in which each node has two daughter nodes leads to a unique price for an option.
- The price of an option in a multi-step tree can be found by stringing the probabilities forward to find the probability of attaining each node in the final layer.
- The price of an option in a multi-step tree can also be found by backwards iteration, i.e. by computing the option price at each node in each layer by starting with the last layer and iterating backwards.
- By fixing the mean and variance over a time interval, and letting the step-size go to zero, we can deduce a model in which the final distribution of the stock is log-normal.
- When pricing options with the limiting process obtained from trees, the real-world drift of the stock plays no role.
- The Black–Scholes price of a call option can be obtained by taking the discounted expected pay-off in a risk-neutral world.

3.11 Further reading

The approach we have developed here is due to Cox, Ross & Rubinstein, [36], and was not the method originally used by Black & Scholes to deduce their famous formula. We will study the method originally used by Black & Scholes in Chapter 5.

A good account of the tree approach covering similar ground can be found in Baxter & Rennie, [12].

68   *Trees and option pricing*

We return to the tree approach at a more theoretical level in Chapter 6. We also return to the practical aspects of pricing on trees at various points in the book, in particular we look at how trees are used in practice in Chapter 7.

3.12 Exercises

Exercise 3.1 Assets A and B are worth 100 today. Asset A will be worth 110 tomorrow with probability 0.9 and 90 otherwise. Asset B will be worth 110 with probability 0.5 and 90 otherwise. Asset C is worth 1 both today and tomorrow. How will the prices of call options on A and B struck at 100 compare?

Exercise 3.2 A stock is worth 200 today and either 190 or 220 tomorrow. There are no interest rates. Price call options struck at 190, 200 and 220.

Exercise 3.3 A stock is worth 100 today. There are no interest rates. It will be worth one of 90, 100 and 110 tomorrow. If the call option stuck at 100 is worth 2, give optimal no-arbitrage bounds on a call option struck at 105.

Exercise 3.4 A stock is worth 100 today. There are no interest rates. It will be worth one of 85, 95, 105 and 115 tomorrow. Give optimal no-arbitrage bounds on a call option struck at 100. If the call option struck at 100 is worth 5, give optimal no-arbitrage bounds on a call option struck at 110.

Exercise 3.5 There are no interest rates. An asset is worth zero today and goes up or down by 1 each day. Find the price of a call option struck at zero as a function of the number of steps to expiry.

Exercise 3.6 A stock is worth 100. Each month its value increases or decreases by precisely 10. The riskless bond is worth e^{rt} at time t years with r equal to 5%. Price a four-month European put option struck at 110. Do the American case too.

Exercise 3.7 A stock is worth 50 today. Interest rates are zero. It is worth 40 or 70 tomorrow. What risk-neutral probabilities should be used to price an option?

Exercise 3.8 A stock is worth 50 today. Interest rates are zero. It is worth 40, 55 or 70 tomorrow. What are the possible risk-neutral probabilities?

Exercise 3.9 Suppose A is worth 100 today and worth 90 or 110 tomorrow. Asset B is worth 100 today and worth 80 or 120 tomorrow. Asset B is worth 80 if and only if A is worth 90. There is no riskless bond. Price a call option on A struck at 100.

3.12 Exercises

Exercise 3.10 For a log-normal Black–Scholes model with spot equals 100, volatility 10%, and interest rates 5%, price a call option struck at 100 with a one-year expiry, using the Black–Scholes formula.

Exercise 3.11 Prove that the price of an American option implied by a tree will always be as much as the price of a European option with the same parameters priced on the same tree.

Exercise 3.12 Prove that the price of a barrier option implied by a tree will always be less than the price of a vanilla option with the same parameters priced on the same tree.

Exercise 3.13 Show that
$$\mathbb{E}\left(e^{\sigma N(0,1)}\right) = e^{\frac{1}{2}\sigma^2}.$$

4

Practicalities

4.1 Introduction

In the last chapter, we developed a model for stock price movements and used this to deduce a unique necessary arbitrage-free price. One might think this was the end of the story for vanilla call and put options – what else is there to say? Note that one consequence of the last chapter's arguments is that by investing the cost of an option, one can reproduce precisely the value of the option at pay-off. Given that this is true, why bother to buy options at all? Instead of purchasing the option, why not just carry out this dynamic replication strategy? In this chapter, we attempt to answer these questions and look at the practicalities of option hedging.

The first thing to observe is that even if our model holds for a bank with its easy access to the markets, it does not follow that it will hold for a general individual, and so one could view buying an option as outsourcing the hedging strategy to an institution better suited to carrying it out. There is also an economy of scale – hedging many options together is not much harder than hedging one as all the Delta hedges will add together and possibly even cancel each other.

4.2 Trading volatility

More fundamentally, it is important to realize that no model is perfect and it will never describe the market perfectly. Where are the imperfections in our model? Given certain parameters, our model produces a price. These parameters are strike, time to maturity, interest rates, current stock price and volatility. One of these parameters is vastly different from the others: all except volatility are either specified by the option contract or are observable in the market. A consequence of this is that if you call a trader and ask for a price on an option, he will not quote you a sum of money, instead he will quote a volatility, or *vol*, as traders typically say. Actually he will quote two vols, the price to buy and the price to sell. The vol used to price is often called the *implied volatility* as it is the volatility implied by the price. A *market maker* is expected to quote two prices or two vols, which are close together so the purchaser can be sure that he is not being cheated. An element of psychology comes in at this point, as the market maker tries to guess whether the purchaser wishes to buy or sell and slants his prices accordingly. Note that the difference between the two quoted prices is where the bank makes its profits. The true price is somewhere in between, so whichever the customer chooses, the bank makes a small profit.

How does the trader estimate volatility? We can measure, and therefore use, the volatility of an asset over any period in the past for which we have market data. The problem is which past period? We could, for example, use a thirty-day average. Although the Black–Scholes model is based on an assumption of constant volatility it is really the average volatility that is important, or more precisely the root-mean-square volatility. A major news event will cause rapid movements in an asset's price and thus a spike in the volatility. We would use a higher value of volatility in the pricing formula if a major news event occurred in the last thirty days than if one did not. This is undesirable because we would end up quoting a much lower vol thirty-one days after a major news event than thirty days after one, despite the fact that very little has changed. One solution to this problem is to use a weighted average with the weight ascribed to a given day decaying as it gets further in the past.

A more subtle issue is that it is not the past volatility that matters. It is the volatility that occurs during the life of the option which will cause hedging costs and the option should be priced thereby. The trader therefore has to estimate the future volatility. This could be based on market prices, past performance and anticipation of future news. It is important to realize that announcements are often expected in advance, and the information they contain will either push the asset up or down. The market knows that the asset price will move but cannot discount the information as it does not know whether it is good or bad. The options trader on the other hand does not care whether the information is good or bad, all he cares about is whether the asset price will move. Thus the anticipation of an announcement will drive estimated vols up.

The trading of options is therefore really about the trading of vol, and the options trader is taking views on the future behaviour of volatility rather than the movement of the underlying asset.

4.3 Smiles

If we collect data on estimated vols from the prices of options and compare these to historical vols, two facts stand out. The first is that vols implied by the market prices of options are higher than historical vols. The second is that two different options on the same underlying with the same expiry date can imply different vols.

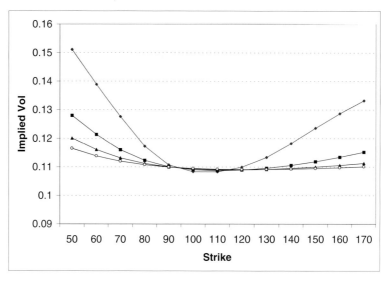

Fig. 4.1. Some possible smiles

Indeed, if one plots the implied volatility as a function of the strike of an option, one obtains a curve which is roughly smile-shaped. The qualitative nature of the curve will vary according to the nature of the underlying but it is very rarely the flat horizontal line which the model would predict. We show some possible smiles in Figure 4.1.

At this point, we would seem to have an arbitrage opportunity. Either the option with higher implied vol is overpriced or the option with lower implied vol is underpriced. We ought to be able to make some money from selling the one with high vol and buying the one with low vol. However, such smiles are a persistent feature of the markets and do not disappear with time, so the arbitrage opportunity is likely to be illusory.

How does the smile arise? Suppose a market maker spends his day buying and selling vanilla options. Each time a client calls him, he quotes a vol for buying and a vol for selling, with a little gap between them. If the first client buys from him then he wants the second client to sell. For if he sells to the first client for $11 and buys from the second from $10, then he has made a riskless profit of $1; he will have to carry out no hedging, and he will be perfectly immune to market changes. He therefore slants his prices to encourage the second client to do the opposite of the first. If more and more clients buy then the price will get higher and higher. If more and more sell, then the price will get lower. Eventually the price will settle down when the number of buyers and sellers are similar.

The crucial point is that the buying and selling behaviour will be different for different values of strike which will drive the volatilities in different directions for

4.3 Smiles

the different strikes. Whilst one can hedge options for different strikes by each other, one is no longer perfectly hedged in a model-free sense, and one is taking on some risk on the basis of the imperfections in the model.

Thus the *smile* expresses the market's view of the imperfections of the Black–Scholes model. There are two obvious criticisms of the model; the first is that the model requires a certain hedging strategy to be carried out which is not practical, and the second is that stock and foreign exchange prices are simply not log-normally distributed.

We address the hedging issues first. The model requires the option seller to hedge his exposure by holding $\frac{\partial O}{\partial S}$ units of the stock, S, at any time. This quantity is known as the *Delta* of the option and the hedging strategy is known as *Delta-hedging*. Therefore, as the stock moves up and down, the option seller has to continuously change his holding to remain Delta-hedged. In the real world this is not practical, as it takes time execute a trade. Thus it is impossible to rehedge truly continuously. As a consequence, the seller will never be perfectly hedged. Another problem is that executing a trade costs money: transaction costs may be low but they will never be non-existent. The more trades made, the more transaction costs mount up and increase the costs of hedging the option.

One interesting feature of the log-normal model of asset prices is that the total amount of wobble is infinite. What is meant by wobble? If we do not let down-moves cancel up-moves but instead measure the total distance the asset has moved during the lifetime of a contract, we obtain infinity. (The wobble is really the variation of the function.) This means that the model requires an infinite amount of rehedging, and thus if we allow transaction costs, the cost of rehedging will be infinite and we are clearly worse off than if we had never hedged at all. The solution is, of course, to hedge discretely rather than continuously. We only rehedge when the hedge required by our model is more than a chosen distance from the hedge we currently hold. Whilst this is effective, we are no longer in the world of perfect hedging and a single necessary price.

One simplification to the trader's position arises from the fact that he will have bought and sold many different options' contracts on the same underlying. Each one of these has a Delta and since the model is linear, the trader can hedge them all simply by adding their Deltas together. Note that the Deltas of long and short positions will have opposite signs and so if the portfolio is a mixture of such positions, the Deltas will at least partially cancel each other. It is only a short conceptual distance now to start thinking about using options to hedge options. Whilst there is not a great deal of point to this if our objective is to Delta-hedge, the use of options allows us more sophisticated methods of hedging.

Recall that the purpose of the Delta-hedge was to eliminate risk from the portfolio, by making the derivative of the portfolio with respect to the stock price zero.

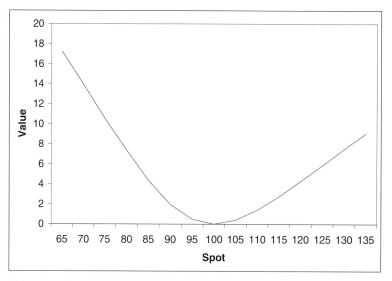

Fig. 4.2. The changes in value of a call option Delta-hedged with spot and strike equal to 100.

A portfolio whose price has zero derivative at a point will still change in value if the asset moves a short distance from that point. However, if the distance is small, the change in value will be proportional to its square, whereas for a non-Delta-hedged portfolio the change will be linear in the distance. As the square of a small number is even smaller, this means the change for small movements is tiny.

If we allow ourselves to use options to hedge, we can do better. The second derivative can also be matched and the portfolio's change in value for small changes in price of the underlying will now be proportional to the cube of the change, which is much smaller again. The second derivative with respect to the spot is called the *Gamma* and the process we have discussed is called *Gamma-hedging*. We illustrate the value changes of hedged portfolios in Figures 4.2, 4.3 and 4.4.

4.4 The Greeks

In the Black–Scholes model, we need to make the distinction between variables and parameters. The spot price is the only variable: it is the only term which is supposed to change within the model. The other terms are parameters: they affect the price but do not change within the model. However, we must remember that the Black–Scholes model is just a model, and the real world is rather different: traders often hedge their exposure to other quantities. Note that we can compute the derivative of the price of a portfolio with respect to any of the underlying parameters, and, as with the stock price, we can buy options which match that derivative. In general,

4.4 The Greeks

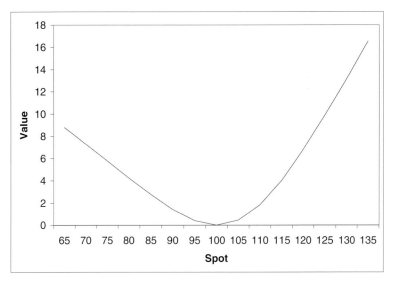

Fig. 4.3. The changes in value of a call option Delta-hedged with spot equal to 100 and strike equal to 110.

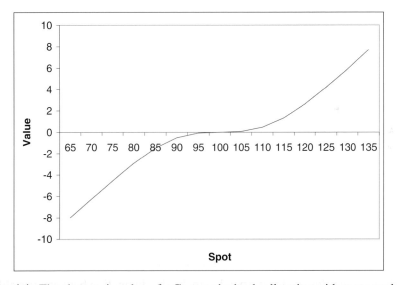

Fig. 4.4. The changes in value of a Gamma-hedged call option with spot equal to 100 and strike 110. Hedging option struck at 100.

if we wish to hedge k of the parameters, we will need $k-1$ different options to carry out the hedging, and we will need to construct the portfolio so as to match all the parameters at once by solving a system of linear equations. If it seems circular to hedge options with options, one should think in terms of hedging complicated

76   *Practicalities*

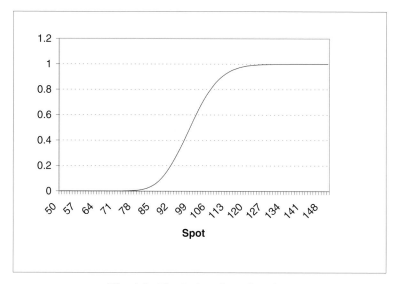

Fig. 4.5. The Delta of a call option

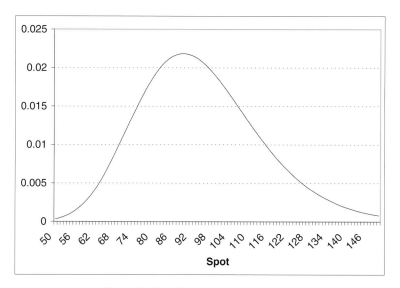

Fig. 4.6. The Gamma of a call option

options with less complicated ones. The complicated option could be an exotic or simply a far out-of-the-money vanilla option.

The derivatives with respect to the various quantities are denoted by Greek letters with initials corresponding to the quantity differentiated, and the derivatives are collectively known as the *Greeks*. In addition, to Delta and Gamma which we have already met, we have Greeks for each of the parameters. See Figures 4.5, 4.6 for illustrations of the Delta and Gamma. The derivative with respect to the short

4.4 The Greeks

rate, r, is called *Rho*. The derivative with respect to time, t, is called *Theta*. The derivative with respect to volatility is known as *Vega* or, by purists, *Kappa*, on the not unreasonable grounds that vega is not a letter in the Greek alphabet. Vega is a very important Greek as the Black–Scholes assumption of constant deterministic volatility is manifestly false, and the trader will wish to reduce his exposure to unexpected changes in volatility. This is called *Vega-hedging*.

Typically, the options trader will monitor all his Greeks but not necessarily continually rehedge them all. Instead he will take a view on which ones are important, and on which risks he wishes to hedge. Equally importantly, he will have a view on which Greeks he does not wish to hedge. This expresses the trader's opinions about which direction market parameters will move, and on which parameters the trader has a firm opinion that he wishes to place money on. For example, if the trader believes that the volatility will increase in the near future, he will go *long Vega*, that is he will ensure that the derivative of his portfolio with respect to volatility is positive. If he believes that vol will decrease he will go *short Vega*.

The Greeks are also important for the risk manager; he assesses the value of the bank's portfolio and tries to estimate the probability of the bank losing a lot of money. The Greeks describe to first order, via Taylor's theorem, the effect of varying the parameters. If $F(S, t, r, \sigma)$ is the value of an option, we have to first order,

$$F(S+\delta S, t+\delta t, r+\delta r, \sigma+\delta\sigma) = F(S, t, r, \sigma) + \delta S \frac{\partial F}{\partial S} + \delta t \frac{\partial F}{\partial t} + \delta r \frac{\partial F}{\partial r} + \delta\sigma \frac{\partial F}{\partial \sigma}. \tag{4.1}$$

Thus for small market changes the Greeks will tell us the portfolio's new value fairly accurately.

The Delta The Delta is the most fundamental Greek. Note from Figure 4.5, how it increases from zero to one as a function of spot. The Gamma of a call option is always positive so its Delta is always an increasing function of spot. At expiry, a call option has pay-off

$$(S - K)_+;$$

for $S < K$ the pay-off has Delta equal to zero. For $S > K$, it has Delta equal to 1. This means that as expiry approaches, the Delta becomes more and more similar in shape to this binary-valued function. Just before expiry it will be almost zero for S more than a little below K and then it will rapidly increase to almost 1 just above K. See Figure 4.7.

Differentiating the Black–Scholes formula, one easily obtains a formula for the Delta of a call option in a Black–Scholes world:

$$\Delta = \frac{\partial C}{\partial S}(S, t) = N(d_1), \tag{4.2}$$

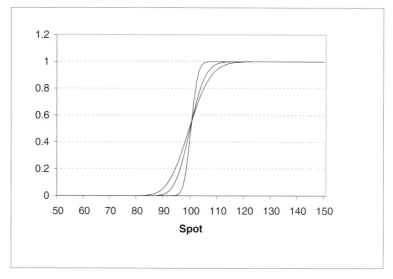

Fig. 4.7. The convergence of the Delta of a call option

where

$$d_1 = \frac{\log(S/K) + \left(r + \frac{1}{2}\sigma^2\right)(T-t)}{\sigma\sqrt{T-t}}. \tag{4.3}$$

The Gamma The Gamma is the derivative of the Delta with respect to spot, or the second derivative of the price. In the Black–Scholes model, it is always positive for calls and puts. The easiest way to see this is by the formula

$$\Gamma = \frac{\partial^2 C}{\partial S^2}(S, t) = \frac{N'(d_1)}{S\sigma\sqrt{T-t}}, \tag{4.4}$$

where $N'(x) = e^{-\frac{x^2}{2}}/\sqrt{2\pi}$. An immediate consequence of this is

Theorem 4.1 *The Black–Scholes price of a call option is a convex function of spot.*

This theorem follows immediately from the fact that any function with strictly positive second derivative is convex. Note that calls and puts have the same Gamma by put-call parity since a forward contract has zero Gamma.

The Gamma is important in that it expresses how much hedging will cost in a small time interval. In particular, if the Gamma of our portfolio is positive, then we will make money by Delta-hedging and if negative we will lose money. If we have sold a call option then we are *short Gamma* and the procedure of hedging will cost us money over the life of the option. If we are *long Gamma* then, over the lifetime of the option, our hedging makes us money realizing the option's value. To see

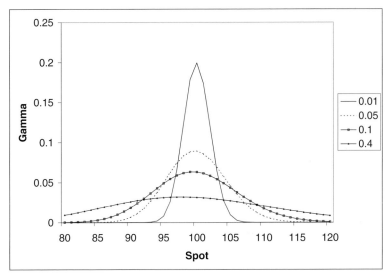

Fig. 4.8. The Gamma of a call option struck at 100 as a function of spot for varying expiries

why this is the case, we expand the Taylor series of the portfolio value, $P(S, t)$, as a function of S, we have

$$P(S + \Delta S, t) = P(S, t) + \frac{\partial P}{\partial S}(S, t)\Delta S + \frac{1}{2}\frac{\partial^2 P}{\partial S^2}(S, t)\Delta S^2 + \mathcal{O}(\Delta S^3). \quad (4.5)$$

If the portfolio is Delta-neutral then the main term is

$$\frac{1}{2}\frac{\partial^2 P}{\partial S^2}(S, t)\Delta S^2,$$

and thus the stock's price variations up and down will cause money to be lost or gained according to the sign of the Gamma.

We saw that as maturity approaches the Delta of a call option behaves more and more like a step function equal to 0 below the strike and 1 above it. This means that the Delta is going from almost 0 to almost 1 in a shorter and shorter interval. Its derivative, the Gamma, must therefore become more and more spiked as maturity approaches. Away from the strike it becomes zero, but at the strike it becomes more and more peaked.

The Vega The Delta and Gamma are Greeks with respect to the spot price, which is expected to move within the model. The Vega, on the other hand, is the derivative with respect to the volatility which is a parameter of the model. As we observed above, the volatility is an uncertain parameter and the trading of vanilla options is largely about correctly estimating it. The Vega expresses the position the trader is

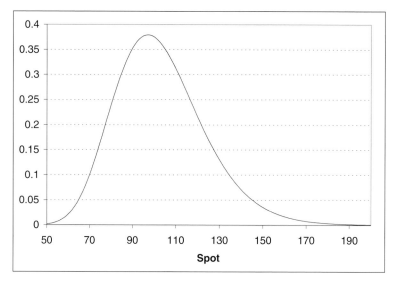

Fig. 4.9. The Vega of a call option

taking on volatility: a positive Vega expresses the opinion that volatilities will go up, and a negative Vega the opinion that they will go down.

The Vega of a call option is given by

$$\frac{\partial C}{\partial \sigma} = S\sqrt{T-t}N'(d_1), \qquad (4.6)$$

with d_1 as in (4.3). As a forward is insensitive to volatility, it follows from put-call parity that the Vega of a put will equal the Vega of the call with the same strike. Note that for a call or put, the Vega is always positive. An immediate consequence is that the map from volatilities to prices is injective – if two volatilities give the same price then they are equal. This is part of the reason that the practice of quoting volatilities instead of prices is popular.

Another important aspect of the Vega is that it gives a natural measure of the uncertainty of the price. Since volatilities are estimated rather than measured, the change in price for changing vol by 1% gives us a good measure of the size of the range the correct price may lie in. The Vega is therefore a good proxy for the bid-offer spread. Note that the size of the Vega can be quite different from the size of the price. For example, a deeply in-the-money put option close to expiry will have very low Vega, as, regardless of volatility, its value will simply be the intrinsic value of the option. An at-the-money put, despite being worth a lot less, will have much higher Vega as the volatility has a real effect on the option's value.

If we consider a digital call option or digital put option instead of a vanilla call or put, then the Vega need no longer be positive. A digital call pays one if spot is above the strike, and zero otherwise; similarly for a digital put. Indeed, if we consider the

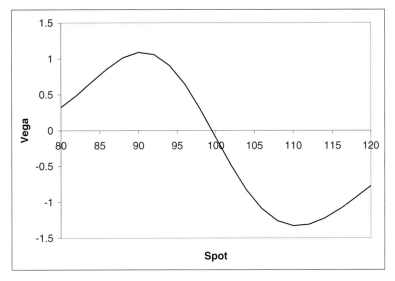

Fig. 4.10. The Vega of a digital call option

portfolio consisting of a digital call and a digital put with the same strike, then the portfolio replicates a zero-coupon bond which has a value independent of volatility. This means that the Vega of a digital put is the negative of the Vega of a digital call, and so we can expect one of them to be negative for any parameter values (unless both are zero).

When a digital option is in-the-money, volatility is bad as it increases the (risk-neutral) probability that the option will finish out-of-the-money without any benefits, so we roughly obtain negative Vega in-the-money and positive Vega out-of-the-money.

4.5 Alternate models

We have seen that the impossibility of perfect hedging takes us away from the Black–Scholes world of perfect arbitrage-free pricing; however there are other criticisms of the Black–Scholes world. The most important of these is simply that stock prices and foreign exchange prices are not log-normally distributed. This failure is manifested in a number of fashions. As we mentioned above, volatility is not even deterministic, let alone constant. Stock and FX prices often do not move continuously: rather they jump. For example, a market crash or a sudden devaluation will move the price quickly with no opportunities for a rehedge. A more subtle criticism is that the logs of asset price changes have *fat tails*.

If one computes the mean and variance of the movements of the logs, and plots the actual distribution against the normal with the same mean and variance, one finds that the density is greater for large (positive and negative) values. Since the

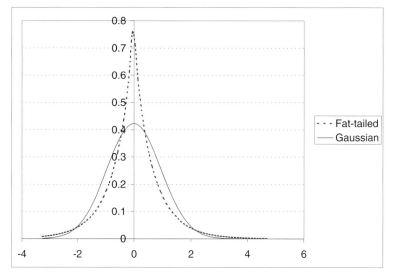

Fig. 4.11. A fat-tailed distribution and a Gaussian with the same mean and variance.

variance is the same this means the distribution is peaked higher in the middle, and is lower in the middle ranges. In statistical language the fourth moment, or *kurtosis,* of the actual distribution is higher. We illustrate this effect in Figure 4.11.

How can the financial mathematician cope with these problems? More sophisticated models are needed. We briefly survey some of these and discuss some of the issues.

4.5.1 Jumps

A very real flaw in the Black–Scholes model is its assumption that the asset price is a continuous function. This is particularly the case with options on equities. The stock market periodically undergoes corrections which involve a rapid downward movement in stock prices. The most famous of these corrections are the crashes of 1929 and 1987. During such crashes, market conditions are anything but normal and continuously rehedging a portfolio as it slides is simply not practical. Indeed it has been suggested that the 1987 crash was exacerbated by derivatives traders trying desperately to sell their hedges in order to remain Delta-hedged as the market tumbled. Conversely, one use of options is the purchase of put options by fund managers to insure against crashes by guaranteeing a price they can sell at. In 1987, there was also a large number of fund managers who had decided that they did not need to buy options, because they could replicate the options themselves – the necessary trading became impossible when the market crashed.

We therefore wish to permit in our model the possibility of a sharp downward move during which rehedging is not permitted. For simplicity, we restrict the jumps

4.5 Alternate models

to be of a particular size, that is we always move the log of the stock price by a fixed amount. For example, we could model jumps which reflect a loss of 25% of the stock's value. As we are assuming the jumps occur too quickly to allow rehedging, this means that in our tree, at each node, we have to allow three possible moves: one a small amount up, a second a small amount down and a third corresponding to the jump down. The problem is that, as we saw before, with three nodes there is no longer a unique arbitrage-free price. Instead one has a continuum of possible prices. How can we choose one? A conservative choice would be to take the price which would allow the option's payoff to be met in all possible worlds. That price would of course be the maximum arbitrage-free price.

A second approach would be to attempt to apply risk-neutral valuation. We saw before that a unique price is still not guaranteed as there are many ways of distributing the probabilities between the three nodes which are consistent with risk-neutrality. One solution is to assume that the market price jumps occur with the real-world probability, and adjust the probabilities of the small moves to obtain the unique risk-neutral probabilities consistent with it. Whilst this certainly allows us to obtain an arbitrage-free price, it is only *an*, not *the* arbitrage-free price. Models of this sort are known as *jump-diffusion models*, a description which expresses the mixture of an underlying diffusive time process consisting of small moves with that of a jumpy process.

Risk-neutral valuation has become such a paradigm in mathematical finance that the continuous-time analogues of such arguments are regularly used in the financial literature. Part of the issue here is the distinction between *risk-neutral* and *real-world* parameters. Given a model, one can produce a black box which takes as inputs certain parameters such as strike, time-to-expiry, spot, volatility and probability of a jump, and outputs a price. One can then start tweaking the unobservable parameters, such as volatility and jump probability, until the price agrees with the market price. Or more generally, one would tweak until prices agreed with all the prices observable in the market. If it proves impossible to find such parameters, one is left with two possibilities: either the model is wrong or the market is wrong. One has to be very confident to be sure of the second and to trade accordingly.

If one has obtained a fit, what do the parameters mean? They do not necessarily reflect anything about the movement of stock prices and are therefore often referred to as being *risk-neutral*, although *market-calibrated* would be a more accurate term. The question which now arises is "what is the good of models?" The perfect reproduction of market prices, whilst seemingly impressive, does not actually tell us very much; after all we already knew the market prices so producing a model that tells us what they are is not overly impressive.

One possible use is to attempt to derive the prices of more complicated, exotic options. Given that our model prices vanilla options correctly, it may be possible to use it to price exotic options either by decomposing them into vanilla options or by using dynamic hedging strategies. These dynamic hedging strategies may involve trading vanilla options and therefore require a good understanding not only of the vanilla options' current prices but also of their expected values at future times. The model can then be used to infer these future prices. However if we were to do so, we would have to be reasonably sure that our model is consistent with pricing at future times as well as the present. For example, some models can be used to fit prices well today but are rather time-inhomogeneous in that they imply that the smile in the future will have a rather different shape that it does today. Given that smiles have persisted in their present form for a number of years, this seems an unreasonable prediction and such models should be treated with care. In general, such models tend to perform poorly in that the hedges have to be rebalanced more often than models which predict sensible future behaviour.

The jump-diffusion model certainly has some validity, in that stock prices are jumpy and that the volatility smiles of equity options are certainly skewed, reflecting the fact that stock prices are much more likely to jump down than up. One could also see this as a reflection of market supply and demand: many fund managers want to buy out-of-the-money put options to protect themselves against the risk of a crash, thereby driving the cost of such options up. We discuss the mathematics of such models in Chapter 15.

4.5.2 Stochastic volatility

A further source of uncertainty in modelling stock prices is volatility. As we mentioned above, stock price distributions tend to have fat tails. One way of modelling this is to make volatility a function of the stock price. As the stock price moves farther away from the initial value, volatility increases making it more likely for the stock price to get even farther away, thus achieving fat tails. The main drawback of this approach is that it results in future smiles that are rather different from present ones. To see this, suppose our original smile had its low point at the current value, which we will take to be 100 for simplicity. We therefore make volatility a function of spot with its lowest value also at 100. The problem is that when the spot moves to 200, the volatility still has its lowest point at 100 rather than at 200, so the shape of the smile has changed and the stock is in a qualitatively different universe. Such a smile is said to be *sticky* as opposed to a *floating* smile which is always qualitatively the same.

A more subtle model would be to make the volatility a random quantity itself. We developed a random process for stock prices movements based on constant volatilities in the last chapter. We could also make the volatility follow such a random process and then feed the random volatility parameter back into the model.

4.5 Alternate models

This would mean that at each stage of our tree, there would first have to be a random draw reflecting the up or down move of the volatility which would determine the magnitude but not the direction of the stock's move, and then a second random draw deciding whether the move would be up or down. The problem left is then that at each time segment the stock can take four possible new values instead of two, and we are back in a world where perfect hedging no longer exists. A risk-neutral price can be developed but should we believe it? These models are known as *stochastic-volatility* models, and we discuss them in Chapter 16.

4.5.3 Random time

Another subtle idea is to make time itself a random process. While this may seem a little artificial, if we think in terms of the variation of the rate of arrival of the information which drives asset movements, then this is not so unreasonable. A small value for the random time will reflect a boring year and a large value an exciting one. Such *Variance Gamma* models can provide good fits to asset price movements. The process arrived at for the stock is then a series of small jumps. The main problem is that, as before, the extra source of randomness removes the possibility of perfect arbitrage-free pricing. We discuss Variance Gamma models in Chapter 17.

4.5.4 Which model?

Which of the above improved approaches is correct? Probably a mixture of all of them is the most accurate model of stock price movements. Which is most useful? This will depend on what sort of option the modeller wishes to hedge; in practice the trader might use multiple models to compare prices and then decides accordingly. The appropriateness of the model will also depend upon the nature of the underlying asset – jumps are a fact of life for stocks but are not so common in foreign exchange markets between major currencies.

The common problem with the more sophisticated models was the impossibility of perfect hedging. How can we cope with this? One solution is to try to divide (in a theoretical rather than practical sense) the asset into two pieces, one of which is hedgeable and the other being totally unhedgeable. The hedgeable portion can then be priced perfectly using risk-neutral or arbitrage-free arguments, as was possible in the simple Black–Scholes world.

The unhedgeable portion is more problematic. It is a risky asset that cannot be hedged. To price it, we have to return to our notions of risk and try to estimate its distribution and then decide how much we are willing to pay for that piece of risk. Since different options can depend on the same underlying, we then have the opportunity to observe the cost of that piece of risk for different options and see whether it varies. If not there is a possibility to hedge one option with another and make a profit. Our no-arbitrage arguments therefore imply that a given piece of risk should have a unique price and this price is referred to as the *market price of risk*.

One consequence of the fact that the price depends on the market price of risk and the impossibility of hedging is that there is considerable resistance in the markets to models incorporating jumps. The great attraction of the Black–Scholes model was that the market price of risk did not enter, and there was a mathematically correct price. The derivatives community is reluctant to give this up despite strong empirical evidence that jumps occur. This is, however, a bit ostrich-like: to use a model because we like its implications rather than because we believe its accuracy is a dangerous path to follow.

Note that most of the alternative models had the property that no-arbitrage arguments did not guarantee a single price. Equivalently there is no method of dynamically reinvesting to replicate an option's payoff with a sum invested today. The markets implied by such models are said to be *incomplete* and there is a theory of no-arbitrage pricing in incomplete markets, which obtains bounds instead of unique prices.

The issues surrounding the pricing of derivatives in a non–Black–Scholes world are still very current and the ideas presented here reflect the debates that are currently going on rather than represent the 'correct' model. A trading bank will typically have a team of research quantitative analysts working purely on the pricing of vanilla options in order to better understand these issues, to which we return in Chapter 18.

4.6 Transaction costs

Although transaction costs are a reality, they tend not to be modelled explicitly when developing pricing models. There is a simple reason for this: transaction costs can never create arbitrages. In other words, if a price cannot be arbitraged in a world free of transaction costs, it cannot be arbitraged in a world with them either.

The proof of this result is very simple. Suppose a price is arbitrageable in the world with transaction costs. Then we can set up a portfolio taking into account transaction costs at zero or negative cost today, which will be of non-negative and possibly positive value in the future. If we neglect to take into account transaction costs then the initial set-up cost of the portfolio will be even lower and thus still be negative or zero. The final value of the portfolio will however be at least as high as there will be no cash drain from any transaction costs during the portfolio's life. We therefore conclude that the portfolio is also an arbitrage portfolio in a world free of transaction costs.

Thus the existence of arbitrage in the world with transaction costs implies arbitrage in a world free of them.

A second reason they tend to be neglected is that hedging is carried out on a portfolio basis. This results in many transactions that would be necessary to hedge a single option, not being necessary because they cancel out with other positions. The precise transaction costs added by a single new trade are therefore a function of the existing positions, and could be effectively negative if a trade offsets existing ones.

4.7 Key points

- The buying and selling of vanilla options is really about the trading of volatility.
- The imperfection of models leads to the *smile* – the practice of using different volatilities to price options at different strikes.
- The derivatives of the price of a portfolio with respect to various parameters are known as the Greeks, and are denoted by Greek letters.
- The derivative with respect to the spot is the Delta.
- The second derivative with respect to the spot is the Gamma.
- The derivative with respect to the volatility is the Vega.
- The derivative with respect to time is the Theta.
- The derivative with respect to interest rates is the Rho.
- The Gamma expresses the amount of money we expect to make or lose from dynamic hedging.
- A possible source of smiles is jumps.
- Another possible source of smiles is stochastic volatility.
- In an *incomplete* market unique prices are no longer guaranteed.

4.8 Further reading

Many of the reasons for smiles, and the whys and wherefores of alternative models are discussed at length in *Volatility and Correlation* by Riccardo Rebonato, [106]. We return to some of the issues in Chapters 15, 16, 17 and 18.

One model for transaction costs is that of Leland, [82], which leads to a Black–Scholes price with a modified volatility. Models for more general cases have been developed by Whalley & Wilmott; see [118] for further discussion and references therein. A rather demoralizing result is due to Soner, Shreve & Cvitanic, [114], who show that if one wishes to be sure of covering the pay-off of a call option at expiry in Black–Scholes world with transaction costs, the best strategy is to buy the stock today and do no further trading. This is the strategy we used to deduce the rational bounds of Chapter 2.

4.9 Exercises

Exercise 4.1 Show that a portfolio of vanilla options with the same expiry is Gamma neutral if and only if it is Vega neutral. Does this result hold if the expiries are not all the same?

Exercise 4.2 How does the graph of the Vega of a call option vary as a function of time to expiry?

Exercise 4.3 If a digital call and a digital put have the same expiry and strike, what relations will their Greeks satisfy?

Exercise 4.4 If a derivative has a negative Vega and volatility increases what happens to the price?

Exercise 4.5 A portfolio consisting of a short position in a call option and a long position in a stock is Delta-neutral. Suppose the stock price jumps; how will the value of the portfolio change if the option is priced according to the Black–Scholes formula before and after the jump?

Exercise 4.6 Derive simple approximations for at-the-money Vega and Theta.

Exercise 4.7 Show that call and put options with the same strike and expiry have the same Vega. Do this without using the Black–Scholes formula.

5

The Ito calculus

5.1 Introduction

We have so far avoided doing any hard mathematics; our objective was to develop the conceptual ideas of mathematical finance in order to provide a motivational framework. However, the time has come where we must start to develop the more complicated tools necessary to manipulate the formulas of mathematical finance. There are two quite different but related approaches to derivatives pricing. The first is to use stochastic calculus to develop a partial differential equation for option prices, and the second is to construct synthetic probability measures which allow option prices to be expressed as expectations. In this chapter, we develop the necessary mathematics to carry out the first of these approaches. For background probability results we refer the author to [53]. For a more rigorous treatment, see [100]. We discuss further reading at the end of the chapter.

5.2 Brownian motion

One of the fundamental tools in option pricing is the theory of stochastic calculus. This theory allows the manipulation of the random processes described in Chapter 3 much as the ordinary differential calculus allows the manipulation of functions. Indeed, Black & Scholes used the Ito calculus to derive their famous equation, and it was several years later that Cox, Ross & Rubinstein developed the more intuitive tree approach. To develop the Ito calculus in a wholly rigorous fashion is quite involved as there are many technical issues, and we therefore skirt over some of them in order to concentrate on the ideas.

To try to motivate better the definitions made in this chapter, we examine in more detail the random processes developed in Chapter 3. There, we divided a time interval into k pieces, and in each piece let the variable move randomly. The move consisted of a fixed deterministic part and a random part consisting of an up or down move. The random variable for each piece was the same. The size of the pieces was then shrunk to zero and the random variables changed in such a way as to make the mean and variance remain constant.

The resultant distribution at the end of our time-frame corresponded to a random variable which was normally distributed with the given mean, μ, and variance, σ^2, and is usually written as $N(\mu, \sigma^2)$, or $\mu + \sigma N(0, 1)$. However, there is nothing special about the end of our interval, and if instead we sum only the variables associated to pieces in the first half of the interval, we obtain the same distribution but with mean $\mu/2$, and variance $\sigma^2/2$ instead. Or more generally, if our initial interval is $[0, T]$ and we consider the subinterval $[a, b]$ we obtain a random variable with mean

$$\frac{\mu(b-a)}{T}$$

and variance

$$\frac{\sigma^2(b-a)}{T} = \left(\frac{\sigma(b-a)^{1/2}}{T^{1/2}}\right)^2.$$

If we now change notation slightly and take the mean to be μT, and the variance to be $\sigma^2 T$, we can construct a collection of random variables, X_t, for $t \leq T$, by taking X_t to be the random variable associated to the interval $[0, t]$. We then have that X_t has mean μt and variance $\sigma^2 t$ and is normally distributed. We can view X_t as a particle which starts at the origin and is displaced the distance X_t at time t. Note for $s < t$ we have that $X_t - X_s$ has the distribution of the random variable associated to the interval $[s, t]$ that is, it is normal with mean $\mu(t-s)$ and variance $\sigma^2(t-s)$.

Each of the small steps we added together was independent of all the others. This means that the steps that go into making $X_t - X_s$ are independent of those that going into making X_s.

It is also independent of the value of X_s, since the random variables we summed were independent of each other. Note that the distribution of X_t given the value of X_s is that determined by that of $X_t - X_s$ and so the behaviour of X_t is totally unaffected by the values of the random variables X_r for r less than s. This is called the *Markov* property.

To summarize, we have constructed a collection of random variables X_t such that $X_t - X_s$ is normally distributed with mean $(t-s)\mu$ and variance $(t-s)\sigma^2$. Such a collection of random variables is called a *Brownian motion* as it models well the random movement of small particles suspended in a fluid. This random motion is caused by jostling by smaller particles, and was first observed by the botanist Robert Brown when observing pollen through a microscope[1]. Note that

---

[1] Whilst this process is named after a physical phenomenon, financial mathematicians love to point out that it was first studied mathematically in connection with the movement of stock prices, (Bachelier 1900).

5.2 Brownian motion

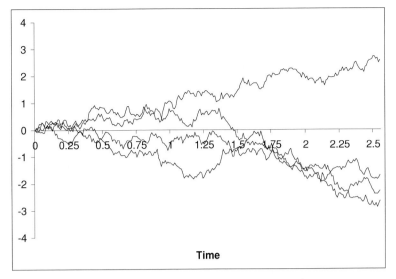

Fig. 5.1. Some paths from a Brownian motion.

the standard deviation of the random variables is $\sigma(t-s)^{1/2}$. This suggests that spreading out happens at a rate that grows with the square root of time. Note how the Markov property expresses very neatly the weak efficiency of markets. The past path of the stock price has no effect on future movements.

Brownian motion has lots of strange properties. If we trace the path followed by a particle moving under a Brownian motion, we find that it is infinitely jagged. It is continuous everywhere and differentiable nowhere. We also find that it has an infinite first variation – the total amount of change in the path is infinite. This means that if one changed all the down movements into up movements, the path would go off to infinity instantly. If it hits a value, then it hits it again an infinite number of times in an arbitrarily short interval afterwards. To be precise these properties do not always hold, but only with probability 1. They are said to hold *almost surely*.

We emphasize that an event occurring with probability 1 is not quite the same as a certain event. For example, consider a uniform random variable, X, on the interval $[0, 1]$. The probability that X lies in an interval $[a, b]$ with $0 \leq a < b \leq 1$ is $b - a$. The probability that X takes on any particular value, x, must therefore be zero, as the probability that X lies in the interval $[x - \epsilon, x + \epsilon]$ is 2ϵ for any $\epsilon > 0$. Yet the random variable must take on some value. So an event of probability zero always occurs. The point is that there is an uncountable infinity of such zero-probability events, and if one takes an uncountable union of sets, probabilities can do strange things.

We make a clear mathematical definition of Brownian motion.

Definition 5.1 We shall say a sequence of random variables, W_t, for $t \geq 0$, is a *Brownian motion* if $W_0 = 0$, and for every t and s, with $s < t$, we have that

$$W_t - W_s \qquad (5.1)$$

is distributed as a normal distribution with variance $t - s$, and the distribution of $W_t - W_s$ is independent of the behaviour of W_r for $r \leq s$.

It is important to realize that the condition that the distribution is normal of variance $t - s$ for every t and s and independent of the path up to time s, is much stronger than requiring W_t to be normally distributed with variance t for every t. For example, if we let

$$Z_t = \sqrt{t} Y, \qquad (5.2)$$

where Y is the same draw from a normal distribution for all t, then we have that Z_t is normally distributed with variance t. However, the paths of Z_t are straight lines. The distribution of $Z_t - Z_s$ is a normal of variance $(\sqrt{t} - \sqrt{s})^2$, and the value of $Z_t - Z_s$ is wholly determined by the value of Z_s.

5.3 Stochastic processes

Whilst Brownian motion is extremely interesting in its own right, it does not make the best model for stock movements since the probability of negative prices is always non-zero. To see this, one simply observes that the Gaussian distribution is non-zero everywhere. We will therefore want to think of the log of the stock as being normally distributed. As our option price will be a function of the stock price, we want to deduce the distribution of the option price from the distribution of the stock price. Our objective is therefore to develop a class of processes which is closed under simple operations. In particular, our principal objective is to achieve a class of processes which is invariant under composition with smooth functions; that is, we want $f(X)$ to be a member of the class if X is, for any smooth function f. Moreover, we want to be able to compute the process for $f(X)$.

For more general processes, we might wish to let μ and σ vary with time or even depend on X_t. The simplest generalization one could try would be to let μ and σ be piecewise constant functions of time.

Suppose we want to know how $X_t - X_s$ is distributed. Suppose the interval $[s, t]$ is divided into $[s, t_1], [t_1, t_2], \ldots, [t_{k-1}, t_k], [t_k, t]$ in such a way that on each of these subintervals, μ and σ are constant. Putting

$$t_0 = s, \quad t_{k+1} = t,$$

let μ_j and σ_j denote the values of μ and σ on the interval $[t_j, t_{j+1}]$. We can then write

$$X_t - X_s = (X_t - X_{t_k}) + (X_{t_k} - X_{t_{k-1}}) + \cdots + (X_{t_2} - X_{t_1}) + (X_{t_1} - X_s).$$

Using the fact that μ and σ are constant on the interval $[t_i, t_{i+1}]$, we have that $X_{t_{i+1}} - X_{t_i}$ is normally distributed, with mean, equal to $\mu_i(t_{i+1} - t_i)$ and variance equal to $\sigma_i^2(t_{i+1} - t_i)$.

The sum of two independent normally distributed random variables is also normally distributed, with mean equal to the sum of the means, and variance equal to the sum of the variances. (This is a very special property of the Gaussian distribution.) We therefore have that $X_{t_2} - X_{t_0}$ is normally distributed with mean

$$\mu_0(t_1 - t_0) + \mu_1(t_2 - t_1)$$

and variance

$$\sigma_0^2(t_1 - t_0) + \sigma_1^2(t_2 - t_1).$$

We deduce, using induction, that $X_{t_{k+1}} - X_{t_0}$ has mean equal to

$$\sum_{j=0}^{k} \mu_j(t_{j+1} - t_j),$$

and variance equal to

$$\sum_{j=0}^{k} \sigma_j^2(t_{j+1} - t_j).$$

Thus for piecewise constant μ and σ we get a mean which is the integral of μ and a variance which is the integral of σ^2. Since every continuous function can be uniformly approximated by piecewise constant functions, it is reasonable to simply define for arbitrary continuous functions μ and σ, that $X_t - X_s$, for all t and s with $s < t$, should be normally distributed with mean equal to

$$\int_s^t \mu(r)dr$$

and variance equal to

$$\int_s^t \sigma^2(r)dr.$$

This definition is consistent because of the additive property of normal distributions mentioned above. In particular, we obtain the same distribution for $X_t - X_s$ if we write it as $(X_t - X_{t'}) + (X_{t'} - X_s)$.

If we take σ to be identically zero we obtain

$$X_t - X_s = \int_s^t \mu(r)dr;$$

that is, X_t is just the integral of $\mu(t)$, or, equivalently that μ is the derivative of X. We can interpret this as saying that our attempts to understand the processes implied by more general μ and σ are really an attempt to generalize calculus to cope with random variables. The reader will recall that in ordinary calculus the function f is said to have derivative equal to $f'(x)$ at the point x if

$$\lim_{h \to 0} \frac{f(x+h) - f(x)}{h} = f'(x).$$

An equivalent way to say this is that,

$$f(x+h) - f(x) - f'(x)h = o(h).$$

(A function, g, is said to be $o(h)$ if it converges to zero faster than h, i.e. $g(h)/h$ converges to zero as h tends to zero.) We recall from high school calculus that one way of looking at this equation is to say that near x, f is well approximated by $f(x) + f'(x)h$. Thus the derivative of f provides a good approximation to f. (In fact, it is the best linear approximation near x.)

We wish to do something similar for random variables. We therefore examine the behaviour of $X_{t+h} - X_t$ as h tends to zero. We can write

$$\mu(r) = \mu(t) + e(r),$$
$$\sigma^2(r) = \sigma^2(t) + f(r)$$

where e and f vanish at $r = t$. We then have that

$$X_{t+h} - X_t = h\mu(t) + \int_t^{t+h} e(r)dr + h^{1/2}\sigma(t)N(0,1) + g(t,h)N(0,1), \quad (5.3)$$

where $g(t,h) = (h\sigma(t)^2 + \int_t^{t+h} f(r)dr)^{1/2} - h^{1/2}\sigma(t)$. The important thing here is that the second and fourth terms are small as h goes to zero. To see this, note that the second term is $o(h)$ since $e(t) = 0$ and e is continuous because μ is. Moreover, we can rewrite g as

$$h^{1/2}\sigma(t)\left(\left(1 + h^{-1}\sigma(t)^{-2}\int_t^{t+h} f(r)dr\right)^{1/2} - 1\right).$$

5.3 Stochastic processes

Recall the binomial expansion which says that for x small

$$(1+x)^{1/2} = 1 + \frac{1}{2}x + \mathcal{O}(x^2), \tag{5.4}$$

where $\mathcal{O}(x^2)$ means that the error is less than Cx^2 for x small and some number C. Let

$$\epsilon_h = h^{-1} \sigma(t)^{-2} \int_t^{t+h} f(r)dr.$$

Since $f(0) = 0$, and f is continuous, we have that $\epsilon_h = o(1)$, i.e.

$$\epsilon_h \to 0 \text{ as } h \to 0.$$

Using the binomial expansion,

$$g = \frac{1}{2}\sigma(t)h^{1/2}\epsilon_h + \sigma(t)h^{1/2}\mathcal{O}(\epsilon_h^2).$$

The variance of g times $N(0, 1)$ will certainly be $o(h)$ as it will be divisible by $h\epsilon_h$. We deduce that

$$X_{t+h} - X_t - h\mu(t) - h^{1/2}\sigma(t)N(0, 1),$$

is a random variable with mean and variance that are both $o(h)$.

We may want to have several random processes all of which are driven by the same random information. We therefore will want the random part of the change, $N(0, 1)$, to be the same for each of them. The crucial example to have in mind is a stock and an option upon it. Changes in the value of the stock will also affect the value of the option. We can achieve this property by requiring the random increments to come from a Brownian motion, and by requiring that the same Brownian motion drive all the random processes. Recall that

$$W_{t+h} - W_t = h^{1/2}N(0, 1), \tag{5.5}$$

in a distributional sense. With this in mind, we define

Definition 5.2 Let W_t be a Brownian motion. We shall say that the family X of random variables X_t satisfies the *stochastic differential equation*,

$$dX_t = \mu(t, X_t)dt + \sigma(t, X_t)dW_t, \tag{5.6}$$

if for any t, we have that

$$X_{t+h} - X_t - h\mu(t, X_t) - \sigma(t, X_t)(W_{t+h} - W_t)$$

is a random variable with mean and variance which are $o(h)$.

We shall call such a family of random variables an *Ito process* or sometimes just a stochastic process. Note that if σ is identically zero, we have that

$$X_{t+h} - X_t - h\mu(t, X_t) \tag{5.7}$$

is of mean and variance $o(h)$. We have thus essentially recovered the differential equation

$$\frac{dX_t}{dt} = \mu(t, X_t). \tag{5.8}$$

The essential aspect of this definition is that if we know X_0 and that X_t satisfies the stochastic differential equation, (5.6), then X_t is fully determined. In other terms, the stochastic differential equation has a unique solution. An important corollary of this is that μ and σ together with X_0 are the only quantities we need to know in order to define a stochastic process. Equally important is the issue of existence – it is not immediately obvious that a family X_t satisfying a given stochastic differential equation exists. Fortunately, under reasonable assumptions on μ and σ, solutions do exist and are unique. Unfortunately, developing the necessary mathematics is beyond the scope of this book.

5.4 Ito's lemma

One of the most important tools for manipulating ordinary differential equations is the chain rule. The chain rule allows us to take the derivative of a function of another function, and simply states that if

$$dX_t = \mu(X_t, t)dt$$

then

$$d(f(X_t)) = f'(X_t)\mu(X_t, t)dt.$$

When f is also a function of t, we obtain

$$d(f(X_t, t)) = \left(\frac{\partial f}{\partial x}(X_t, t)\mu(X_t, t) + \frac{\partial f}{\partial t}(X_t, t)\right)dt.$$

Our objective is to develop a generalization of the chain rule which holds for functions of random processes.

An obvious first guess is that if

$$dX_t = \mu(X_t, t)dt + \sigma(X_t, t)dW_t, \tag{5.9}$$

then

$$d(f(X_t)) = f'(X_t)dX_t = f'(X_t)\mu(X_t, t)dt + f'(X_t)\sigma(X_t, t)dW_t. \tag{5.10}$$

5.4 Ito's lemma

Unfortunately, this first guess is wrong. To see this, consider the simple function

$$f(x) = x^2, \tag{5.11}$$

applied to Brownian motion. Since $f'(x) = 2x$, we would obtain

$$d(W_t^2) = 2W_t dW_t. \tag{5.12}$$

This means that the stochastic differential equation for W_t^2 would have no drift: up-moves are as likely as down-moves. However, we know that W_t has mean zero and variance t. As the mean is zero, the expected value of the square is just the variance. So the expected value of W_t^2 is t.

This suggests that there is a missing drift term: W_t^2 is drifting away from the origin at a constant rate of 1. A guess for the stochastic differential equation is therefore

$$d(W_t^2) = dt + 2W_t dW_t. \tag{5.13}$$

Our objective in this section is to understand where the dt comes from.

Let's examine the behaviour of

$$W_{t+h}^2 - W_t^2$$

for h small. We can write

$$\begin{aligned} W_{t+h}^2 - W_t^2 &= (W_{t+h} - W_t)(W_{t+h} + W_t) \\ &= 2W_t(W_{t+h} - W_t) + (W_{t+h} - W_t)^2. \end{aligned} \tag{5.14}$$

The first term is what we expect from $2W_t dW_t$. The second term is new. As $W_{t+h} - W_t$ has variance h and mean zero, we conclude that $(W_{t+h} - W_t)^2$ has mean h. In other words,

$$(W_{t+h} - W_t)^2 - h$$

has mean zero. Our definition of a stochastic differential equation (SDE) requires us to find terms μ and σ such that

$$(W_{t+h}^2 - W_t^2) - \mu h - \sigma(W_{t+h} - W_t)$$

has both mean and variance which are $o(h)$. Our candidates are

$$\mu = 1, \tag{5.15}$$
$$\sigma = 2W_t. \tag{5.16}$$

Our computation above shows that this expression above evaluates to

$$(W_{t+h} - W_t)^2 - h,$$

which we have already shown to have mean zero. It remains to show that this term has small variance. We can write

$$W_{t+h} - W_t = h^{1/2}Z,$$

with Z a standard normal Gaussian variable. (It would actually be a different identically distributed variable for each h but this is unimportant.) We deduce that

$$((W_{t+h} - W_t)^2 - h)^2 = (hZ^2 - h)^2 = h^2(Z^2 - 1).$$

The variance is therefore

$$h^2 \text{Var}(Z^2 - 1),$$

which is certainly $o(h)$. In conclusion, we have shown that W_t^2 is such that

$$d(W_t^2) = dt + 2W_t dW_t. \tag{5.17}$$

Our objective in the rest of this section is to generalize this argument to apply to any smooth function of any solution of a stochastic differential equation. As we defined SDEs in terms of $X_{t+h} - X_t$ for h small, we look at $f(X_{t+h}) - f(X_t)$. The key to examining this difference is Taylor's theorem. Taylor's theorem expresses the local behaviour of a function in terms of its derivatives. It implies that if f is smooth function then for y close to z, we have

$$\left| f(y) - \left(f(z) + (y-z) f'(z) + \frac{(y-z)^2}{2} f''(z) \right) \right| \leq C |y-z|^3,$$

for some constant C. Note that we have gone one step further than the definition of the derivative of a function; we have the best parabolic approximation instead of the best linear approximation. Taylor's theorem will apply equally well when y and z are stochastic.

Let X_t satisfy (5.6). If we put $y = X_{t+h}$, $z = X_t$, we deduce that

$$f(X_{t+h}) - f(X_t) = f'(X_t)(X_{t+h} - X_t) + \frac{f''(X_t)}{2}(X_{t+h} - X_t)^2 + e, \tag{5.18}$$

where $|e(X_{t+h}, X_t)| \leq C|X_{t+h} - X_t|^3$. Recalling the definition of a stochastic differential equation, we have that $X_{t+h} - X_t$ is equal to

$$\mu(X_t, t)h + h^{1/2}\sigma(X_t, t)N(0, 1)$$

plus an error term. Remember in what follows that the $N(0, 1)$ term comes from the Brownian motion and we can always substitute

$$h^{-1/2}(W_{t+h} - W_t)$$

for it.

5.4 Ito's lemma

If we now substitute into (5.18), we obtain for the right-hand side,

$$f'(X_t)\mu(X_t,t)h + h^{1/2}\sigma(X_t,t)f'(X_t)N(0,1)$$
$$+ \frac{f''(X_t)}{2}(\mu(X_t,t)h + h^{1/2}\sigma(X_t,t)N(0,1))^2 \quad (5.19)$$

plus error terms, which are all of mean and variance at least $o(h)$. If we now expand the square, it can be rewritten as

$$h\sigma^2 + h\sigma^2(N(0,1)^2 - 1) + 2h^{3/2}\mu\sigma N(0,1) + h^2\mu^2, \quad (5.20)$$

where we have dropped the dependence of σ and μ on X_t and t for tidiness. The last two terms are trivially of $o(h)$ in both mean and variance. Recall that the normal distribution is of mean 0 and variance 1, so the definition of variance implies that the mean value of $N(0,1)^2 - 1$ is zero. This means that the second term has mean zero and the coefficient of h guarantees it has variance which is $o(h)$. We conclude, modulo terms that are of mean and variance $o(h)$, that

$$f(X_{t+h}) - f(X_t) = f'(X_t)\mu(X_t,t)h + \frac{f''(X_t)}{2}\sigma(X_t,t)^2 h$$
$$+ h^{1/2}\sigma(X_t,t)f'(X_t)N(0,1), \quad (5.21)$$

plus terms of mean and variance $o(h)$. Substituting $h^{-1/2}(W_{t+h} - W_t)$ for $N(0,1)$, we conclude that $f(X_t)$ satisfies the stochastic differential equation

$$d(f(X_t)) = \left(f'(X_t)\mu(X_t,t) + \frac{f''(X_t)}{2}\sigma(X_t,t)^2\right)dt + \sigma(X_t,t)f'(X_t)dW_t. \quad (5.22)$$

This is the chain rule for stochastic calculus and is almost the same as the chain rule for ordinary calculus, except that the additional term involving the second derivative appears in the dt term. This rule is known as *Ito's lemma* and the extra term is sometimes called the *Ito term*.

If we allow f to be a function of time as well as x a simple extension of the argument above yields

$$d(f(X_t,t)) = \left(\frac{\partial f}{\partial t}(X_t,t) + \frac{\partial f}{\partial x}(X_t,t)\mu(X_t,t) + \frac{1}{2}\frac{\partial^2 f}{\partial x^2}\sigma(X_t,t)^2\right)dt$$
$$+ \sigma(X_t,t)\frac{\partial f}{\partial x}(X_t,t)dW_t. \quad (5.23)$$

Ito's lemma is the fundamental tool in stochastic calculus and in its applications to finance. It is probably most easily remembered as follows.

Theorem 5.1 (Ito's Lemma) *Let X_t be an Ito process satisfying*

$$dX_t = \mu(X_t,t)dt + \sigma(X_t,t)dW_t, \quad (5.24)$$

and let $f(x,t)$ be a twice-differentiable function; then we have that $f(X_t,t)$ is an Ito process, and that

$$d(f(X_t,t)) = \frac{\partial f}{\partial t}(X_t,t)dt + f'(X_t,t)dX + \frac{1}{2}f''(X_t,t)dX_t^2 \quad (5.25)$$

where dX_t^2 is defined by

$$dt^2 = 0, \quad (5.26)$$
$$dt\,dW_t = 0, \quad (5.27)$$
$$dW_t^2 = dt. \quad (5.28)$$

Note that the final multiplication rule is the crucial one which gives the extra term.

A similar argument gives us a rule when we have several Ito processes based on the same Brownian motion.

Theorem 5.2 (Ito's Lemma) *Let $X_t^{(j)}$ be an Ito process for each j satisfying*

$$dX_t^{(j)} = \mu_j(t, X_t)dt + \sigma_j(t, X_t)dW_t, \quad (5.29)$$

and let $f(t, x_1, \ldots, x_n)$ be a twice-differentiable function; then we have that $f(t, X_t^{(1)}, X_t^{(2)}, \ldots, X_t^{(n)})$ is an Ito process, and that

$$d\left(f\left(t, X_t^{(1)}, X_t^{(2)}\right)\right) = \frac{\partial f}{\partial t}dt + \sum_{j=1}^n \frac{\partial f}{\partial x_j}dX_t^{(j)} + \frac{1}{2}\sum_{j,k=1}^n \frac{\partial^2 f}{\partial x_j \partial x_k}dX_t^{(j)}dX_t^{(k)}, \quad (5.30)$$

where $dX_t^{(j)}dX_t^{(k)}$ is defined by

$$dt^2 = 0, \quad (5.31)$$
$$dt\,dW_t = 0, \quad (5.32)$$
$$dW_t^2 = dt. \quad (5.33)$$

Note that all our Ito processes here are defined by the same Brownian motion. Later on we will want to consider stocks driven by different Brownian motions.

One important consequence of Ito's lemma is a product rule for Ito processes; if we let $f(x, y) = xy$, then we have

Proposition 5.1 *If X_t and Y_t are Ito processes then*

$$d(X_t Y_t) = X_t dY_t + Y_t dX_t + dX_t dY_t. \quad (5.34)$$

Note that this generalizes the Leibniz rule from ordinary calculus by involving a third term.

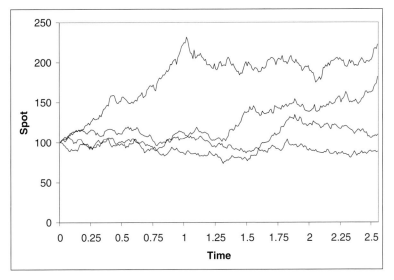

Fig. 5.2. Some paths for a stock following a log-normal process.

5.5 Applying Ito's lemma

As this may all have been a bit abstract, we now do some concrete examples. A standard model for the evolution of stock prices is geometric Brownian motion, that is

$$dS_t = \mu S_t dt + \sigma S_t dW_t, \qquad (5.35)$$

with μ and σ both constant. This is often written as

$$\frac{dS_t}{S_t} = \mu dt + \sigma dW_t. \qquad (5.36)$$

The idea here is that movements in a stock's value ought to be proportional to its current value, as it is percentage movements that matter not absolute ones. We give some example paths in Figure 5.2. For example, for a stock worth $1000 losing $10 is inconsequential, whereas for one worth $20 losing $10 has lost half its value. The term μ is called the *drift* of the stock since it expresses the trend of the stock's movements, and σ is called the *volatility* of the stock as it expresses how much the price wobbles up and down, or equivalently how risky it is. Since investors generally expect greater yields in return for greater uncertainty, we expect that the drift will be higher for stocks with high volatility. If our risk-free money market account follows the process

$$dB_t = rB_t dt, \qquad (5.37)$$

which is equivalent to $B_t = B_0 e^{rt}$, then the difference $\mu - r$ expresses the size of the risk premium. It is the amount of extra growth investors demand to compensate for extra risk introduced by the Brownian motion. Since we expect the premium to increase with volatility the ratio

$$\lambda = \frac{\mu - r}{\sigma}, \qquad (5.38)$$

is often useful, and is called *the market price of risk*. We stress that λ is only the market price of this specific piece of risk and other unrelated pieces of risk may well have different market prices. One might, however, use the market price of risk as a guide to which stocks are good value in the sense of giving good returns at low risk. (Note that this ignores the difference between diversifiable risk and systemic risk discussed in Chapter 1.)

There can be only one growth rate for portfolios which have no random part; for suppose we have that

$$dB_t = rB_t dt, \qquad (5.39)$$
$$dB'_t = r'B'_t dt, \qquad (5.40)$$

with $r < r'$. Without loss of generality, we suppose $B_0 = B'_0 = 1$ (otherwise, take a linear multiple). The portfolio consisting of $B' - B$ will have zero value initially and value

$$e^{r't} - e^{rt} > 0$$

for all $t > 0$. The portfolio therefore constitutes an arbitrage and we conclude that no-arbitrage requires that r is equal to r'.

This model is often called the log-normal model of stock price evolution as the log of the stock price follows a normal distribution. To see this, we use Ito's lemma and some stochastic calculus; suppose we put $S_t = e^{Y_t}$ or $Y_t = \log S_t$, what SDE does Y_t satisfy? By Ito's lemma

$$dY_t = d(\log S_t) = (\log S_t)' \mu S_t dt + (\log S_t)' \sigma S_t dW_t + \frac{1}{2} (\log S_t)'' \sigma^2 S_t^2 dt. \qquad (5.41)$$

As $\log S'_t = S_t^{-1}$ and $\log S''_t = -S_t^{-2}$, we obtain

$$dY_t = \left(\mu - \frac{1}{2}\sigma^2\right) dt + \sigma dW_t. \qquad (5.42)$$

We conclude that $\log S$ is a simple Brownian motion with drift. Recalling our derivations above, we have that

$$Y_t - Y_0 = \left(\mu - \frac{1}{2}\sigma^2\right) t + \sigma \sqrt{t} N(0, 1). \qquad (5.43)$$

5.5 Applying Ito's lemma

Now, as $S_t = e^{Y_t}$, we conclude that

$$S_t = S_0 e^{(\mu - \frac{1}{2}\sigma^2)t + \sigma\sqrt{t}N(0,1)}. \tag{5.44}$$

We have thus solved the stochastic differential equation (5.36). Unfortunately, this is one of the very few stochastic differential equations that have explicit solutions. In general, one can only write down facts *about* the solution. Fortunately, we shall not need the solutions of other SDEs, and in fact most of our work shall consist of manipulating stochastic differential equations in such a way as to eliminate randomness; that is, we turn stochastic differential equations into partial differential equations by mixing quantities judiciously in such a way as to eliminate the dW_t term that is the source of the randomness.

If we wish to interpret these ideas in terms of the market, we can regard dW_t as modelling the arrival of information which may be good or bad and therefore drives the stock price up or down. An option on a stock will be driven by the same information and therefore its stochastic differential equation will be driven by the same dW_t, so if we combine the option and the stock judiciously we ought to be able to eliminate the randomness. This observation is at the heart of the Black–Scholes approach to pricing options.

Before proceeding to the derivation of the Black–Scholes equation, we look at a further example of the application of Ito's lemma. Suppose our stock movements were not strictly proportional to level but instead obeyed a power law:

$$dS_t = S_t^\alpha \mu dt + S_t^\beta \sigma dW_t, \tag{5.45}$$

with $\beta \neq 0, 1$. Such a process is called a *constant elasticity of variance* process or a *CEV* process. In order to solve the SDE we would like to make the process constant coefficient. If we take $d(f(s))$ for some smooth function f then the volatility term of the new process will be, from Ito's lemma,

$$f'(S)S^\beta \sigma.$$

For the volatility term to be constant coefficient, we need

$$f'(S) = S^{-\beta}. \tag{5.46}$$

Thus, we let

$$f(S) = \frac{S^{1-\beta}}{1-\beta}. \tag{5.47}$$

Note that

$$f''(S) = -\beta S^{-\beta-1}. \tag{5.48}$$

Applying Ito's lemma, we have

$$d(f(S)) = \left(S^{\alpha-\beta}\mu - \frac{\beta}{2}S^{\beta-1}\sigma^2\right)dt + \sigma dW. \tag{5.49}$$

For general σ and μ this will be constant coefficient only if α and β are 1 – that is, we are back in the log-normal case which we had already ruled out. (We had assumed as well that they were not zero which would also give us a constant coefficient process.)

This example illustrates the fact that solving SDEs is difficult; the existence of two interacting terms generally makes it impossible to simplify via a change of coordinates.

5.6 An informal derivation of the Black–Scholes equation

Suppose we wish to price a call option, C, on a stock, S, with expiry T and strike K. We assume that S follows a geometric Brownian motion with drift μ and volatility σ. We take the risk-free money-market account to be continuously compounding at a rate r. The value of C at a time $t < T$ will depend on the value of S and the value of t, so we write $C(S, t)$. Note that writing C as $C(S, t)$ implicitly assumes that there is a unique well-defined price for the option. We shall eventually justify this assumption.

All other parameters are fixed throughout so we need not explicitly consider dependence upon them. We can apply Ito's lemma to deduce the stochastic differential equation for $C(S, t)$:

$$dC = \frac{\partial C}{\partial t}dt + \frac{\partial C}{\partial S}(S, t)dS + \frac{1}{2}\frac{\partial^2 C}{\partial S^2}(S, t)dS^2. \tag{5.50}$$

Expanding dS, we obtain

$$dC = \left(\frac{\partial C}{\partial t}(S, t) + \mu S\frac{\partial C}{\partial S}(S, t) + \frac{1}{2}\sigma^2 S^2\frac{\partial^2 C}{\partial S^2}(S, t)\right)dt + \sigma S\frac{\partial C}{\partial S}(S, t)dW_t. \tag{5.51}$$

This equation contains, of course, the derivatives of C with respect to t and S and if we know them, then simply by integrating we can find the original function. Nevertheless, we can manipulate it in a useful way. If we consider the portfolio consisting of the option and α stocks, we obtain from (5.51)

$$d(C + \alpha S) = \left(\frac{\partial C}{\partial t}(S, t) + \mu S\frac{\partial C}{\partial S}(S, t) + \frac{1}{2}\sigma^2 S^2\frac{\partial^2 C}{\partial S^2}(S, t) + \alpha\mu S\right)dt$$
$$+ \sigma S\left(\frac{\partial C}{\partial S} + \alpha\right)dW_t. \tag{5.52}$$

If we set α equal to $-\frac{\partial C}{\partial S}(S,t)$, we obtain

$$d(C + \alpha S) = \left(\frac{\partial C}{\partial t}(S,t) + \frac{1}{2}\sigma^2 S^2 \frac{\partial^2 C}{\partial S^2}(S,t) \right) dt; \qquad (5.53)$$

actually we do not obtain this, as we are ignoring the derivative of α which we cannot do – we will present a more rigorous argument in the next section. Our choice of α means that we are carrying out Delta-hedging: we now have a portfolio which is deterministic; that is, it has no random component. Since a risk-free portfolio must grow at the risk-free rate, we conclude that the drift of $C + \alpha S$ must be equal to $r(C + \alpha S)$. We therefore conclude that

$$\frac{\partial C}{\partial t}(S,t) + \frac{1}{2}\sigma^2 S^2 \frac{\partial^2 C}{\partial S^2}(S,t) = r\left(C - S\frac{\partial C}{\partial S}\right). \qquad (5.54)$$

Upon rearranging we have

$$\frac{\partial C}{\partial t}(S,t) + rS\frac{\partial C}{\partial S} + \frac{1}{2}\sigma^2 S^2 \frac{\partial^2 C}{\partial S^2}(S,t) = rC. \qquad (5.55)$$

We have thus deduced that the value of a call option satisfies a second-order partial differential equation. This equation is called the *Black–Scholes equation* after its inventors.

We do not yet have quite enough information to solve the equation since a second-order linear partial differential equation has many solutions. However, we certainly know what the value of our option will be at expiry, so for $t = T$ we have

$$C(S,T) = \max(S - K, 0), \qquad (5.56)$$

since the option would be exercised if and only if S was bigger than K. Note that it is only at this point that the fact that C is a call option comes in. We could apply this analysis to any European contingent claim, i.e. a derivative which pays off a function, f, of S at time T, simply by putting

$$C(S,T) = f(S). \qquad (5.57)$$

With this final condition, we have enough data to find a unique solution.

5.7 Justifying the derivation

Before proceeding to the solution, we prove that the solution to the equation is, in fact, the unique arbitrage-free price for the option. The argument we have given made a couple of dubious assumptions – α is a function of S so why can we ignore its derivative? Also we assumed that C is a well-defined function of S and t which is really part of what we are trying to prove.

We prove the validity by a replication argument. In particular, we show that if $C(S_0, 0)$ is the solution of the Black–Scholes equation at time zero and with spot equal to S_0, today's spot, then it is possible to execute a trading strategy which results in having precisely $\max(S_T - K, 0)$ pounds at time T. This trading guarantees that we have the option's pay-off no matter what happens in between and what the value of S_T is.

Once we have proven that we can do the replication, then the price of the option follows by an arbitrage argument. If the price of the option were greater than $C(S_0, 0)$ then one could find an arbitrage by selling the option and executing this trading strategy to cover the cost of its exercise at time T. If the price of the option were less than $C(S_0, 0)$, one would buy the option and execute the negative of the trading strategy and once again make a risk-free profit no matter what happened: we would realize an arbitrage opportunity.

To describe our trading strategy, we need the concept of a *self-financing portfolio*. This is a portfolio which is set up at time 0, i.e. today, and to which no cash is injected or extracted during the lifetime of the contract. However, selling some assets to buy new assets is permitted. This means that all changes in value of the portfolio come from the changes in value of the underlying assets. In our case, we are interested in a portfolio of risk-free bonds and the underlying stock. If we write

$$P = \alpha S + \beta B, \tag{5.58}$$

then our self-financing condition becomes

$$dP = \alpha dS + \beta dB, \tag{5.59}$$

thus expressing the notion that all changes come from the changes in S and B rather than from those in α and β. It is important to realize that equation (5.59) is not just the linearity of differentiation as α and β will generally not be constants. Indeed, we could let α and β at time t depend on the entire path of S up to time t. We would not, however, let α and β depend on the value of S after time t, as this would involve in some sense seeing the future; we will return to this point in the next chapter. In the derivation that follows we will take α and β to be functions of $S(t)$ and t only. Note that as the path of S is continuous, the value of $S(t)$, and hence α and β is determined in some sense by information from the past.

The self-financing condition expresses the simple financial idea that money is neither added to nor subtracted from the portfolio. As we wish to execute trades within the portfolio, but cannot add or subtract cash, there are clearly constraints on the values the pair (α, β) can take. If we increase α through trading, β must decrease and vice versa. This means that if we specify one of α and β and the other is then determined. Mathematically, we have

5.7 Justifying the derivation

Theorem 5.3 *Let S be a stock following geometric Brownian motion and let B be the money-market account continuously compounding at rate r. Then given a smooth function $\alpha(S, t)$ and an initial value P_0 then there is a unique smooth function $\beta(S, t)$ such that*

$$P = \alpha S + \beta B, \tag{5.60}$$

is a self-financing portfolio with initial value P_0.

We now have the tools to prove our main result that $C(S_0, 0)$ is the unique arbitrage-free price of the option. Letting $C(S, t)$ be the solution of the Black–Scholes equation, we set up a portfolio P with initial value $C(S_0, 0)$ and, following our heuristic argument above, we set

$$\alpha(S, t) = \frac{\partial C}{\partial S}.$$

Remember that this was the hedge that made the portfolio instantaneously riskless. As we are now trying to replicate C rather than hedge a long position in C, the sign of our hedge has changed.

We show that

$$P(S, t) = C(S, t) \text{ for } 0 < t \le T.$$

In particular, it will follow that

$$P(S, T) = \max(S_T - K, 0)$$

which is another way of saying that the portfolio P will precisely reproduce the option's pay-off as desired.

To show that $P(S, t) = C(S, t)$, we consider the difference. This will initially be zero, by construction, and we have

$$\begin{aligned}
d(P(S, t) - C(S, t)) &= dP - dC, \\
&= \frac{\partial C}{\partial S} dS + \beta(t) dB - \frac{\partial C}{\partial t} dt \\
&\quad - \frac{\partial C}{\partial S} dS - \frac{1}{2} \frac{\partial^2 C}{\partial S^2} dS^2, \\
&= \beta(t) r B dt - \frac{\partial C}{\partial t} dt - \frac{1}{2} \frac{\partial^2 C}{\partial S^2} \sigma^2 S^2 dt. \quad (5.61)
\end{aligned}$$

Recalling that $P = \beta B + S \frac{\partial C}{\partial S}$, and that by the Black–Scholes equation,

$$\frac{\partial C}{\partial t} + \frac{1}{2} \sigma^2 S^2 \frac{\partial^2 C}{\partial S^2} = r \left(C - S \frac{\partial C}{\partial S} \right),$$

it follows that

$$d(P - C) = r\left(P - S\frac{\partial C}{\partial S}\right)dt - r\left(C - S\frac{\partial C}{\partial S}\right)dt. \qquad (5.62)$$

We therefore find that

$$d(P - C) = r(P - C)dt. \qquad (5.63)$$

As we have the initial condition

$$(P - C)(S, 0) = 0, \qquad (5.64)$$

we conclude that the unique solution of this differential equation is identically zero. To see this simply observe that zero is a solution and so by uniqueness is the only solution. Thus we have that

$$P(S, t) = C(S, t) \text{ for } 0 \leq t \leq T, \qquad (5.65)$$

as required.

We have shown that by investing $C(S, 0)$ at time zero and carrying out Delta hedging, we have precisely the value of the payoff of the option at time T, no matter what happens. That is we have replicated the option's pay-off on a path by path basis. It therefore follows that the option must be worth $C(S_0, 0)$ at time 0, and indeed $C(S_t, t)$ at any time in between, by the same argument. We stress once again that there is nothing probabilistic in this conclusion – there is no sense in which we are pricing on average, we have eliminated all risk via our hedging strategy. Note also that our derivation shows that the option has a unique well-defined price – we have not assumed that the price exists other than in the heuristic part of our derivation. Indeed, the fact that the option has a well-defined price depending only on the variables S and t, and the parameters r and σ is the most important aspect of the result. The reason this is so important is that it is so unexpected; without the possibility of perfect hedging one would expect a role for the investors' risk preferences and hence the impossibility of a unique price.

We also emphasize that the price is independent of μ – the drift of the stock is irrelevant. This means that two investors with totally different opinions of the value of μ can agree on the price of the option provided they agree on the volatility. This surprising fact simply reflects the fact that the hedging strategy ensures that the underlying drift of the stock is balanced against the drift of the option. The conceptual reason that the drifts are balanced is that the drift reflects the risk premium demanded by investors to account for an uncertainty and that uncertainty has been hedged away.

5.8 Solving the Black–Scholes equation

We thus have a partial differential equation that the price of an option satisfies, and, of course, we want to solve it. The surprising thing about the Black–Scholes equation is that it is fairly easy to write down the solution. Indeed the Black–Scholes equation is really just the one-dimensional heat equation in disguise. To see this, we can rewrite it as

$$\frac{\partial C}{\partial t} + \left(r - \frac{1}{2}\sigma^2\right) S \frac{\partial C}{\partial S} + \frac{1}{2}\sigma^2 \left(S \frac{\partial}{\partial S}\right)^2 C = rC. \tag{5.66}$$

Recalling that the stochastic differential equation for the stock S was much simpler when expressed in terms of $\log S$, we can try the same approach here. Let $S = e^Z$, that is $Z = \log S$. The equation then becomes

$$\frac{\partial C}{\partial t} + \left(r - \frac{1}{2}\sigma^2\right) \frac{\partial C}{\partial Z} + \frac{1}{2}\sigma^2 \frac{\partial^2 C}{\partial Z^2} = rC, \tag{5.67}$$

which is constant coefficient.

To simplify the equation further, consider that it is not really the current time that will affect the price of an option but rather the amount of time to go. Putting $\tau = T - t$, we obtain

$$\frac{\partial C}{\partial \tau} - \left(r - \frac{1}{2}\sigma^2\right) \frac{\partial C}{\partial Z} - \frac{1}{2}\sigma^2 \frac{\partial^2 C}{\partial Z^2} = -rC. \tag{5.68}$$

As we are trying to price the value of a possible cash flow in the future, we can write $C = e^{-r\tau} D$, expressing the notion that we are discounting the possible future cash flow D to the current time. We then obtain

$$\frac{\partial D}{\partial \tau} - \left(r - \frac{1}{2}\sigma^2\right) \frac{\partial D}{\partial Z} - \frac{1}{2}\sigma^2 \frac{\partial^2 D}{\partial Z^2} = 0. \tag{5.69}$$

Next, we eliminate the first-order term. Now the mean value of Z at time t is $Z(0) + \left(r - \frac{1}{2}\sigma^2\right)t$. It is therefore reasonable to shift coordinates to take this into account. We therefore let

$$y = Z + \left(r - \frac{1}{2}\sigma^2\right)\tau$$

and our equation becomes

$$\frac{\partial D}{\partial \tau} = \frac{1}{2}\sigma^2 \frac{\partial^2 D}{\partial y^2}. \tag{5.70}$$

This is the one-dimensional heat-equation. Thus to solve the Black–Scholes equation we simply transform the boundary condition, solve the heat equation and transform back. We leave this as an exercise for the enthusiastic reader and simply state the solution for a call option

$$C(S, t) = SN(d_1) - Ke^{-r(T-t)}N(d_2), \qquad (5.71)$$

with $N(x)$ denoting the cumulative normal distribution, $\frac{1}{\sqrt{2\pi}}\int_{-\infty}^{x} e^{-s^2/2}ds$, and

$$d_1 = \frac{\log(S/K) + (r + \frac{1}{2}\sigma^2)(T-t)}{\sigma\sqrt{T-t}}, \qquad (5.72)$$

$$d_2 = \frac{\log(S/K) + (r - \frac{1}{2}\sigma^2)(T-t)}{\sigma\sqrt{T-t}}. \qquad (5.73)$$

In Chapter 6, we give an alternative method of deriving the solution depending upon probabilistic ideas which makes the provenance of the terms d_1 and d_2 a lot clearer.

That this is a solution can be verified by direct substitution. Under fairly mild conditions, for example positivity or exponential boundedness, solutions of the heat equation are unique, so we can be sure that this solution is the correct one. Remember that the price of an option is always positive, since there is always some possibility that it can make the holder some money in the future.

The fact that the Black–Scholes equation is the heat equation in disguise has some interesting consequences. The solution operator for the heat equation is a *smoothing operator.* This means that even for a complicated final pay-off with lots of spikes and jumps, the value of the derivative is always a smooth function of (S, t) in the sense that it can be differentiated infinitely often.

An important property of heat flows is their asymmetry. If one attempts to flow heat backwards in time, one does not obtain a solution of the heat equation. This is unlike the wave equation which does not see the direction of time. The idea is that heat flows destroy information – everything blurs together. This means that the prices of options which satisfy the time-reversed heat equation, can only ever flow backwards in time, i.e. we can price an option in the future, and as the expiry of the option approaches, the solution becomes less blurred as it converges towards the pay-off; see Figure 5.3.

The formula we have deduced for the price of a call option is, of course, the same as the one that we deduced in Chapter 3; the two very different approaches lead to the same answer. Of course, it would be worrying if they did not. The approach in this chapter is much closer to that of the original paper by Black & Scholes, whereas the tree approach was introduced later by Cox, Ross & Rubinstein. Both techniques have their uses; one's choice depends on the specific properties of the option being studied. When studying exotic options, we will examine the advantages and disadvantages in several cases.

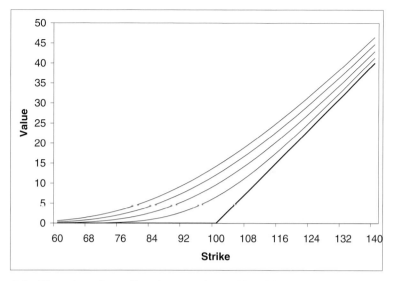

Fig. 5.3. The price of a call option struck at 100, with vol 30%, with times 0, 0.25, 0.5, 0.75 and 1 year remaining. As time to expiry goes to zero, the graphs becomes progressively more like the final pay-off.

5.9 Dividend-paying assets

The equation we have derived is only valid for an option on a non-dividend paying stock. In practice, many stocks do pay dividends; moreover we may want to consider options on foreign currencies, since a foreign currency holding will grow at the risk-free rate in that currency, currencies can also be considered to pay dividends. We therefore need to bring this into our model. We will model the dividend yield on a stock as if it were a currency; that is, the number of stocks held will grow at a continuously compounding constant rate. In order to motivate what follows, let us examine a little more carefully what happens when we hedge a stock. We hold at any given time $\frac{\partial C}{\partial S}$ stocks. We know that as $t \to T-$, C converges to $\max(S - K, 0)$ and, away from $S = K$, differentiating this implies that $\frac{\partial C}{\partial S}$ converges to 1 for $S > K$, and 0 for $S < K$. (For those worried about the interchange of limits here, this follows from properties of the heat equation.)

At the pay-off time, the hedger therefore holds an empty portfolio if S is less than K, and otherwise holds a portfolio consisting of one stock and $-\pounds K$.

Ultimately, when carrying out a replicating strategy, our objective is to hold this portfolio at the expiry time; how we achieve this portfolio is not important. In particular, suppose that instead of hedging with the stock, we hedge with contracts that involve payment for the stock today but delivery of the stock at time T. For a non-dividend-paying stock, the cost of such a contract will be identical to that of a stock. To see this, just observe that in any state of the world at time T, the contract and the stock will have identical worth as both just involve the investor holding precisely one stock. For a dividend-paying stock things are slightly different; the investor who buys the stock at time t will hold $e^{d(T-t)}$ stocks at time T, whereas the investor who buys the delivery contract at time t will hold just one stock at time T. We therefore conclude that buying $e^{-d(T-t)}$ stocks today is equivalent to buying one delivery contract. This means that the price of the delivery contract, X_t, must be $e^{-d(T-t)}S_t$. Note that as the delivery contracts are replicable by a trading strategy, the fact that they do not exist in the market is irrelevant. All that matters is the fact that they can be replicated.

What does this buy us? An option on X_t with expiry T must have the same value as an option on S_t, since S_T equals X_T in all world states. The asset X_t is however non-dividend-paying so we can apply the Black–Scholes analysis to it directly. Note that X_t is described by the process

$$dX_t = (\mu + d)X_t dt + X_t \sigma dW_t, \tag{5.74}$$

which is just geometric Brownian motion with a different drift. But the drift does not affect the Black–Scholes price. We therefore conclude that the price of the option C satisfies

$$\frac{\partial C}{\partial t}(t, X) + \frac{1}{2}\sigma^2 X^2 \frac{\partial^2 C}{\partial X^2}(t, X) + rX\frac{\partial C}{\partial X}(t, X) - rC(t, X) = 0. \tag{5.75}$$

Transforming to an equation in S we obtain

$$\frac{\partial C}{\partial t}(t, S) + \frac{1}{2}\sigma^2 S^2 \frac{\partial^2 C}{\partial S^2}(t, S) + (r - d)S\frac{\partial C}{\partial S}(t, S) - rC(t, S) = 0. \tag{5.76}$$

We can now solve this with the same boundary condition as before, or, more simply, just substitute into the original solution of the Black–Scholes equation to obtain

$$C(S, t) = Se^{-d(T-t)}N(d_1) - Ke^{-r(T-t)}N(d_2) \tag{5.77}$$

where

$$d_1 = \frac{\log(S/K) + \left(r - d + \frac{1}{2}\sigma^2\right)(T - t)}{\sigma\sqrt{T - t}}, \tag{5.78}$$

$$d_2 = \frac{\log(S/K) + \left(r - d - \frac{1}{2}\sigma^2\right)(T - t)}{\sigma\sqrt{T - t}}. \tag{5.79}$$

A similar analysis can be applied to options on commodities. The essential difference between money and commodities is that holdings of money grow as interest is paid on them whereas holdings of commodities cost money just to hold since warehousing must be paid for. This money is known as *cost of carry* and can be represented very simply as a negative dividend. One therefore obtains the same equations but with $d = -q$, where q is the cost of carry. In commodities markets, it is sometimes very important to actually hold the physical asset in case one actually wants to use it. Commodities are therefore often modelled with an additional positive dividend-type process called the *convenience yield*, y. One then simply sets

$$d = y - q.$$

Note that d could be either positive or negative.

In our analysis of a dividend-paying asset, we hedged the option with contracts involving payment today but delivery at the expiry of the option. We could equally well have hedged with forward contracts, that is with contracts that involve payment and delivery at time T but with the size of the payment fixed today. This has certain advantages and we will return to this point when we have developed more theory.

5.10 Key points

In this chapter we have covered a lot of ground. We introduced the concept of an Ito process and developed a calculus for manipulating them. We used that calculus to deduce a necessary price for a call option and extended the model to cope with dividend-paying assets.

- A Brownian motion is a process W_t such that for any $t > s$, $W_t - W_s$ is normally distributed with variance $t - s$ and mean zero and is independent of W_s.
- Stochastic calculus generalizes ordinary calculus by letting the derivative have a random component coming from a Brownian motion.
- Stochastic calculus deals with Ito processes, in which the random variable's derivative has both a deterministic linear part and a random part which is normally distributed.
- The fundamental tool of stochastic calculus is Ito's lemma which says that if

$$dX_t = \mu(X_t, t)dt + \sigma(X_t, t)dW_t$$

then

$$df(X_t, t) = \frac{\partial f}{\partial t}(X_t, t)dt + \frac{\partial f}{\partial x}(X_t, t)dX_t + \frac{1}{2}\frac{\partial^2 f}{\partial x^2}(X_t, t)dX_t^2,$$

with the multiplication rules

$$dt^2 = 0,$$
$$dt\,dW_t = 0,$$
$$dW_t\,dW_t = dt.$$

- A standard model for stock evolution is geometric Brownian motion:
$$dS_t = S_t\mu dt + S_t\sigma dW_t.$$
- The SDE for geometric Brownian motion is solved by
$$S(t) = S(0)e^{(\mu-\frac{1}{2}\sigma^2)t+\sigma\sqrt{t}N(0,1)}.$$
- For a stock following geometric Brownian motion the quantity
$$\lambda = \frac{\mu-r}{\sigma}$$
is called the market price of risk.
- Risk can be eliminated by holding a portfolio in which the random parts of two different assets cancel each other.
- No-arbitrage implies that a portfolio from which risk has been eliminated must grow at the riskless rate.
- A self-financing portfolio is a portfolio in which assets are bought and sold but no money is taken in or out.
- A European option's payoff can replicated by a self-financing portfolio consisting of dynamic trading in stock and riskless bond.
- No-arbitrage guarantees that the price of a European option must satisfy the Black–Scholes equation.

5.11 Further reading

There are many books on stochastic calculus. One that is very popular and reasonably accessible is *Stochastic Differential Equations: an Introduction with Applications*, by Bernt Oksendal, [100].

A more rigorous book on stochastic calculus written for the pure mathematician which requires hard work, but is worthwhile for those who are not willing to take results on faith is Karatzas & Shreve's *Brownian Motion and Stochastic Calculus*, [78].

The PDE approach to mathematical finance developed in this chapter has now been superseded by the martingale approach we describe in the next chapter. However, the approach can be pushed quite a long way and two accessible books following this approach are Wilmott, Howison & Dewynne, [119], and Wilmott, [118].

5.12 Exercises

Exercise 5.1 If
$$dX_t = \mu X_t dt + \sigma X_t dW_t,$$
what process does X_t^k follow?

5.12 Exercises

Exercise 5.2 If

$$dX_t = \mu(t, X_t)dt + \sigma(X_t)dW_t,$$

with σ positive, show there exists a function f such that

$$d(f(X_t)) = v(t, X_t)dt + VdW_t$$

where V is a constant. How unique is f?

Exercise 5.3 Suppose S_t is the price of a non-dividend paying stock following geometric Brownian motion. Let F_t for $t < 1$ be the forward price of the stock for a contract expiring at time 1. What process does F_t follow?

Exercise 5.4 If S_t follows geometric Brownian motion, what process does S_t^{-1} follow?

Exercise 5.5 What sort of qualitative behaviour does a stock following a process of the form

$$dS_t = \alpha(\mu - S_t)dt + S_t\sigma dW_t,$$

exhibit? What qualitative effects do altering μ and α have? What effects do they have on the price of a call option?

Exercise 5.6 Show that S_t is a solution of the Black–Scholes equation. Why should this be so?

Exercise 5.7 Show that Ae^{rt} is a solution of the Black–Scholes equation. Why should this be so?

Exercise 5.8 Show that the solution of the Black–Scholes equation for a call option is bounded by 0 and S.

Exercise 5.9 Show that if $f(S, t)$ and $g(S, t)$ are solutions of the Black–Scholes equation and $f(S, T) \leq g(S, T)$ for all S and some T then $f \leq g$ for $t \leq T$.

Exercise 5.10 If

$$dX_t^{(j)} = \mu_j dt + \sigma dW_t$$

and $\mu_1 < \mu_2$, with $X_0^{(0)} = X_0^{(1)}$ show that for $t > 0$,

$$X_t^{(1)} < X_t^{(2)}.$$

Exercise 5.11 Suppose an asset follows Brownian motion instead of geometric Brownian motion. Find the analogue of the Black–Scholes equation.

Exercise 5.12 Suppose we have a call option on the square of the stock price. That is the pay-off is $(S_T^2 - K)_+$. What equation does the price of the call option at time t satisfy? What is the solution to this equation?

6

Risk neutrality and martingale measures

6.1 Plan

In this chapter, we introduce and study the theory of pricing using martingales. This is a difficult and complicated topic with many aspects. Our plan of attack is as follows:

- We review pricing on trees from a slightly different viewpoint.
- We show that under very slight assumptions vanilla option prices for a single time horizon are given by an expectation under an appropriate probability density.
- We show how to compute this density and observe that it implies that the stock grows at rate r.
- In the Black-Scholes model, this density is that implied by giving a stock a drift equal to r instead of μ.
- We return to multiple time horizons and view stochastic processes as the slow revelation of a single path drawn in advance.
- The concept of information is examined in this context, and defined using filtrations.
- In the discrete setting, we define expectations conditioned on information.
- A martingale is defined to be a process such that its value is always equal to its expected value.
- It is shown that martingale pricing implies absence of arbitrage in the discrete setting.
- The basic properties of conditioning on information are surveyed in the continuous setting.
- Continuous martingales are identified to be processes with zero drift (up to technical conditions.)
- Pricing with martingales in the continuous setting is introduced.
- The Black-Scholes equation is derived using martingale techniques.

- We study how to hedge using martingale techniques.
- The Black-Scholes model with time-dependent parameters is studied.
- The concept of completeness, the perfect replication of all options, is introduced and its connections with uniqueness of martingale measures are explored.
- Numeraires are introduced and it is shown that the derivation of the Black-Scholes formula can be greatly simplified by using the change of numeraire technique.
- We extend the martingale pricing theory to cover dividend-paying stocks.
- We look at the implications of regarding the forward price as the underlying instead of the stock.

6.2 Introduction

We have presented two different approaches to deriving the Black–Scholes equation. The first approach relied on approximating Brownian motion by a discrete process in which the value at each node of a tree was determined by no-arbitrage arguments. The second approach relied on deriving a partial differential equation for the price of an option. Our purpose in this chapter is to derive a third, more fundamental, approach that relies on the concept of a martingale measure which we will introduce.

Let us assume there are no interest rates. Recall that when studying risk-neutral valuation on trees, we saw that the method worked because once the risk-neutral probabilities had been chosen, the expectation of every portfolio's value in the future was equal to today's value. This meant that once a portfolio had been created it could not have zero value today and positive value in the future with positive probability without also having non-zero probability of negative value. This meant that under the risk-neutral probabilities it was impossible for arbitrages to occur. And as the existence of arbitrage only depends on the sets of non-zero probability, the existence of arbitrage was only possible with real-world probabilities if it was possible with risk-neutral ones and so could not occur.

In the tree, we required the probabilities to be risk-neutral at every point in the tree: at any point in the tree all portfolios must have expectation equal to today's value. Why do we need this property? Why is it not enough just to have that all portfolios have expectation equal to today's value at the start. The reason is that we have to be careful what we mean by a portfolio. If we include in our set of portfolios all self-financing portfolios, then it is enough to require the property holds today, as then we cannot be arbitraged by any trading strategy having put our expectation condition on all possible portfolios that might arbitrage us. On the other hand, if we only place our condition on static portfolios today, which is equivalent to placing the condition on single assets, then we have made a much weaker condition, one which is not sufficient.

To see this, consider a simple example. There is a riskless bond of constant value 1 and an asset of initial value 0. On the first day, a coin is flipped, if the value comes up heads the asset is worth 1, otherwise it is worth −1. On the second day the value of the asset doubles.

The expected value of the asset at time 1 is zero, and at time 2 it is zero also. This means that any static portfolio has value at time 2 equal to its value at time zero, and we conclude that no static portfolio is an arbitrage.

However, if we are allowed to trade at time 1 the situation changes. We set up an initially empty portfolio. If the coin comes up heads, then at time 1 we borrow a bond and buy the asset, otherwise we do nothing. Our portfolio is then worth 1 at time 2 if the coin came up heads and zero otherwise. We have constructed an arbitrage.

In this example, our self-financing portfolio did not have the property that its future expectation was equal to today's value. This was caused by the fact that the asset at time 1 did not have future expectation equal to its value at time 1.

In conclusion, if we want to avoid arbitrage via trading strategies then we need to construct probabilities such that, if at any trading time and state of the world we take the expectation of any asset's future value, then it will be equal to the value it has in that state of the world at that trading time. This is called the *martingale* property.

For a tree, it is reasonably clear what this means: we simply go through each node of the tree assigning probabilities so that the expected value at each node is the expectation of its values on the next time slice. In continuous time, however, life is much more complicated. If we can trade at any time then we need the expectation property to hold at every time. We no longer have small time steps to work with, and so we must consider many times at once. We must also think a little about what we are assigning probabilities to: there is no longer any simple concept of up and down moves.

Our purpose in this chapter is to explore all these issues and use them to develop a powerful pricing theory. We start by examining the case of a single time horizon in more detail.

6.3 The existence of risk-neutral measures

One of the surprising facts of mathematical finance is that option prices actually define probability measures. These measures, however, do not make statements about the probability distribution of the asset's future price movements, but instead always imply a 'risk-neutral' evolution where the asset's rate of growth is the risk-free interest rate. Here we show how one can construct this risk-neutral distribution for a single maturity from option prices and examine how it relates to the option prices.

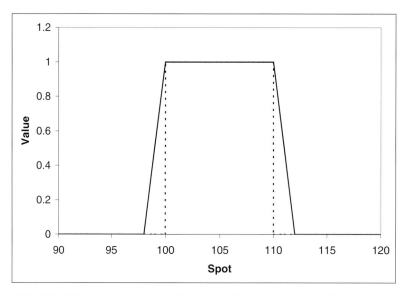

Fig. 6.1. The approximation of a double digital option by call options.

Suppose we can observe the price in the market for options of all strikes for a single maturity T. Let O_K be the price at time 0 of a call option struck at K on the underlying S. Let B_T be the price of a zero-coupon bond expiring at time T. We consider the price ratios $C_K = O_K/B_T$. In what follows, we implicitly assume that there are no interest rates, that is $B_T = 1$. However, this is merely for clarity and all the arguments work with little modification in the general case.

We know that C_K must be a decreasing function of K in an arbitrage-free market. We also know that any linear combination of different C_K's which lead to a non-negative final pay-off must be of positive value. Let D_I denote an option which pays 1 if spot ends in the interval I and zero otherwise. The interval could be of the form (K_1, K_2) or $[K_1, K_2]$. We can approximate such options by a portfolio of call options.

In particular, if we take $L_\epsilon = \epsilon^{-1}(C_{K_1-\epsilon} - C_{K_1} - C_{K_2} + C_{K_2+\epsilon})$, then we have pay-off as follows

$$0 \quad \text{for} \quad S_T < K_1 - \epsilon, \tag{6.1}$$

$$1 \quad \text{for} \quad S_T \in [K_1, K_2], \tag{6.2}$$

$$0 \quad \text{for} \quad S_T > K_2 + \epsilon, \tag{6.3}$$

and the pay-off varies linearly between 0 and 1 in the omitted intervals. See Figure 6.1.

6.3 The existence of risk-neutral measures

As the pay-off is at most 1, and can be 0, we conclude that the value of this portfolio is between 0 and 1. The value must be a decreasing function of ϵ as for $\epsilon < \epsilon'$ we have for the pay-offs

$$L_\epsilon(S, T) \leq L_{\epsilon'}(S, T)$$

with strict inequality for some values of S. The value is thus a strictly decreasing function of ϵ and non-negative. If we let ϵ tend to zero, it must therefore converge to a non-negative number less than 1.

We certainly have

$$\lim_{\epsilon \to 0+} L_\epsilon(S_0, 0) \geq D_{[K_1, K_2]}(S_0, 0).$$

We can do a similar approximation for the interval (K_1, K_2), by taking the portfolio

$$L'_\epsilon = \epsilon^{-1}(C_{K_1} - C_{K_1+\epsilon} - C_{K_2-\epsilon} + C_{K_2}), \tag{6.4}$$

and letting ϵ tend to zero. Plainly, one could also approximate $[K_1, K_2)$ and $(K_1, K_2]$ similarly. Will these four options always have the same value? Clearly, we have that $L_\epsilon(S_t, t) > L'_\epsilon(S_t, t)$ for all $\epsilon > 0$. This means that we have

$$\lim_{\epsilon \to 0+} L_\epsilon(S_0, 0) \geq D_{[K_1, K_2]}(S_0, 0) \geq D_{(K_1, K_2)}(S_0, 0) \geq \lim_{\epsilon \to 0+} L'_\epsilon(S_0, 0). \tag{6.5}$$

Intuitively, we would expect these limits to be the same. If the probability of the spot finishing on one of the two values K_1 and K_2 precisely is zero, then a portfolio consisting of being long $D_{[K_1, K_2]}$ and short $D_{(K_1, K_2)}$ has zero value at time T except on a set of zero probability.

Recall that an arbitrage portfolio is a portfolio of zero value which has non-negative value with probability 1 and positive value with non-zero probability. If $D_{[K_1, K_2]}$ and $D_{(K_1, K_2)}$ do not have the same value, we go long $D_{(K_1, K_2)}$, short $D_{[K_1, K_2]}$ and put the proceeds in riskless bonds. At expiry time, our portfolio will consist solely of the riskless bonds unless spot lands on K_1 or K_2 which is a zero probability event. Thus an arbitrage would exist and we conclude that the two options have the same value at time zero.

However, we still want to show that

$$\lim_{\epsilon \to 0+} L_\epsilon(S_0, 0) = D_{[K_1, K_2]}(S_0, 0). \tag{6.6}$$

In fact, no-arbitrage is too weak a condition to force equality. However, if the limit were higher we could do extremely well by shorting the portfolio L_ϵ for a very small ϵ and buying the option $D_{[K_1, K_2]}$.

We then make the sum $L_\epsilon(S_0, 0) - D_{[K_1, K_2]}(S_0, 0)$ which is of course greater than

$$x = \lim_{\epsilon \to 0+} L_\epsilon(S_0, 0) - D_{[K_1, K_2]}(S_0, 0).$$

So for any ϵ greater than 0 we make at least x unless the final spot lands in either of the intervals $[K_1 - \epsilon, K_1)$ and $(K_2, K_2 + \epsilon]$. The probability that the final spot lands in either of these intervals can be made arbitrarily small by taking ϵ arbitrarily small, and the maximum we can lose in that case is $1 - x$. Note that the variance of the difference portfolios will go to zero also. We can therefore make x by taking on an arbitrarily small amount of risk.

This does not imply an arbitrage: for an arbitrage we have to make money with no risk, here we are only making money with an arbitrarily small amount of risk. We could call such a situation an *arbitrarily good deal*. A good model will not allow such deals; we therefore need an extra condition to outlaw them.

Definition 6.1 We shall say that a market admits free lunches with vanishing risk if there exists a sequence of portfolios ϕ_n such that

(i) the expected value of ϕ_n is bounded below by x greater than zero independent of n, for n large,
(ii) the set-up cost of ϕ_n is less than or equal to zero for n large,
(iii) the variance of ϕ_n tends to zero as $n \to \infty$.

If a market does not admit free lunches with vanishing risk, we shall say that it satisfies the no free lunch with vanishing risk (NFLWVR) condition.

The NFLWVR condition is really a continuity condition on the pricing function; it implies that if a sequence of portfolios approximates a portfolio arbitrarily well, then the prices must converge to the limiting portfolio's price.

If we now impose the NFLWVR condition, we can conclude that

$$\lim_{\epsilon \to 0+} L_\epsilon(S_0, 0) = D_{[K_1, K_2]}(S_0, 0) = D_{(K_1, K_2)}(S_0, 0) = \lim_{\epsilon \to 0+} L'_\epsilon(S_0, 0). \quad (6.7)$$

Let $P(I)$ denote the value of D_I at $(S_0, 0)$. We now know that we can deduce this price from the price of the traded vanilla call options under the no-arbitrage and NFLWVR conditions.

What properties do the prices $P(I)$ have? Take a sequence of intervals

$$I_n = [a_n, a_{n+1})$$

such that $a_0 = 0$ and $a_n \to \infty$ as $n \to \infty$; then if we consider a portfolio consisting of all these options it will pay 1 whatever happens. So the initial portfolio is of value 1 too. This means that

$$\sum_{n=0}^{\infty} P(I_n) = 1. \quad (6.8)$$

6.3 The existence of risk-neutral measures

This is slightly suspect in that we have introduced a portfolio consisting of an infinite number of options. However, if we consider the portfolio Q_n consisting of the first n options then we can use the NFLWVR condition to show that the value of Q_n must converge to 1 as $n \to \infty$. To see this consider the portfolio, R_n, consisting of Q_n and going short a zero-coupon bond of notional 1.

The expected value of R_n is the negative of the probability that the spot ends above a_{n+1}. This clearly goes to zero as $n \to \infty$. The variance of R_n will be equal to the expectation as the pay-off is zero or 1. If the value of Q_n does not converge to 1, then it must converge to x less than 1 since it is bounded above by 1 and increasing with n. The setup cost of R_n is then at most $x - 1$. The sequence of portfolios consisting of $1 - x$ bonds together with R_n is now a free lunch with vanishing risk.

This means that under the NFLWVR assumption, we have assigned a number to every interval between zero and 1 such that the infinite interval $[0, \infty)$ receives the number 1. For disjoint intervals, A and B, we also have that

$$P(A \cup B) = P(A) + P(B).$$

We can regard $P(I)$ as the probability that the stock is in the interval I, at time T. However this is not a real-world probability but a synthetic probability – we can regard it as the probability the market is choosing to price with. We can extend P to any set which is a countable disjoint union of intervals (which includes any open set) by summing the individual probabilities. These sums are bounded above by 1, so they must converge.

Note that if the real-world probability that the stock landed in an interval I was zero then the value of the digital D_I must be zero also; otherwise, selling D_I is an arbitrage. Conversely, if the value of D_I is zero then the real-world probability of landing in the interval I must be zero or buying D_I (for nothing) is an arbitrage.

This means that the two probability measures have the same sets of probability zero. Whilst this relationship is quite weak, it is important and we will repeatedly return to this property:

Definition 6.2 Given two probability measures P, Q, on a sample space Ω which assign probabilities to the same collections of events, $\mathcal{F}$, then we shall say P and Q are equivalent if they have the same sets of zero measure.

We have constructed a probability measure P from the option prices under some mild assumptions. Suppose we are given P but not the option prices. Can we get the option prices back? We prove that we can under the mild additional assumption that P is given by the integral of a continuous density function, g. In practice, this would generally be the case. Thus we have

$$P(I) = \int_a^b g(x)dx. \tag{6.9}$$

First, by definition, $P(I)$ was the price of an option paying 1 if spot is in the interval I at time T and zero otherwise. We therefore certainly have the prices of all the digital options.

We can use no-arbitrage to deduce the price of any option with a (measurable) compactly supported pay-off, f – i.e. a pay-off which is zero outside a continuous set and not too unreasonably behaved. We take f to be piecewise continuous for simplicity. Thus suppose f is continuous on each of a finite set of intervals

$$I_n = (a_n, b_n), \tag{6.10}$$

is zero outside $\bigcup [a_b, b_n]$, and has a continuous extension to $[a_n, b_n]$ for each n. Thus it is allowed to jump at the end-points of intervals but cannot have worse discontinuities.

We show that the value of an option paying f is its expectation under the measure P. We do so by showing that it can be approximated by digital options to arbitrary accuracy. By linearity, it is enough to consider the case where f is continuous on the interval $[a, b]$ and zero otherwise. (The values of f at a and b will not affect the price as we have assumed the probability of spot finishing at a single given point is zero.) The expectation of f will just be the integral

$$\int f(x)g(x)dx.$$

As f is continuous on the compact interval $[a, b]$, it is uniformly continuous, (see for example [112].) That is, given $\epsilon > 0$, there exists $\delta > 0$ such that if $x, y \in [a, b]$ and $|x - y| < \delta$, then $|f(x) - f(y)| < \epsilon$. Uniformity means that δ does not depend on x and y.

Now suppose we take $\epsilon > 0$, and choose N such that $N\delta > b - a$. We divide the interval $[a, b]$ into N pieces of equal size, J_l, of length $(b - a)/N$. For x and y inside one of these pieces, we have $|x - y| < \delta$ and thus $|f(x) - f(y)| < \epsilon$. For each piece J_l, we therefore have that

$$\max_{x \in J_L} f(x) - \min_{x \in J_L} f(x) < 2\epsilon. \tag{6.11}$$

If we let

$$\phi_\epsilon(x) = \max_{x \in J_L} f(x) \text{ for } x \in J_l. \tag{6.12}$$

$$\psi_\epsilon(x) = \min_{x \in J_L} f(x) \text{ for } x \in J_l, \tag{6.13}$$

6.3 The existence of risk-neutral measures

then at all points, we have

$$0 \leq \phi_\epsilon - \psi_\epsilon < 2\epsilon, \tag{6.14}$$

and

$$\psi_\epsilon \leq f \leq \phi_\epsilon. \tag{6.15}$$

(The values at the ends of the intervals J_l are not well-defined but they are not relevant as the value at a single point does not affect an integral nor an expectation against a continuous density.)

Both ϕ_ϵ and ψ_ϵ are step functions by linearity, that is they are a linear combination of digital functions. The market price of options paying ϕ_ϵ and ψ_ϵ must therefore be equal to their expectations under P. We also have that the price of an option paying f must be between the prices of options paying ϕ_ϵ and ψ_ϵ by no-arbitrage and (6.15). The condition (6.15) also implies that the expectation of f lies between the two expectations.

We also have by (6.14) that the difference in expectations (and therefore prices) of options paying ϕ_ϵ and ψ_ϵ is less than $2\epsilon(b-a)$.

To summarize we have shown that given any $\epsilon > 0$, both the price of an option paying f and the expectation under P of the pay-off lie in a single fixed interval of size at most $2\epsilon(b-a)$. As ϵ was arbitrary this means that the price and expectation must be equal.

We have thus proved

Lemma 6.1 *If there are zero interest rates and the synthetic measure P is given by the integral of a continuous function, then the price of a European option paying a compactly supported continuous function f of spot at time T is equal to its expectation under P.*

We want to extend this result to more general options such as forwards and call options. Note that the pay-off of a put is already accounted for in the statement of the lemma. To encompass such infinitely supported options, we need something more than no-arbitrage as we can only approximate on finite sets. We therefore use the NFLWVR condition. Note however that the probability of a stock lying above 10^{100} is effectively zero, and so while we need to carry out this argument for mathematical rigour, it is perfectly reasonable from a financial standpoint to assume the stock price is constrained to lie in a bounded interval.

Thus suppose we have an option C paying f, a piecewise continuous function of spot, at time T. Let C_n denote the option paying $f(S)$ if spot finished below n at time T and zero otherwise. We know from our lemma that the price of C_n is its expectation under P. Without loss of generality, let's suppose that f is non-negative. (Just decompose as a difference of two non-negative functions.) We then have that the value of C_n must increase with n as the pay-off can only improve.

The option $C - C_n$ then has a non-negative pay-off. As n gets large, the value of $C - C_n$ gets progressively smaller and the probability of $C - C_n$ paying off goes to zero as the probability of spot ending above n will go to zero. This is true in both the real-world measure and the synthetic measure. If our function f is unbounded, which it certainly is for call options, we need some decay conditions on the probability measures to go further.

We assume that both real-world and synthetic probability measures are given by rapidly decaying density functions h and g, and that the pay-off of f is polynomially bounded. By rapidly decaying, we mean that any polynomial times h or g is bounded. These assumptions are an overkill; however they are satisfied by standard models such as the Black–Scholes model, and the pay-off functions of all traded options. In equation terms, our assumption is that for any k there exists D, such that

$$0 \leq g(x) \leq D(1+x)^{-k}, \qquad (6.16)$$

$$0 \leq h(x) \leq D(1+x)^{-k}, \qquad (6.17)$$

and there exists N such that

$$0 \leq f(x) \leq D'(1+x)^N. \qquad (6.18)$$

Without loss of generality, we can take $D = D'$. The expected value of $C - C_n$ under P is

$$\int_n^\infty g(x) f(x) dx \leq D^2 \int_n^\infty (1+x)^{N-k} dx. \qquad (6.19)$$

Picking k sufficiently large ensures that this goes to zero as n goes to ∞. The real-world expectation goes to zero by the same argument, with g replaced by h. The expected value of C under P is therefore the limit of the prices of C_n.

The real-world variance of the pay-off of $C - C_n$ is equal to

$$\int_n^\infty g(x) f(x)^2 dx \leq D^3 \int_n^\infty (1+x)^{2N-k} dx, \qquad (6.20)$$

minus the square of the left hand side of (6.19). Both the terms clearly go to zero as k tends to infinity.

The price of C_n must be less than that of C by no-arbitrage as C always pays at least as much as C_n. By the same token, the price of C_n must be increasing. We want to show that it converges to the price of C. By the uniqueness of limits this would imply that the price of C equals the expected value of its pay-off under P. Suppose the prices do not converge to the price of C; then they must converge to a lower number, x. The sequence of portfolios, E_n, consisting of being short C

and long C_n then has the property of having variance going to zero, expected value going to zero and with a set-up cost less than x minus the set-up cost of C for all n. The set-up cost is therefore less than $y < 0$ for all n. It follows immediately from the NFLWVR condition that such a sequence cannot exist. We conclude that the market prices of C_n do converge to the price of C, and deduce that the price of C equals the expected value of its payoff under P.

Forwards and risk-neutrality

Suppose we apply pricing by expectation to a forward contract. If spot is S_0 and interest rates are zero, we have already seen that by buying the underlying today we can enforce a no-arbitrage price of $S_0 - K$ for a forward struck at K. We therefore deduce that

$$\mathbb{E}(S_T - K) = S_0 - K, \tag{6.21}$$

where $\mathbb{E}$ denotes expectation under P, and as K is constant, this implies that

$$\mathbb{E}(S_T) = S_0. \tag{6.22}$$

This means that our constructed probability measure contains no allowances for risk-premia. The expected value of the asset is today's value. Such a measure is said to be risk-neutral.

Thus the mild assumption that it is possible to buy the underlying today implies that all the call options must be priced by a measure which is risk-neutral. Note that if we had an option on something which was not tradable, for example a contract paying off according to the temperature on a given day, then this argument would no longer work and the pricing density could have any mean.

Computing the density directly

Our construction of the risk-neutral measure is somewhat opaque. We can use the fact that the measure must price the call options correctly to deduce its value in a much more transparent fashion. We must have

$$\mathbb{E}((S_T - K)_+) = C_K. \tag{6.23}$$

If we make the mild assumption that the probability measure is given by a continuous density $p(s)$, then we can write this as

$$C_K = \int_K^\infty (s - K) p(s) ds. \tag{6.24}$$

This must hold valid for all K so if we differentiate with respect to K, we retain a true relation, and we have

$$\frac{\partial C_K}{\partial K} = -\int_K^\infty p(s)ds. \tag{6.25}$$

If we differentiate again, we obtain

$$p(K) = \frac{\partial^2 C_K}{\partial K^2}. \tag{6.26}$$

(This will hold true in a distributional sense if p is not a continuous function.) This is simple formula allows immediate computation of p from the call prices. Recall that Theorem 2.10 stated that the call option price is a convex function of strike. For a twice-differentiable function, convexity is equivalent to the second derivative being non-negative, and so no-arbitrage brings us back to the fact that probability densities are non-negative.

Putting interest rates back in

How do things change if we allow non-zero interest rates? As long as we take care of discounting, there is little difference. The value of an option paying 1 at time T will of course be the price of a zero-coupon bond (ZCB) maturing at time T. This means that if we divide the interval into digital options then the total value of these digitals will not be 1 but the price of the ZCB instead.

The solution is to assign to each interval a probability equal to the price of the digital option associated to it, divided by the price of the ZCB. Let $Z(t)$ denote the price at time t of a zero-coupon bond expiring at time T. The probabilities then add up to 1 and we have a synthetic density, p, as before. We can make similar arguments for the pricing of options, and an option with pay-off f will have price equal to

$$C(0) = Z(0) \int_0^\infty p(S) f(S) dS. \tag{6.27}$$

If our option is C, then using the fact that $Z(T) = 1$, we can write this as

$$\frac{C(0)}{Z(0)} = \mathbb{E}\left(\frac{C(T)}{Z(T)}\right), \tag{6.28}$$

as $C(T)$ will be equal to $f(S)$. This simple formula is the most important equation in mathematical finance so study it carefully! (Yes, even more important than the Black–Scholes equation.) We will use it and variants of it time and time again.

We can also deduce an expression for p in terms of the derivatives of K, similar to (6.26),

$$p(K) = Z(0) \frac{\partial^2 C_K}{\partial K^2}, \qquad (6.29)$$

where $C(K)$ is, as before, the price of a call option struck at K.

The value of a forward contract struck at K is $S_0 - Z(0)K$. We therefore have that

$$\frac{S_0 - Z(0)K}{Z(0)} = \mathbb{E}(S_T - K), \qquad (6.30)$$

using the fact that K is constant the forward contract is priced correctly if and only if

$$\mathbb{E}(S_T) = Z(0)^{-1} S_0. \qquad (6.31)$$

The average growth of the stock, in the synthetic measure, is therefore precisely equal to the growth in value of a zero-coupon bond. Just as in the case of zero interest rates, the fact that we can hedge a forward contract precisely guarantees that the synthetic measure is risk-neutral, and the stock grows at the riskless rate on average.

The risk-neutral density in the Black–Scholes world

If we return to the Black–Scholes world, we already know the price of a call option on a stock with drift μ, volatility σ and with risk-free rate r. We can therefore deduce the pricing measure implied by the Black–Scholes price just by differentiating twice. However, recall that the Black–Scholes price does not contain μ so the risk-neutral density must be independent of μ. As the differentiation is more tedious than illuminating we do not carry it out. However, if we think back to chapter 4, the answer becomes clear. We showed there that the price of an option expiring at time T is equal to its discounted expectation when the log of the stock is given by the distribution of

$$\left(r - \frac{1}{2}\sigma^2\right)T + \sigma\sqrt{T}N(0,1)$$

with $N(0,1)$ a standard normal distribution.

However, as we saw in Chapter 5, this is the same as the density obtained by evolving an asset under the process

$$\frac{dS}{S} = r\,dt + \sigma\,dW_t. \qquad (6.32)$$

We have therefore shown that the Black–Scholes price can be obtained for any European option, simply by taking the expected discounted pay-off of the option under the density obtained by letting the asset grow at the risk-free rate rather than at the real-world rate μ. This observation is the heart of the idea of risk-neutral pricing and can be applied to any derivative asset not just European options.

To summarize, we have shown that a probability density for the evolution of the stock at each time is implied by the observable call prices in the market and that this probability density is not the real-world density but instead a density which implies no compensation for risk-taking. The price of any European option is then its discounted expectation under this measure. In the specific case of the Black–Scholes model the density is that implied by taking the growth rate of the stock to be r instead of μ.

Why should this be the case? Essentially, the reason is that in the Black–Scholes model, we can eliminate all risk by continuous trading. Once the risk has been eliminated, there can be no risk premium. We must therefore price everything without a risk premium, that is as if everything grows at the risk-free rate.

Multiple time horizons

Now suppose that we know nothing about the price process of the asset but that rather than just having the value of the call options for one maturity, we have them for all maturities. We then have a synthetic density for each maturity. One might be tempted to conclude that the densities are enough to give us a price process for the asset, not the real world one of course but instead an implied risk-neutral one. However, this is not the case. This is essentially because we only have information about the assets distributions at single time frames. We have no information about how the value of the asset at time T_1 affects its value at time T_2 for $T_2 > T_1$. A derivative product affected by the values at both time T_1 and T_2 might give us some of that information, but we do not have it from the call options.

We shall return to this point once we have developed more ideas on what a process is and have learnt how to express the evolution across several time spans.

6.4 The concept of information

We have developed two different approaches to the arbitrage pricing of options. Both of these approaches led to a single unique price which was not affected by the risk-preferences of the investor. We can regard either of these approaches as being aspects of a single more fundamental one: risk-neutral pricing. As we know that investors' risk-preferences do not affect the price, we may as well assume that they have none, that is that they are risk-neutral. If all investors are risk-neutral then a stock will grow at the same rate as a riskless bond, so stocks following a geometric Brownian motion will have drift r rather than μ. We can now take a naive expectation of the option's value with this growth rate for the stock. Rather surprisingly, this leads to the Black–Scholes price. We therefore have a very powerful alternative method for pricing options. Justifying this procedure requires an excursion into some deep and powerful mathematics.

Before we can proceed to a better understanding of option pricing, we need a better understanding of the nature of stochastic processes. In particular, we need to think a little more deeply about what a stochastic process is. We have talked about a continuous family of processes, X_t, such that $X_t - X_s$ has a certain distribution. As long as we only look at a finite number of values of t and s this is conceptually fairly clear, but once we start looking at all values at once it as a lot less obvious what these statements mean.

One way out is to take the view that each random variable X_t displays some aspect of a single more fundamental variable. Instead of considering our asset price or particle moving through time via the process X_t which gives us a random path, we stand outside time and draw the entire path, ω, at once using a random variable. The random variable X_t is then the point $\omega(t)$. As time progresses, more of the path ω is revealed.

We can think of the goddess of probability living in eternity outside of the ephemeral world of an options trader. She draws an entire stock price path from a jar containing all possible stock price paths. The god of time stops us from looking into the future and slowly reveals the stock price path to us, second by second. The moral is that although the stock price is determined in one go, we have to trade as if it were not; we can only trade on the information available at the time of trading. Our objective in this section is to make this idea mathematical.

The path ω will have to be drawn at random from the space of all paths or, rather, all continuous paths. To do this will require a measure on the space of paths to determine the distribution of ω. That is we will need a map from a set of subsets of the space of continuous paths, C, to $[0, 1]$ which makes it into a probability space. The measure assigns to each subset the probability that the path is in that subset. This therefore expresses the probability of certain events occurring. Subsets could be

$$\{\omega : \omega(0) < \omega(1)\}$$

or

$$\{\omega : \omega(1) > 1\}$$

or any condition on ω one likes. For technical reasons we shall not explore, it is impossible to assign a probability to every subset of the space of paths but we shall not worry too much about this. In any case, constructing sets which cannot be assigned probabilities is actually quite hard work and so any set we come across naturally will not be a problem. The important fact for us is that it is possible to assign a probability measure to the space of continuous paths on the interval $[0, T]$ in such a way that for every t and s, $X_t - X_s$ is normally distributed with mean 0 and variance $t - s$. This measure is known as *Wiener measure,* and ensures that Brownian motion actually exists in a mathematical sense. One curious aspect of the theory is that it is the existence of the measure that is important rather than its precise values. We will often want to prove that the measure has certain properties, but it is important to realize that we will never actually compute with it directly.

In this set-up, we should really think of X_t as being a function from the space of paths to the real line. For X_t the function is very simple: it is just the evaluation map

$$\theta_t : \omega \mapsto \omega(t). \tag{6.33}$$

So X_t is a family of maps from C to $\mathbb{R}$. The distribution of X_t is then given by

$$P(X_t \in A) = P\left(\omega \in \theta_t^{-1}(A)\right) = P(\omega \in A_t), \tag{6.34}$$

where A_t is the subset of the space of paths consisting of ω such that $\omega(t) \in A$, that is $\theta_t^{-1}(A)$. Note that the measure on C wholly determines the distribution of the random variables X_t, since the probability that X_t is in any set A is determined by the measure on C.

The joint distributions of the random variables X_t are also determined by the measure on C. We have that if A is a subset of $\mathbb{R}^n$ then

$$P((X_{t_1}, X_{t_2}, \ldots, X_{t_n}) \in A) = P((\omega(t_1), \ldots, \omega(t_n)) \in A) \tag{6.35}$$

for any $t_1, t_2, \ldots, t_n \in [0, T]$. In fact, the converse is also roughly true: if one knows all the finite-dimensional distributions of the variables X_t, then this determines a measure on C. This, however, is a deep fact we cannot address here, and is at the heart of the proof of existence of Wiener measure, that is the measure on the space of paths which yields Brownian motion. We refer the reader to [78] for discussion of these and other technical points in this chapter.

We have been discussing only the one-dimensional case, but if we were considering two assets then we would need to think of the path as being a path in two-dimensional space and each random variable being the value of one of the coordinates at time t. More generally, if we were considering a market then we would need a dimension for each asset, and we might well need to consider paths in a thousand-dimensional space with each stock described by a random variable given by one of the coordinates. The market index would be a weighted average of all the coordinates. Note that any reasonable function of the path ω will define a random variable. For example, the first time that the path ω reaches 100 is a random variable, as is $\omega(1)/\omega(0)$.

6.4 The concept of information

For the following discussion, we return to the one-dimensional case. As we have mentioned above, we need some concept of what information is available at a given time. If we are observing the random variable X_t, then at time t_1 we know only about the values of X_t for $t \leq t_1$; this means that the events in the space of paths that are determined at time t_1, are precisely those which are definable in terms of $\omega(t)$ for $t \leq t_1$.

What does this mean? Let A be a subset of the space of paths which is determined at time t_1, suppose a path ω is in the set A. If we deform ω by changing its values after time t_1, then it must still be in the set A as it implies precisely the same values for X_t for $t \leq t_1$.

This means that an event A is determined at time t_1 if A is invariant under the operation of replacing a path ω by another path ω' such that

$$\omega(t) = \omega'(t)$$

for $t \leq t_1$. The space of events determined at time t is denoted $\mathcal{F}_t$. The set of these spaces is called the *filtration* of information. Clearly, we have

$$\mathcal{F}_{t_1} \subset \mathcal{F}_{t_2} \quad \text{for} \quad t_1 < t_2. \tag{6.36}$$

For example, the event $\{X_t > \alpha\}$ is in $\mathcal{F}_t$, but is not in $\mathcal{F}_s$ for $s < t$. It is, of course, in $\mathcal{F}_r$ for $r > t$. Thinking in terms of a stock, all this really says is that at time t, we know the prices of the stock for times up to time t but not for times after time t. We cannot hedge on the basis of whether the stock will finish in the money but instead only on the basis of where it is today.

A closely related concept is that of a *stopping time*. We can define a *random time* to be a function from the space of paths to the positive numbers. However, we wish to distinguish those random times which are practical in the sense that the information available at a given time determines whether the time has been reached. A stopping time is a random variable which gives a random time with the critical property that the information available at a time determines whether or not the stopping time has been passed. Thus the first time t such that a stock price reaches 100 is a stopping time, but the first time t such that the stock price is above 100 at time $t+1$ is not a stopping time as it involves knowing information about a time that has not yet been reached.

The technical definition is that the random time τ is a stopping time if for any t the event $\tau < t$ is in the set $\mathcal{F}_t$. This simply says that the event that time τ has passed at time t is in the set of information available at time t. We can write this as

$$\{\tau < t\} \in \mathcal{F}_t \tag{6.37}$$

for all t.

The concept of information is important for hedging strategies as our hedge at time t should be based purely on the information available. We cannot hedge on the basis of the future values. To illustrate this, recall from Chapter 1, the *stop-loss* hedging strategy. The stock starts at 95 and moves continuously. We have sold a call option struck at 100 and wish to hedge our risk. We assume zero interest rates. When the stock crosses 100 we buy a unit of stock with borrowed money and when it crosses back down we sell the stock again paying off our loan. If the stock ends in-the-money, we are holding the stock and the exercise price pays off our loan, if it ends out-of-the-money then we hold nothing and owe nothing. So we have hedged our risk at zero cost whatever happens.

The flaw in this argument is that it involves seeing the future. The first time that the stock reaches 100 is a stopping time, but the first time that the stock crosses 100 is not. To see this observe that the concept of crossing is crucially dependent on what happens next, the stock could just go to 100 and then go back down, as easily as going to 100 and going on past it. Which of these happens is only known a little time into the future.

The concept of a stopping time is also crucial when considering options with American-style early exercise features. Recall that an American option is an option that can be exercised at any time before expiry rather than only at expiry. The holder is then faced with the problem of deciding what the optimal exercise time is. The decision whether to exercise at time t must be made on the basis of information available at time t so an exercise strategy is really just a stopping time.

To illustrate some of these ideas further, we return to the discrete setting. We suppose our experiment consists of tossing a coin ten times, and after each coin toss the value of the stock is the number of heads so far minus the number of tails so far. We take our space of paths to be all the different possible strings of heads and tails. This space has $2^{10} = 1024$ elements since order matters. We assign a probability measure which makes all paths equally likely, thus each path occurs with probability $1/1024$.

The probability of the first coin toss being a heads is the probability of the event that the first element of the string is heads. Now as precisely half of the paths start with a heads and all strings are equally likely, the probability of the first path being a heads is just 0.5, as we might expect. There is nothing special about the first coin toss, and we equally have that half of the paths have any particular coin toss being heads. We therefore conclude that the probability of any individual coin toss being heads is 0.5. A little further thought shows that all the different tosses are independent of each other; if the first toss is heads then we know the path is one of the 512 paths which start with a heads but for each subsequent toss precisely half the paths express a heads and so the probability is 0.5.

After each toss, the stock moves up or down one according to the coin toss. After k tosses, the path of the stock up to time k tells us whether each of the first k coins tosses were heads or tails. We can therefore make decisions at time k purely on the values of the first k tosses as one might expect.

As up and down moves are equally likely, the stock, X_k, has an interesting property. No matter what its current worth and no matter what time it is, its expected value at any future time is always its current value. This is called the *martingale property*. To express this notion mathematically requires the concept of conditioning on information. At time 0 we do not know what the value of X_k will be; however we do know what information will be available at time k. The expected value of X_r for $r > k$ at time k can only depend on the information available at time k, and so we want to define it to be a random variable defined by the information available at that time. Let me emphasize that the expectation of X_r at time k will itself be a random variable. It will, however, be a random variable that depends on a smaller amount of information than X_r does. In particular, it will be possible to determine its value from information available at time k, whereas X_r is determined by information only available at time r. We will denote this random variable $\mathbb{E}(X_r | \mathcal{F}_k)$.

What properties should this expectation have? The first is that the expectation should only depend on information available at time k so the event

$$A = \{\omega : \mathbb{E}(X_r | \mathcal{F}_k) \in I\}$$

should be in $\mathcal{F}_k$ for any subset I of $\mathbb{R}$. The second property is complementary and expresses the idea that taking the expectation with respect to $\mathcal{F}_k$ should throw away just enough information to obtain a random variable determined by $\mathcal{F}_k$ and no more. We therefore require that for any event, A, in $\mathcal{F}_k$, that the expectation of $\mathbb{E}(X_r | \mathcal{F}_k)$ over the set A should be equal to the expectation of X_r over A. Since a random variable is a function from the sample space of paths to the real numbers, this says that the sum over paths, ω, in A, weighted by probabilities of the function $X_r(\omega)$ should equal the sum of $\mathbb{E}(X_r | \mathcal{F}_k)$ with the same weightings.

That is we require

$$\sum_{\omega \in A} p(\omega) X_r(\omega) = \sum_{\omega \in A} p(\omega) \mathbb{E}(X_r | \mathcal{F}_k)(\omega), \tag{6.38}$$

for all A in $\mathcal{F}_k$. These two properties are enough to determine $\mathbb{E}(X_r | \mathcal{F}_k)$ uniquely. Of course, we have written a sum here but in the continuous setting, we would have an integral over a subset of the space of paths.

6.5 Discrete martingale pricing

We now specialize to the discrete setting as the concepts of martingale pricing are less obscured by technical details in that case. We return to the continuous setting later in the chapter. In fact, in the discrete setting there is an easy way to construct $\mathbb{E}(X_r | \mathcal{F}_k)$; this method works via decomposing every set in $\mathcal{F}_k$ into a union of elementary sets. The elementary sets will have the properties that the intersection of any two is empty and that every set in $\mathcal{F}_k$ is a union of them.

An elementary set will just be a set of paths which agree up to time k and as the set has to be in $\mathcal{F}_k$, once one path is in a set, all paths which agree with it up to time k will have to be in it. Agreement up to time k is clearly an equivalence relation on the set of paths, so we partition the entire sample space into a union of disjoint subsets. As any set in $\mathcal{F}_k$ is defined purely in terms of properties of the path up to time k, the elements of $\mathcal{F}_k$ will just be unions of these elementary sets.

We now define $\mathbb{E}(X_r|\mathcal{F}_k)(\omega)$ to be the expectation of X_r over the elementary set containing ω. As this expectation only depends on the first k steps in ω it satisfies our first property. It satisfies the second property for any elementary set by definition, and a simple summing shows that it satisfies that property in general simply by decomposing the set into elementary sets.

Returning to our example, we have that

$$\mathbb{E}(X_r|\mathcal{F}_k) = X_k. \tag{6.39}$$

To see this, observe that for any path ω the left-hand side will only depend on the first k steps in ω, and that for the remaining steps there will always be another path with the heads and tails reversed that will cancel their values when taking the expectation.

In general, we shall say that a process X_j is a *discrete martingale* if for all k and r such that $r > k$, equation (6.39) holds. An important point here is that it is possible for a process X_j to satisfy

$$\mathbb{E}(X_r) = \mathbb{E}(X_r|\mathcal{F}_0) = X_0, \tag{6.40}$$

for all r, without being a martingale. For example, if we define $X_0 = 0$, and let X_1 take values 1 and -1 with equal probability, and then let $X_j = jX_1$, then we do not have a martingale. Indeed our sample space has only two paths which have non-zero probability. The first is the ascending sequence $(0, 1, 2, 3, 4, 5, \ldots)$, and the second is its negative. We therefore have that

$$\mathbb{E}(X_2|\mathcal{F}_1) = 2X_1, \tag{6.41}$$

rather than X_1. The point here is that the expected value in the future is no longer today's value at any time except zero.

The importance of martingales lies in their relationship with the condition of no arbitrage. Suppose we are working in a zero interest-rate environment and that every tradable asset is a martingale. Suppose additionally that we have a portfolio, P, of zero value today and that it is possible with non-zero probability for it to have positive value at time T in the future. We have that the expected value of the portfolio at time T is equal to today's value, zero, by the martingale property. For

6.5 Discrete martingale pricing

the expected value to be possibly positive and the expectation zero, there has to be a positive probability of the portfolio being negative.

This means that an arbitrage is impossible as we have shown that a portfolio which can take positive values with non-zero probability cannot be positive in all possible worlds. Recall that the definition of arbitrage allows us to ignore zero probability events. It may be possible to set up a portfolio which is always non-negative and which is positive on a non-empty set which occurs with probability zero but we shall not regard such portfolios as true arbitrages. In any case, in the discrete setting there is no real difference.

In one special case, we have seen that the martingale property leads to the impossibility of arbitrage. However, this special case is not particularly useful because there certainly will be interest rates. We therefore need to work with discounted prices instead of real prices. If we are working with a constant interest rate, r, which continuously compounds, then we can consider all asset prices to be multiplied by e^{-rt} to discount the future prices to today. If we write B_t to be the value of £1 invested in a riskless bond which is continuously compounding, then we are requiring X_t/B_t to be a martingale. The argument we gave above still works; if a portfolio is of zero value and can be positive with positive probability tomorrow then to get the expectation to be zero, there must be a positive probability of negative value tomorrow. Hence, as before arbitrage is impossible.

This is still not particularly useful however, as we know that a risky asset will in general grow faster than a riskless bond on average due to the risk aversion of market participants. To get round this problem, we ask what the rate of growth means for a stochastic process. The stochastic process is determined by a probability measure on the sample space which is the space of paths. However, the definition of an arbitrage barely mentions the probability measure. All it says is that it is impossible to set up a portfolio with zero value today which has a positive probability of being of positive value in the future, and a zero probability of being of negative value. The actual magnitude of the positive probability is not mentioned. In particular, it could be very small or close to 1. A consequence of this is that if we take a second probability measure, Q, and if it has the same space of events as the original measure, P, and the same set of events of probability zero then a portfolio creates an arbitrage under Q if and only if it creates an arbitrage under P. Note how weak the condition of arbitrage is from a probabilistic point of view; we can change the probability measure in a massive way and the condition of arbitrage does not notice.

We recall the definition of equivalent probability measures:

Definition 6.3 Two probability measures P and Q on a sample space Ω with event spaces $\mathcal{F}_P$ and $\mathcal{F}_Q$ are equivalent if $\mathcal{F}_P = \mathcal{F}_Q$, and for all events E in $\mathcal{F}_P$, we have that $P(E) = 0$ if and only if $Q(E) = 0$.

One simple consequence is that for any event, E, $P(E)=1$ if and only if $Q(E)=1$. This follows from considering the complementary event E^c which will be of probability 0 if and only if E is of probability 1.

We can now broaden our result greatly: instead of requiring our asset prices divided by the riskless bond to be martingales under the original, 'real-world', measure, we can conclude that there is no arbitrage if there exists an equivalent measure under which the discounted asset prices are all martingales. This condition is weak enough to be useful. This new measure is called the *risk-neutral measure*, or the *martingale measure*, as it implies that all assets grow at the same rate, that is no risk premium is demanded for the risky assets.

To illustrate the condition, we return to a two-state model. The stock price S takes value 100 today. Tomorrow it takes the value 120 with probability p and 100 with probability $1 - p$. The bond B is worth 100 today and tomorrow takes value 110 with probability 1. There are only two elements in our sample space, the path (100, 120) which occurs with probability p and the path (100, 100) with probability $1 - p$.

The expected value of S_1/B_1 is therefore,

$$\frac{1}{110}(120p + 100(1-p)) = \frac{1}{110}(100 + 20p). \tag{6.42}$$

As S_0/B_0 is equal to 1, we conclude that S/B is a martingale with respect to the probability measure obtained by taking $p = 0.5$. We can therefore conclude that there is no arbitrage provided the original measure had the same null sets as the $p = 0.5$ measure. This will be the case unless the original measure had $p = 0$ or $p = 1$, in which case only one path could occur and the other defined a null set. This is the result we want, as the probabilities 0 and 1 imply arbitrages. If $p = 1$ then the portfolio $S - B$ defines an arbitrage, and if $p = 0$ then $B - S$ defines one. We thus see in this special case that an equivalent martingale measure exists if and only if the original measure permits no arbitrage.

The reason this is so powerful is that it allows us to price options. Suppose we have an option to buy at 110 on day one. That is, it is an asset worth 10 if the stock is 120 and zero otherwise. We want to find the price on day zero that causes no arbitrage. Suppose there is such a price, call it C_0 and denote the value on day one by C_1.

Suppose a risk-neutral martingale measure for the three assets B, S and C exists. We have already seen the existence and uniqueness for the pair B and S, so there is only one possible candidate for the trio B, S and C, that is we must have $p = 0.5$. If this is to be a martingale measure then we must have

$$\frac{C_0}{B_0} = \mathbb{E}\left[\frac{C_1}{B_1}\right], \tag{6.43}$$

6.5 Discrete martingale pricing

where the expectation is taken in the martingale measure, that is with $p = 0.5$. We thus have

$$C_0 = B_0 \mathbb{E}\left[\frac{C_1}{B_1}\right]. \tag{6.44}$$

The value C_0 is thus $\frac{100}{110} (0.5 \times 10 + 0.5 \times 0) = \frac{50}{11}$. We have thus shown that for a risk-neutral measure to exist, C_0 must have a specific computable value; our computation also shows that this is the only case in which a risk-neutral measure exists. Since we have shown that the existence of a risk-neutral measure implies the absence of arbitrage, we conclude that we have computed an arbitrage-free price for the option. The reader is encouraged to compare the price computed here with that obtained from the arguments in Chapter 3.

As well as having computed *an* arbitrage-free price for the option, it is important to consider whether we have computed the *only* arbitrage-free price for the option. We have shown that the price computed is the only price that leads to the existence of a risk-neutral measure. So our question is really whether the absence of arbitrage implies that a risk-neutral measure exists. If it does we conclude that the computed price is the unique arbitrage-free price for the option. Fortunately, a deep theorem due to Harrison & Kreps, says precisely that the existence of a risk-neutral measure in the discrete setting is equivalent to the absence of arbitrage.

Of course, a one-period two-branch model is not very useful for option pricing. However, as we did in Chapter 3, we can now apply our techniques to multi-period settings with a branching at each stage. This is really just a reinterpretation of the arguments of Chapter 3 but is nonetheless useful. Thus suppose we have a tree of asset prices and a riskless bond B. Suppose the tree has n periods. At each node in the tree the stock can move up or down to the next node, each move having some non-zero probability. Each move is independent of how the stock arrived at that node. We let our sample space be the space of paths of the stock through the tree. The probability of a given path in the sample space is just the product of the probability that at each stage the stock moves up or down in the way the path specifies. The probability of the path is therefore obtained by taking the product of probabilities of the moves along the path: this is implied by the independence of the moves. As we have assumed that each move has non-zero probability this means that each path has non-zero probability also, and thus that the only event of probability zero is the empty event.

Since the asset price has to be a martingale under the risk-neutral measure, the expected value at the next time has to be equal to the current price after division by the riskless bond B. Thus if the price at a node N is S_N, the price after an up-move is S_{Nu}, and after a down-move is S_{Nd}, then we must have that

$$\frac{S_N}{B_t} = p \frac{S_{Nu}}{B_{t+1}} + (1-p) \frac{S_{Nd}}{B_{t+1}}; \qquad (6.45)$$

this clearly has a unique solution p provided $S_{Nu} \neq S_{Nd}$, and the solution will be between 0 and 1 if the discounted price at time t is between the discounted prices at time $t+1$. Note that if the discounted price were not in between the evolved prices, there would be an arbitrage opportunity, as the stock would have to be either worth more than the riskless bond in all possible worlds, or worth less than the riskless bond in all possible worlds.

Having computed the risk-neutral probability at each node, we can then assign a probability to each path by stringing together the probabilities along the path. As the risk-neutral probabilities are non-zero, the probability of every path is non-zero, and the new measure is equivalent to the old. We wish to price our European option which has a determined value at each node on the final layer. Let B_T denote the price of the riskless bond on the final layer. By construction, the probability measure we have constructed on the tree is the unique martingale one, and we have to assign the value, C_N, at the node N at time t in such a way that we always have that

$$\frac{C_N}{B_t} = \mathbb{E}\left(\frac{C_T}{B_T}\right), \qquad (6.46)$$

where the expectation is taken over paths passing through N. As the only indeterminate quantity in this equation is C_N, it is directly determined and the option is priced. To actually carry out the procedure in practice, we would compute the price at each node in the second last layer by taking the risk-neutral expectation at the nodes in the last layer, which would just be the weighted average

$$\frac{B_{T-1}}{B_T}(pC_{\text{up}} + (1-p)C_{\text{down}}) \qquad (6.47)$$

where p is the probability of an up move, and C_{up} and C_{down} are the prices after up and down moves. This process would then be iterated back to the start to give the price today.

An important issue here is uniqueness, we have given a method of determining an arbitrage-free price for the option. First construct a risk-neutral measure on the space of paths of the stock and then price according to the risk-neutral expectation of the option. To what extent is this price unique? If the risk-neutral measure is unique then a unique price will be implied. The problem occurs when the risk-neutral measure is not unique. In the examples above, we carefully stuck to the case where there were precisely two branches emanating from each node as this guaranteed uniqueness. However, if we allow more branches this is no longer the

6.5 Discrete martingale pricing

case. We return to an example from Chapter 3; suppose on day zero the stock is price 100 and on day one it can take the values 90, 100 or 110. For simplicity, we take zero interest rates so the riskless bond is of value 1 on both days. A measure will be determined by assigning the probability of an up move, p_u, and a down move p_d. Under such a measure, we have that

$$\mathbb{E}(S_1) = 110 p_u + 100(1 - p_u - p_d) + 90 p_d, \tag{6.48}$$

this will be equal to 100 if and only if $p_u = p_d$. We therefore have an infinity of risk-neutral measures and consequently an infinity of possible prices for an option. In particular, a call option struck at 100 will be worth $(110 - 100) \times p_u = 10 p_u$.

We have seen previously that if an option can be replicated then the cost of setting up the replicating portfolio is the only possible price for the option. However, multiple risk-neutral measures suggest multiple prices so we deduce that there must be a connection between the uniqueness of risk-neutral measures and the existence of replicating portfolios. To illustrate this point, it is helpful to consider a very specific instrument which while unlikely to exist in the real world, is very useful from a conceptual point of view. Suppose we have T time steps and on the final layer there are k possible values of the stock price. Let the security δ_j pay out one riskless bond, B_T, in state j at time T and zero in all other states. Such a security is known as an Arrow–Debreu security and is essentially a delta function on a given state.

What is the price of this security? Well, we take the discounted risk-neutral expectation

$$B_T^{-1} \mathbb{E}_{RN}(\delta_j B_T) = \mathbb{E}_{RN}(\delta_j); \tag{6.49}$$

however the right-hand expectation is simply the probability that the stock is in state j at time T, as an expectation is just the sum of values in states times the probabilities of states. The price of the security δ_j at time 0 is just the probability (in the risk-neutral measure) that the stock will end up in state j at time T. If it is possible to replicate the security δ_j then this price will be determined by the setup cost of the replicating portfolio and thus the risk-neutral probability is determined by the replication. If all the Arrow–Debreu securities at time T are replicable then we can read off the probabilities of all the states at time T. That is we can read off the distribution of the stock at time T in the risk-neutral measure. Thus the risk-neutral measure is constrained by the prices of the Arrow–Debreu securities if they are replicable.

Of course, knowing the final distribution of the stock price is not enough to determine the risk-neutral measure completely. As the risk-neutral measure is a measure on the space of paths, we consider a security associated to a path, ω, of discrete prices which pays a unit of riskless bond at time T if the path ω is realized and zero otherwise. Letting this security be δ_ω, the price of this security in a risk-neutral measure will simply be the probability of it occurring. Thus if it is replicable, its price will be determined and then the risk-neutral probability of occurrence is determined. Thus if all securities were replicable then all prices and hence all probabilities would be determined. But in a discrete setting, if we know the probability of each path then we know the probability of every subset of the set of paths and hence we know the probability measure. Thus the probability measure is fully determined and therefore unique if every possible security can be replicated.

What do we mean by a security here? We really mean an asset that pays off at some time an amount determined by the behaviour of the stock up to that time. We are therefore really studying a claim which is contingent on the behaviour of the stock. We will therefore often talk about contingent claims. A market in which every possible *contingent claim* can be replicated is said to be *complete*.

Of course, the concept of completeness would not be useful unless natural examples were complete. The simplest example of a complete market is a binary tree. We have already seen that any security paying off at a single time as a function of the state at that time is replicable. This means that the Arrow–Debreu securities are in particular replicable. In fact, in any discrete market we can build up a general claim from Arrow–Debreu securities. Thus the replicability of Arrow–Debreu securities implies the replicability of a general claim, and hence the uniqueness of the risk-neutral measure.

How do we build up a general claim from Arrow–Debreu securities? It is enough to consider a claim that pays a fixed sum in a single state at some time T conditional on the realization of certain previous states, as any claim can be decomposed into a sum of such claims. By linearity, we need only consider the case where the sum to be paid is one unit of riskless bond. We prove the existence of such a decomposition via induction. If the pay-off only depends on the value of the stock at the time of payoff then we are back in the case of an Arrow–Debreu security. If a claim, C, pays off at a time T depending upon being in a state j at that time, and upon being in a subset E at time $T-1$, then the first element of our decomposition is the Arrow–Debreu security, A, that pays one bond in state j at time T. At time $T-1$, either a state in subset E is realized in which case C and A agree at both time T and time $T-1$, or a state outside E is realized in which case C is worth zero but A may not be. However for each state outside E at time $T-1$ there is an Arrow–Debreu security which we can use to cancel the value of A to match the value of C. Note that as the Arrow–Debreu security is replicable we know its value not just at time 0 but in any state – we simply roll the replicating portfolio forward to that state and see what its value is.

Our replicating strategy for a two-stage claim is now clear. We buy the Arrow–Debreu security A, and value it in every state outside E at time $T-1$ and go short

6.5 Discrete martingale pricing

multiples of Arrow–Debreu securities to cancel these values. At time $T - 1$, if a state in E is realized, then all the securities other than A in the portfolio are of zero value, and A is of the same value as C whatever happens at time T which means that C has been replicated. If a state outside E is realized, then C is valueless and A is not, but precisely one of the other securities pays off with negative the value of A so we just sell A and we have an empty portfolio of zero value which replicates C once again.

Note that whilst we have talked of buying and selling Arrow–Debreu securities for this replication, we could equally well buy and sell the portfolios which replicate them. Thus we do not need to assume the tradability of Arrow–Debreu securities for these arguments to work, merely their replicability in terms of the underlying.

For a three-period claim, D, that pays off at some state j at time T, depending on states having been realized at times $T - 1$ and $T - 2$, we proceed similarly. Let E_l be the set of states which must be realized at time $T - l$ for the payoff to occur. We can define a modified claim D' to pay off at time T in state j provided a state in E_1 is realized. The claim D' is replicable as we have shown that all two-period securities are replicable. At time $T - 2$, D and D' agree in the set of states E_2 and disagree outside it. However, as in the two-period case, we are able to cancel the values outside E_2 by using linear multiples of Arrow–Debreu securities. Thus any three-period claim is replicable.

The deduction of the N-period case from the $N - 1$-case is essentially the same and we leave it to the reader to check the details. To conclude, we have shown that the replicability of Arrow–Debreu securities implies that any claim is replicable. We have also shown that the replicability of a general claim implies that there is at most one risk-neutral measure. We conclude that for a binary tree, the risk-neutral measure is unique.

The only example of a complete discrete market we have seen so far is the binary tree. There is a simple reason for this: we have been working with two assets – the stock and the riskless bond, and with two assets we can only replicate values in two arbitrary states. To see this regard the value of each asset tomorrow as a vector. Each entry in the vector represents the value of the asset in some state. The possible values one can replicate will be the linear span of the assets' associated vectors. With two assets, the replicable set will therefore be two-dimensional (assuming linear independence of the asset prices which for stock and bond will hold.) However the dimensionality of the set of possible securities will be equal to the number of branches. Thus the two sets will be equal if and only if the number of branches is two.

Binary trees are neither as restrictive as they might first appear nor as necessary as the above completeness argument might suggest. They are not as restrictive as they might appear because one can always subdivide into lots of little time steps which then give the same terminal distributions as we did in Chapter 3. They are not as necessary as they might appear, in that they are really only important because they provide a good approximation to the continuous case. If one justifies the market completeness in the continuous case and one then passes to a discretization of the risk-neutral measure, the completeness is not important as the measure is already determined. However, this is getting ahead of ourselves as we still need to understand martingales in the continuous setting.

The reason binary trees are not the only complete discrete market is the availability of other hedging instruments. If we return to our simple example of a stock that can take the values of 90, 100 and 110 tomorrow in a world with no interest rates, where there are many risk-neutral measures, we can make the market complete by adding one extra asset – the call option struck at 100. In a risk-neutral measure such that the probability of an up-move is p_u, we saw that the price of this call option was $10p_u$, so once we add it in as an extra tradable, the risk-neutral measure is fixed by its price. We will, however, have to specify its price at every node in a multi-step tree, not just its initial price and payoff. The price vectors of the stock, bond and option are independent so all possible claims are replicable and the market is complete. Of course, there is nothing special about the option struck at 100: we could use any option struck between 80 and 120 as our new hedging instrument and then the prices of all other options would be determined. Why not use an option struck outside the range 80 to 120? The option then loses its non-linearity and can be replicated as a linear multiple of stock and bond which of course then implies that it cannot be used to replicate any new options. The same arguments will apply to any market in which the asset can move to three possible states. We need simply take an option struck in between the two extreme states to make the market complete.

We can similarly see that if we have k new possible states then we need k linearly independent instruments to hedge with. These observations are mainly of use when we wish to hedge complicated options in incomplete markets. If we develop a model of how vanilla options change in value, then we can use them to hedge the exotic option perfectly although the market consisting of stock and bond is incomplete if the market consisting of stock, bond and all vanilla options is complete.

6.6 Continuous martingales and filtrations

We return to the continuous setting; however the discrete setting should be borne carefully in mind remembering that a binary tree with a million nodes is a good approximation to Brownian motion. The principal reason for the ubiquity of martingales in probability theory is that Brownian motion is a martingale; indeed in a certain sense it is the archetypal continuous martingale.

We wish to carry out the same procedures in the continuous case that we did in the discrete case. This will require us to identify the Ito processes which are martingales and understand how to carry out a change of measure on the space of paths. We will also need to comprehend the effect of a change of measure on an Ito process.

First, we revisit conditional expectations. We sketched out the definition of $\mathbb{E}(X|\mathcal{F}_s)$ above. In what follows, it has three important properties and as long we bear them in mind everything follows easily.

The first property is

$$\mathbb{E}(X|\mathcal{F}_0) = \mathbb{E}(X). \tag{6.50}$$

This simply says that if we take a conditional expectation based on no information, we get the ordinary expectation.

The second property is that if we have $s < t$ and we first take the conditional expectation at time t followed by the conditional expectation at time s then this is the same as taking the conditional expectation at time s. This is called the *Tower Law*. We can write the Tower Law as

$$\mathbb{E}(\mathbb{E}(X|\mathcal{F}_t) \mid \mathcal{F}_s) = \mathbb{E}(X|\mathcal{F}_s). \tag{6.51}$$

An immediate consequence of the second property is that we now have a very easy method of creating martingales. Recall that a martingale is a random process, X_r, such that

$$\mathbb{E}(X_r|\mathcal{F}_s) = X_s, \tag{6.52}$$

for any $s < r$. It is immediate from the Tower Law that if we put

$$X_r = \mathbb{E}(X|\mathcal{F}_r), \tag{6.53}$$

then X_r is a martingale. This observation is the key to martingale pricing.

The third property of conditional expectations is that if we condition on information which is independent of the value of the random variable then we get the same value as conditioning on no information. What does it mean for the random variable to be independent of the information? A heuristic definition we will use is that if changing the path up to time s does not affect the value of the random variable then the random variable is independent of $\mathcal{F}_s$. So if X is independent of $\mathcal{F}_s$ we have

$$\mathbb{E}(X|\mathcal{F}_s) = \mathbb{E}(X). \tag{6.54}$$

As usual, we have glossed over a lot of technical points in this section and we refer the reader to [78] for a fully rigorous discussion.

6.7 Identifying continuous martingales

With the properties of conditional expectation established, we can now turn to identifying martingales. We first show that Brownian motion is a martingale, we need to check that

$$\mathbb{E}(W_t|\mathcal{F}_s) = W_s \tag{6.55}$$

for s less than t. We can write $W_t = W_s + (W_t - W_s)$. Using the linearity of expectation, we have

$$\mathbb{E}(W_t|\mathcal{F}_s) = \mathbb{E}(W_s|\mathcal{F}_s) + \mathbb{E}(W_t - W_s|\mathcal{F}_s). \tag{6.56}$$

The first term on the right-hand side is clearly W_s since W_s contains no information not contained in $\mathcal{F}_s$. It remains to show that

$$\mathbb{E}(W_t - W_s|\mathcal{F}_s)$$

is equal to zero. Recall that $W_t - W_s$ is distributed as a normal variable with mean zero and variance equal to $t - s$. We defined the value of $W_t - W_s$ to be independent of the value of W_s and the path up to time s, so when we condition on information available at time s we are effectively conditioning on no information. This means that the conditional expectation of $W_t - W_s$ with respect to $\mathcal{F}_s$ is simply the ordinary expectation which is zero. Thus we conclude that W_t is a martingale. It is important to realize that nothing mysterious is happening here; all we are saying is that at any time the expected future value of W_t is its current value.

We wish to identify which processes, X_t, of the form

$$dX_t = \mu(X_t, t)dt + \sigma(X_t, t)dW_t \tag{6.57}$$

are martingales. As we require $\mathbb{E}(X_t|\mathcal{F}_s) = X_s$ for all s and t, we in particular require it for $t = s + \epsilon$ with ϵ very small. We have defined X_t is such a way that

$$X_t - X_s = \mu(X_s, s)(t - s) + \sigma(X_s, s)(t - s)^{1/2} N(0, 1) + \text{error}, \tag{6.58}$$

where the error term is small for small $t - s$. Hence unless μ is zero there will be a bias upwards or downwards preventing the martingale property.

We therefore require

$$dX_t = \sigma(X_t, t)dW_t, \tag{6.59}$$

that is, we require the drift of the process to be zero. Subject to technical conditions, this is equivalent to X_t being a martingale. The technical conditions ensure that the conditional expectation $\mathbb{E}(X_t|\mathcal{F}_s)$ exists by stopping X_t from blowing up too quickly. When the expectation fails to exist but the process is essentially a martingale, it is said to be a *local martingale*.

6.8 Continuous martingale pricing

We have completed the first stage of martingale pricing – identifying the processes which are martingales. We are interested in the case where the stock, S_t, is following geometric Brownian motion and the bond, B_t, is continuously compounding at the risk-free rate. Following the procedure in the discrete case, we want S_t/B_t to be a martingale. We have

$$dS_t = \mu S_t dt + \sigma S_t dW_t, \tag{6.60}$$

$$dB_t = r B_t dt \tag{6.61}$$

and therefore

$$d\left(\frac{S_t}{B_t}\right) = \frac{dS_t}{B_t} + S_t d\left(\frac{1}{B_t}\right), \tag{6.62}$$

with no additional Ito-type cross terms since B_t is deterministic. Since

$$B_t = B_0 e^{rt},$$

we have

$$d(B_t^{-1}) = -r B_0^{-1} e^{-rt} dt = -r B_t^{-1} dt.$$

Thus

$$d\left(\frac{S_t}{B_t}\right) = (\mu - r)\frac{S_t}{B_t} dt + \sigma \frac{S_t}{B_t} dW_t. \tag{6.63}$$

This will be driftless if and only if $\mu = r$, that is, if and only if the stock grows at a risk-neutral rate.

Therefore, as in the discrete case, the fact that investors are not in general risk-neutral means that our discounted asset prices are not martingales in the 'real-world' measure. We must therefore employ a change of measure to make the discounted prices into martingales. This will, of course, be a change in the measure on the space of paths which underlies the Brownian motion driving the stock-price movements. Such a measure change must preserve probability 1 and probability 0 events. This means that the measure-changed paths will still be continuous and infinitely jagged (i.e. they will still have infinite first variation and finite non-zero second variation).

It turns out that a measure change on the space of Brownian paths is equivalent to changing the drift of the Brownian motion process. In other words, by changing the measure we can make a Brownian motion have the drift of our choice. The mathematics behind this is quite complex, and we do not attempt to address the details here. However, I remark that the amazing thing about the measure of Brownian motion is not that one may change the drift but the fact that one cannot do anything else. We shall return to the issue of measure changes in Chapter 8 where we will need more explicit knowledge of the measure change. Here we quote

Theorem 6.1 Girsanov's Theorem: *Let W_t be a Brownian motion with sample space Ω and measure $\mathbb{P}$. If ν is a reasonable function then there exists an equivalent measure $\mathbb{Q}$ on Ω such that $\tilde{W}_t = W_t - \nu t$ is a Brownian motion.*

We can equivalently write this as

$$d\tilde{W}_t = dW_t - \nu dt, \tag{6.64}$$

with $\tilde{W}_t$ a Brownian motion. Roughly, what is happening here is that we are giving extra weight to paths that move in a certain direction and less weight to paths that move in the opposite way.

We can now apply Girsanov's theorem to (6.63) and we obtain

$$d\left(\frac{S_t}{B_t}\right) = (\mu - r)\frac{S_t}{B_t}dt + \sigma\frac{S_t}{B_t}d\tilde{W}_t + \nu\sigma\frac{S_t}{B_t}dt. \tag{6.65}$$

We solve for ν to make the process driftless, that is we put

$$\nu = -\frac{\mu - r}{\sigma}, \tag{6.66}$$

which is of course just the negative of the market price of risk. Having made S_t/B_t driftless, it is, under the mild technical conditions we are ignoring, a martingale.

We have shown that there exists a unique ν which makes the discounted-stock-price process driftless. This is equivalent to saying that there is a unique change of measure on the space of paths which makes the discounted-stock-price process a martingale.

We can proceed as in the discrete case to pricing via risk-neutral expectations, and, because there is a unique martingale measure, we are guaranteed that the prices are both unique and arbitrage-free. This is better than we did with the PDE approach as that simply showed us the only possible arbitrage-free price, not that the price was arbitrage-free. On the other hand, it did show us how to trade in such a fashion as to enforce the arbitrage-free price, which our argument here has not. We shall return to the issue of how to hedge but now show how to use martingale pricing in a practical way.

We have changed measure to make the discounted-price process a martingale by changing drift. In the new measure, we have

$$d\left(\frac{S_t}{B_t}\right) = \frac{S_t}{B_t}d\tilde{W}_t, \tag{6.67}$$

or equivalently, recalling that $dB_t = rB_t dt$, that

$$dS_t = rSdt + \sigma S d\tilde{W}_t. \tag{6.68}$$

Here we use the fact that as B_t is purely deterministic, a change of probability measure has absolutely no effect on it.

The interesting thing about (6.68) is that it is the evolution equation for a stock-price under geometric Brownian motion with drift the risk-free rate. Our switch to an equivalent martingale measure has the same effect as pretending investors are risk-neutral – the market price of risk can be taken to be zero. To price an option on S we simply take its discounted price to be a martingale under the risk-neutral measure. Thus we define the price of an option O at time s to satisfy

$$\frac{O_s}{B_s} = \mathbb{E}\left(\frac{O_T}{B_T} \Big| \mathcal{F}_s\right), \tag{6.69}$$

where, for a vanilla option, T is the time of expiry. It is immediate from this definition that O_s/B_s is a martingale. The value of O_0 implied by (6.69) is therefore necessarily arbitrage-free. As in the discrete case, the fact that O_t/B_t is a martingale means that it cannot have a non-zero probability of increase without a non-zero probability of decrease. This applies to all portfolios containing O, S and B: a non-zero possibility of increase implies a non-zero possibility of decrease and hence we have no arbitrage.

How do we actually use (6.69)? For a call option with strike K, we know the value of O_T at expiry as a function of spot and we know the terminal distribution of spot in the risk-neutral measure, so the expectation is directly evaluable, and we have

$$O_0 = \frac{B_0}{B_t} \mathbb{E}\left((S_t - K)_+\right). \tag{6.70}$$

In the risk-neutral world we have that

$$S_t = S_0 \exp\left(rt - \frac{1}{2}\sigma^2 t + \sigma\sqrt{t} N(0, 1)\right), \tag{6.71}$$

where $N(0, 1)$ denotes a draw from a standard Gaussian distribution with mean 0 and variance 1. The value of our option is

$$\frac{B_0}{B_t} \mathbb{E}\left(\left(S_0 \exp\left(rt - \frac{1}{2}\sigma^2 t + \sigma\sqrt{t} N(0, 1)\right) - K\right)_+\right). \tag{6.72}$$

Recalling that the density of $N(0, 1)$ is $\frac{1}{\sqrt{2\pi}} e^{-x^2/2}$, we can write this as

$$\frac{e^{-rt}}{\sqrt{2\pi}} \int e^{-\frac{x^2}{2}} \left(S_0 \exp\left(rt - \frac{1}{2}\sigma^2 t + \sigma\sqrt{t} x\right) - K\right)_+ dx. \tag{6.73}$$

The integrand is non-zero if and only if

$$S_0 \exp\left(rt - \frac{1}{2}\sigma^2 t + \sigma\sqrt{t} x\right) \geq K \tag{6.74}$$

which is of course equivalent to

$$rt - \frac{1}{2}\sigma^2 t + \sigma\sqrt{t}x \geq \log(K/S_0). \tag{6.75}$$

Thus the integral must be taken over

$$x \geq \frac{\log(K/S_0) + \frac{1}{2}\sigma^2 t - rt}{\sigma\sqrt{t}}. \tag{6.76}$$

Denote the right-hand side of this equation by l. Our integral now has two terms; the second simple term is just

$$\frac{e^{-rt}}{\sqrt{2\pi}} \int_l^\infty e^{-\frac{x^2}{2}} K\,dx. \tag{6.77}$$

The integral of the normal density from l to ∞ is equal to $N(-l)$, which is its integral from $-\infty$ to $-l$ by evenness. The second term is therefore equal to

$$e^{-rt} KN\left(\frac{\log(S_0/K) - \frac{1}{2}\sigma^2 t + rt}{\sigma\sqrt{t}}\right). \tag{6.78}$$

This leaves us with the first term,

$$\frac{e^{-rt}}{\sqrt{2\pi}} \int_l^\infty e^{-\frac{x^2}{2}} \left(S_0 \exp\left(rt - \frac{1}{2}\sigma^2 t + \sigma\sqrt{t}x\right)\right) dx. \tag{6.79}$$

Performing the change of variables $x = \bar{x} + \sigma\sqrt{t}$, this becomes

$$\frac{e^{-rt}}{\sqrt{2\pi}} \int_{l-\sigma\sqrt{t}}^\infty e^{-\frac{\bar{x}^2}{2}} S_0 e^{rt} d\bar{x}. \tag{6.80}$$

Proceeding as for the second term, it follows that the price of a vanilla call option is

$$S_0 N(d_1) - Ke^{-rt} N(d_2), \tag{6.81}$$

where

$$d_1 = \frac{\log(S_0/K) + \left(r + \frac{1}{2}\sigma^2\right)t}{\sigma\sqrt{t}}, \tag{6.82}$$

$$d_2 = \frac{\log(S_0/K) + \left(r - \frac{1}{2}\sigma^2\right)t}{\sigma\sqrt{t}}. \tag{6.83}$$

This is, of course, the solution to the Black–Scholes equation so our two quite different methods result in the same price, which is a relief.

6.9 Equivalence to the PDE method

One neat consequence of the martingale method is a very short easy derivation of the Black–Scholes equation. In any case, we want to be sure that the prices obtained by both methods are equal in general and not just for a vanilla call option. Let C_t be a derivative product which we know will continue to exist for a small amount of time. All we are assuming is that it is not going to transform into something else by virtue of expiry or hitting a knock-out barrier or whatever.

Let B_t be the continuously compounding money market account. We then have in the risk-neutral measure that C_t/B_t is a martingale. As the price of C_t is defined in terms of a risk-neutral expectation, we can write C_t as a function of S and t only. (Of course, it will depend parametrically on r and σ, but they are constants.)

We can therefore apply Ito's lemma to $C(S, t)$ to obtain

$$dC = \frac{\partial C}{\partial t}dt + \frac{\partial C}{\partial S}dS + \frac{1}{2}\frac{\partial^2 C}{\partial S^2}dS^2; \qquad (6.84)$$

as $dS = rSdt + \sigma S dW$ in the risk-neutral world, we have

$$dC = \left(\frac{\partial C}{\partial t} + \frac{\partial C}{\partial S}rS + \frac{1}{2}\frac{\partial^2 C}{\partial S^2}\sigma^2 S^2\right)dt + \sigma S\frac{\partial C}{\partial S}dW. \qquad (6.85)$$

We can compute

$$d\left(\frac{C}{B}\right) = \frac{1}{B}\left(\frac{\partial C}{\partial t} + \frac{\partial C}{\partial S}rS + \frac{1}{2}\frac{\partial^2 C}{\partial S^2}\sigma^2 S^2 - rC\right)dt + \sigma \frac{S}{B}\frac{\partial C}{\partial S}dW. \qquad (6.86)$$

However, in the risk-neutral world C/B is a martingale. It therefore has zero drift. So, we obtain

$$\frac{\partial C}{\partial t} + \frac{\partial C}{\partial S}rS + \frac{1}{2}\frac{\partial^2 C}{\partial S^2}\sigma^2 S^2 - rC = 0, \qquad (6.87)$$

which is of course the Black–Scholes equation.

Thus the two methods lead to the same PDE for a general option. For both methods, we will want to impose boundary conditions depending on the properties of the option. For example, for a European option, it is simply that the value at time T is equal to a prescribed function of spot. This is an initial condition in the PDE case, and in the martingale case it is a profile over which to take an expectation. This equivalence of methods reflects a deep theorem from PDE theory – the Feynman–Kac theorem which states that certain diffusion equations can be solved by taking expectations of diffusion processes. Indeed, one approach to mathematical finance is to use the Feynman–Kac theorem to deduce the existence of risk-neutral measures from the Black–Scholes equation, [17].

We shall see many other examples of options which can be priced by both methods. For example, a knockout call option can never be exercised if spot falls below

152   *Risk neutrality and martingale measures*

a certain pre-specified value called the barrier. If the barrier is at B, then for the PDE approach we have to impose an extra boundary equation on the level $S = B$, which is that the option price vanishes there. In the martingale approach, we would instead compute the joint distribution of the minimum of the stock price and the terminal value of the stock price, and write the value as expectation against this distribution. We discuss this example in detail in Chapter 8.

6.10 Hedging

The PDE method did not just show us how to find the arbitrage-free price, it also showed us how to enforce that price. The enforcement came from a hedging strategy that involved continuous trading in the underlying. In particular, the strategy was to Delta hedge; that is for an option with price $C(S, t)$ to hold $\frac{\partial C}{\partial S}$ units of the underlying at any time. This allowed us to construct the payoff of the option precisely by starting with the initial value of the option units of cash. So far our treatment of the martingale theory has not mentioned hedging, and we would be laughed at by a trader who would not care about an arbitrage-free price unless it could be enforced.

The key to hedging for the martingale approach is the martingale representation theorem which roughly says that Brownian motion is the archetypal continuous martingale. Any other martingale will have a rate of change which is some varying multiple of that of the underlying Brownian motion. To state our theorem precisely

Theorem 6.2 *Let W_t be a Brownian motion with associated filtration $\mathcal{F}_t$. Suppose that M_t is a continuous martingale with respect to $\mathcal{F}_t$. Then there exists a predictable function ϕ such that*

$$dM_t = \phi dW_t. \tag{6.88}$$

Of course, we need the meaning of *predictable* here. Predictable means that the function's value for time s is always known from the behaviour of W_t for $t < s$. So it can only depend on the path followed by W_t for $t < s$. In practice, for the martingales we are interested in, ϕ will be even more benign. Indeed it will be a continuous function of the form $\phi(W_t, t)$, or equivalently $\phi(S_t, t)$. Note that a continuous function of this form is predictable because W_t is continuous. Knowing its value for $t < s$ fixes its value for $t = s$, as the continuity implies that it is the limit of W_t as t tends to s from below.

The crucial point about the martingale representation theorem is that M_t is a martingale with respect to the filtration generated by the Brownian motion. This means that the information which determines the movement of M_t is contained in the behaviour of W_t. It is therefore not so surprising that its instantaneous movements have to be multiples of that of the underlying Brownian motion.

6.10 Hedging

If C_t is a derivative and B_t is the money market account then

$$M_t = \frac{C_t}{B_t}$$

is a martingale in the risk-neutral measure, so there exists ϕ such that

$$d\left(\frac{C_t}{B_t}\right) = \phi dW_t. \tag{6.89}$$

We want to identify ϕ. In fact, we can compute it using Ito's lemma; we already did this in equation (6.86). Taking the dW_t term we have

$$d\left(\frac{C_t}{B_t}\right) = \sigma \frac{S_t}{B_t} \frac{\partial C_t}{\partial S_t} dW_t. \tag{6.90}$$

We also have

$$d\left(\frac{S_t}{B_t}\right) = \sigma \frac{S_t}{B_t} dW_t. \tag{6.91}$$

Combining the last two equations, we obtain

$$d\left(\frac{C_t}{B_t}\right) = \frac{\partial C}{\partial S}(S_t, t) d\left(\frac{S_t}{B_t}\right). \tag{6.92}$$

Since B_t is deterministic, this immediately implies the random part of dC_t is equal to the product of $\partial C/\partial S$ and the random part of dS. That is, we can Delta-hedge C by holding $\partial C/\partial S$ units of dS. We have thus recovered the idea of Delta-hedging from the martingale theory.

The other part of the PDE theory was that not only could we hedge the instantaneous changes in value of C by holding units of S, but that we could synthesize the value of C by setting up a self-financing portfolio consisting of units of S and B, together with a trading strategy. How do we see this is in the martingale world?

We start off with the amount of cash $C(0, S_0)$ and we need to turn this into a self-financing portfolio which will replicate the value of C at all times, no matter what spot does. We have already decided that we ought to hold $\frac{\partial C}{\partial S}(t, S_t)$ units of the stock at time t if the spot is S_t, so we are left with $C(t, S_t) - \frac{\partial C}{\partial S}(t, S_t)S_t$ units of cash to hold bonds with. We do so. Thus our self-financing portfolio is to hold α_t bonds and β_t stocks which is worth

$$\alpha_t B_t + \beta_t S_t$$

and

$$\alpha_t = \frac{C_t}{B_t} - \frac{\partial C}{\partial S}(t, S_t)\frac{S_t}{B_t}, \tag{6.93}$$

$$\beta_t = \frac{\partial C}{\partial S}(t, S_t) \tag{6.94}$$

By construction, the value of this portfolio will be C_t. We need to check that the portfolio is self-financing, that is, we need to prove that

$$dC_t = \alpha_t dB_t + \beta_t dS_t. \tag{6.95}$$

From the variant of the Leibniz product rule for the Ito calculus, we have that

$$dC_t = d\left(\frac{C_t}{B_t}\right)B_t + \left(\frac{C_t}{B_t}\right)dB_t + d\left(\frac{C_t}{B_t}\right)dB_t. \tag{6.96}$$

The fact that B_t is deterministic makes the final term disappear. Thus we have

$$dC_t = B_t \frac{\partial C_t}{\partial S} d\left(\frac{S_t}{B_t}\right) + \frac{C_t}{B_t} dB_t. \tag{6.97}$$

Using the determinism of B_t again, we obtain

$$d\left(\frac{S_t}{B_t}\right) = \frac{1}{B_t} dS_t - \frac{S_t}{B_t^2} dB_t. \tag{6.98}$$

Combining these, we have

$$dC_t = \frac{\partial C_t}{\partial S} dS_t + \left(\frac{C_t - S_t \frac{\partial C_t}{\partial S}}{B_t}\right) dB_t, \tag{6.99}$$

which is precisely the self-financing condition.

We have shown that the pay-off of an option is replicable by a self-financing portfolio if we are working in a Black–Scholes world, and this once again guarantees the arbitrage-free price.

6.11 Time-dependent parameters

In the perfect Black–Scholes world we have so far inhabited, volatility and interest rates have always been constant. It is possible, and indeed quite reasonable, to make them follow random processes in their own rights and later in the book we shall do just that. However, the simplest generalization is just to let them be deterministic functions of time. Thus suppose the volatility $\sigma(t)$ is not constant but that r is still constant. What difference does this make to pricing an option?

We can follow through the entire Black–Scholes argument from Chapter 5 and conclude that the Black–Scholes equation still holds – the only difference is that σ is a function. We have the same final boundary conditions for European options. We therefore need to solve

$$\frac{\partial C}{\partial t}(S, t) + rS \frac{\partial C}{\partial S}(S, t) + \frac{1}{2} \sigma(t)^2 S^2 \frac{\partial^2 C}{\partial S^2}(S, t) = rC, \tag{6.100}$$

with the usual boundary conditions. It is not so clear how to proceed. We can reduce to the heat equation but that would still leave us with a time-dependent coefficient.

6.11 Time-dependent parameters

If instead we use the martingale approach, life becomes simple. Passing to the risk-neutral measure, the spot follows the process

$$dS = rS\,dt + \sigma(t)S\,dW_t. \tag{6.101}$$

We have that the log of the spot follows the process

$$d(\log S) = \left(r - \frac{1}{2}\sigma(t)^2\right)dt + \sigma\,dW_t. \tag{6.102}$$

To evaluate the expectation $\mathbb{E}(B_T^{-1}C(S_T, T))$, we need the distribution of S_T, or equivalently that of $\log S_T$. Remembering our definition of an Ito process from Chapter 5 (see Section 5.3), we have that

$$\log S_T - \log S_0 = rT - \frac{1}{2}\int_0^T \sigma(s)^2\,ds + \sqrt{\int_0^T \sigma(s)^2\,ds}\,N(0,1), \tag{6.103}$$

where $N(0, 1)$ denotes, as usual, a draw from a normal distribution with mean 0 and variance 1.

If we let $\bar{\sigma}$ denote the root-mean-square value of $\sigma(t)$, across the interval $[0, T]$, i.e.

$$\bar{\sigma} = \sqrt{\frac{1}{T}\int_0^T \sigma^2(s)\,ds},$$

then we can write (6.103) as

$$\log S_T - \log S_0 = \left(r - \frac{1}{2}\bar{\sigma}^2\right)T + \bar{\sigma}\sqrt{T}N(0,1). \tag{6.104}$$

But this is the distribution at time T for the log of a geometric Brownian motion with constant volatility $\bar{\sigma}$. We therefore have immediately that the price of an option for a stock with variable volatility $\sigma(t)$ is given by the Black–Scholes formula with volatility $\bar{\sigma}$.

How do we hedge? We simply hold $\frac{\partial C}{\partial S}$ stocks at a time, as in the constant volatility case. The only complication is what value of volatility to use. At any given time, $C(S_t, t)$ is the Black–Scholes price for an option with volatility given by the root-mean-square volatility over the interval $[t, T]$ – there was nothing special about 0 in the argument above. Differentiating with respect to S has no effect on the root-mean-square volatility. Hence at time t our hedge is the Delta of the Black–Scholes price with root-mean-square volatility over the period $[t, T]$, and using any other value for volatility will lead to an incorrect hedge and thus imperfect replication.

6.12 Completeness and the uniqueness of the risk-neutral measure

Let's pause and recap what we have shown about martingale pricing. First, we showed that the existence of a measure in which the ratio of any tradable asset to the riskless bond is a martingale implies that there is no arbitrage. Second, we used Girsanov's theorem to show that there existed a unique change of measure which made the ratio of asset price to bond price a martingale. We then defined the price of every other tradable to be the price implied by the expectation of its price ratio in the risk-neutral measure. This made the price of every other tradable a martingale which guaranteed the absence of arbitrage, and they had to take this price or there would not exist any measure which made all asset price ratios martingales. Finally, we showed that every claim could be replicated by a self-financing portfolio involving only the trading of the bond and the asset.

We thus have two ways of seeing the uniqueness of the price of an option: the first is that the measure making the asset price to bond price ratio is unique – this involves the fact from Girsanov's theorem that all measure changes are drift changes. Thus if we believe the deep result that the existence of a risk-neutral measure is implied by absence of arbitrage then the price of an option can be nothing other than its discounted expectation in the risk-neutral measure.

The second way of seeing the uniqueness of the price is that we have shown that the market is *complete* – every claim can be replicated by a self-financing trading strategy. Since the claim can be replicated, its value must be precisely the cost of setting up a trading strategy or we clearly have an arbitrage.

We can hazard that there must be a connection between market completeness and the uniqueness of the risk-neutral measure. We show that uniqueness of the risk-neutral measure is implied by market completeness.

We first prove that the risk-neutral probability of any event is determined in a complete market. Let A be an event that is determined at time T, that is we have that $A \in \mathcal{F}_T$. We can define a contingent claim D that pays 1 unit of the money market account at time T if the event A has occurred and 0 otherwise. The value of A is then precisely the probability of A occurring in the risk-neutral measure. If the market is complete then A can be replicated, and its value is determined by the cost of setting up the self-financing portfolio. We therefore have that the value of A is determined, and thus that for any risk-neutral measure the probability of A occurring is determined. This means that the two risk-neutral measures are the same and we conclude that the risk-neutral measure is unique.

Note that this argument, that completeness implies uniqueness of the risk-neutral measure, did not depend on the fact we are working in a Black–Scholes world. However, most alternative models are in fact incomplete, and there is then the issue of which risk-neutral measure to choose. We have not proven the converse result that uniqueness implies completeness which is true under certain additional assumptions but beyond our scope. We refer the reader to the original papers of Harrison & Kreps, and Harrison & Pliska, [57], [58, 59].

6.13 Changing numeraire

We have developed the martingale approach to pricing in the context of constant deterministic interest rates. This meant we could take a riskless bond, B, such that

$$B_t = B_0 \exp(rt)$$

as our unit of account, with which to do all our discounting. In practice, interest rates are not even deterministic, let alone constant. However, by a slight shift in viewpoint, one can see that option pricing can still be carried out in a similar fashion. The key to seeing this is the powerful technique known as *change of numeraire*.

The numeraire is the unit of account. We have so far used the riskless bond or money market account as our unit of account. What special properties does the numeraire have? It is deterministic and it is always positive. It turns out that only the second of these properties is important. We recall our argument that led to risk-neutral pricing. Suppose N is our numeraire and we have a probability measure such that $R = A/N$ is martingale for every asset A.

We conclude that no arbitrages can exist if a portfolio of zero set-up cost today must have expected value of R equal to zero. This means that R must have non-zero probability of negative value if it has non-zero probability of positive value. As N is strictly positive, this is equivalent to saying the portfolio must have non-zero probability of negative value if it has non-zero probability of positive value. In other words, no portfolio can be an arbitrage portfolio. As we have the martingale property, this applies equally well to trading strategies.

Example 6.1 Suppose in a simple Black–Scholes world we take the stock as numeraire. In the real world, we have as usual

$$dS = \mu S dt + \sigma S dW_t \tag{6.105}$$

$$dB = rB dt. \tag{6.106}$$

The condition that S/S should be a martingale is, of course, vacuous as it is the constant 1. However, we now have the new condition that B/S should be a martingale which has replaced the previously vacuous martingale condition for B/B.

We compute via Ito's lemma:

$$d\left(\frac{B}{S}\right) = \frac{dB}{S} - \frac{B}{S^2} dS + \frac{1}{2} \frac{2}{S^3} B dS^2, \tag{6.107}$$

which reduces to

$$d\left(\frac{B}{S}\right) = \frac{(r+\sigma^2-\mu)B}{S}dt - \sigma\frac{B}{S}dW. \quad (6.108)$$

This is a martingale if and only if $\mu = r + \sigma^2$. Invoking Girsanov's theorem we can change the drift of S to be $r + \sigma^2$. We conclude that the dynamics of B and S in the martingale measure associated to the numeraire S are

$$dS = (r+\sigma^2)S dt + \sigma S dW_t, \quad (6.109)$$
$$dB = rB dt. \quad (6.110)$$

Suppose we want to price a European option using this numeraire. Let the option, C, pay f at time T. We have that

$$\frac{C(0)}{S(0)} = \mathbb{E}\left(\frac{C(T)}{S_T}\right) = \mathbb{E}\left(\frac{f(S_T)}{S_T}\right). \quad (6.111)$$

We rewrite a call option's payoff as

$$S_T I_{S_T \geq K} - K I_{S_T \geq K},$$

where $I_{S_T \geq K}$ is 1 for $S_T \geq K$ and 0 otherwise, and apply (6.111) to the first term. That is, let D pay

$$S_T I_{S_T \geq K},$$

at time T. We then have

$$\frac{D(0)}{S(0)} = \mathbb{E}\left(\frac{D(T)}{S_T}\right) = \mathbb{E}\left(I_{S_T \geq K}\right). \quad (6.112)$$

The final expectation is just the probability that S_T is greater than K in the S-numeraire martingale measure. Using the solution of the SDE for a Brownian motion with drift, this is equal to the probability that

$$S_0 e^{(r+\sigma^2/2)T + \sigma\sqrt{T}N(0,1)} \geq K. \quad (6.113)$$

A straightforward computation gives us the first term in the Black–Scholes formula.

To get the second term in the Black–Scholes formula, it is easier to use B as numeraire. Note that this neatly explains the division of the Black–Scholes formula into two terms with coefficients S_0 and $e^{-rT}K$. Note also that the computation of the first term is made substantially easier by the use of the correct numeraire.

6.14 Dividend-paying assets

The PDE approach lets us deduce the Black–Scholes equation for an option on a dividend-paying asset quite easily from the non-dividend-paying case. We similarly wish to understand the behaviour of a dividend-paying asset in the martingale measure.

We can use the same trick. Let S_t be the price of a dividend-paying stock which grows with continuous dividend rate d, and such that the real-world price process is geometric Brownian motion with drift μ and volatility σ. We wish to know how S_t behaves in the risk-neutral measure. Suppose we work over a finite time horizon $[0, T]$. As before, we let X_t be the price of a contract which involves the delivery of one unit of S_t at time T. The value of X_t at time t will be $e^{-d(T-t)}S_t$.

Our equivalent martingale measure analysis applies directly to X_t as it is a non-dividend-paying asset. We therefore have that in the risk-neutral measure

$$dX_t = rX_t dt + \sigma X_t dW_t.$$

We simply compute the process for S_t:

$$\begin{aligned} dS_t &= d\left(X_t e^{d(T-t)}\right) \\ &= dX_t e^{d(T-t)} - X_t d e^{d(T-t)} dt \\ &= (r-d)S_t dt + \sigma S_t dW_t. \end{aligned} \quad (6.114)$$

Thus our process is simply adjusted so that the growth rate is $r - d$ instead of r, to compensate for the fact that the number of units of the asset held will be growing at rate d.

We can now proceed as before. The price of an option will just be the discounted risk-neutral expectation. Note that we have not changed the bond price process so we will be discounting according to e^{-rT} as before. We can now trace through the derivation of the Black–Scholes price in Section 6.8 with the modified drift and the result is an alternate derivation of the formula obtained in Section 5.9.

6.15 Working with the forward

We have so far regarded the underlying as the fundamental quantity, but one could equally well replace it by the forward price at a vanilla option's time of expiry. Thus suppose we have a call option expiring at time T with strike K. At time T, we have that the forward price, $F_T(T)$, for transacting at time T is equal to the spot price, S_T. This means that the pay-off of the call option is equal to

$$\max(F_T(T) - K, 0) = (F_T(T) - K)_+.$$

We can therefore equally regard the call option as a derivative on the forward price as on the spot price.

How would we price? If we have a constant continuously compounding interest rate, r, and a dividend rate d, then the forward price at time t is equal to

$$F_T(t) = e^{(r-d)(T-t)} S_t.$$

Taking the riskless bond as numeraire, we have

$$dS_t = (r-d)S_t dt + \sigma S_t dW_t, \qquad (6.115)$$

which immediately yields

$$dF_T(t) = F_T(t)\sigma dW_t. \qquad (6.116)$$

This means that the forward price is driftless and is a martingale.

The value of our call option is equal to

$$e^{-rT}\mathbb{E}((F_T(T) - K)_+) = e^{-rT}\mathbb{E}((F_T(0)e^{-\frac{\sigma^2}{2} + \sigma\sqrt{T}X} - K)_+) \qquad (6.117)$$

where is X is an $N(0, 1)$ variable. We can evaluate this expectation to get

$$e^{-rT}\left(F_T(0)N(h_1) - KN(h_2)\right), \qquad (6.118)$$

where

$$h_j = \frac{\log\left(\frac{F_T(0)}{K} + (-1)^{(j-1)}\frac{\sigma^2}{2}T\right)}{\sigma\sqrt{T}}. \qquad (6.119)$$

This expression is called the Black formula. It was originally derived in the context of options on futures, [18]. Note that if we substitute the $S_0 e^{(r-d)(T)}$, for the forward price, we immediately get back to the Black–Scholes formula.

However, the Black formula is in some ways neater. The discounting appears purely in the global multiplier e^{-rT} and the effects of interest rates and dividend rates on the risk-neutral growth rate are carefully hidden in the forward price. The Black formula is particularly important in the context of interest-rate derivatives which are often naturally defined in terms of options on rates rather than options on assets.

If we want to price options on stocks in a world with stochastic interest rates (e.g. the real world) then the Black formula is very useful. With stochastic interest rates the processes for the forward price and the spot price are no longer so simply related, and instead will depend upon the process chosen for the interest rates and the correlation between the interest rate process and the spot price process.

However, if instead of trying to produce the forward price process from the stock price process, we regard the former as the fundamental quantity and forget about the spot price then we can still price options. We therefore suppose that in the real-world measure the forward price process is geometric Brownian motion with some drift. The choice of numeraire is now a trickier issue since in a stochastic

6.15 Working with the forward

interest rate world, we can choose any of a continuum of riskless bonds as the numeraire. There is however a natural choice: there is one bond whose value we can be absolutely sure of at time T, namely the zero-coupon bond with maturity T. We denote the value of this bond at time t by $P(t, T)$.

When working with the forward, we have to be more careful in certain ways. In the spot world, we were working with the price of a tradable asset which means that the ratio of the spot price to the numeraire must always be a martingale. This is not so obvious for the forward price and indeed we saw above that in a special case the forward price itself was the martingale rather than the ratio to the numeraire. In order to understand the dynamics of the forward price, we need to work with a closely related proxy: the forward contract. Recall that the forward contract is the agreement to buy or sell the asset at a price K and it carries both right and obligation. The forward price, by definition, is the strike which makes the contract valueless. Suppose the contract is to buy at K and expires at time T, and the forward price is $F_T(t)$ at time T. We can enter into a second contract to sell for $F_T(t)$ at time T. This means that at time T, we will receive and sell the asset so our asset position does not change. We also receive £$F_T(t)$ and pay £K. This means that a forward contract to buy, struck at K, is equivalent to the right to receive

$$£(F_T(t) - K)$$

at time T. In other words the contract has the value

$$£(F_T(t) - K)P(t, T).$$

The contract *is* a traded asset so the ratio of its value to the numeraire must be a martingale in the numeraire's martingale measure. We therefore conclude that on taking $P(t, T)$ as numeraire, we have

$$\frac{(F_T(t) - K)P(t, T)}{P(t, T)} = F_T(t) - K \qquad (6.120)$$

is a martingale for any K. In particular, $F_T(t)$ is a martingale. The martingale measure associated to the bond $P(T)$ is therefore sometimes called the *forward measure* associated to time T.

As changes of measure can only change drift, and as $F_T(t)$ followed a geometric Brownian motion, we conclude that in the forward measure

$$dF_T(t) = F_T(t)\sigma dW_t. \qquad (6.121)$$

We can now price our call option. We take $P(T)$ as numeraire and conclude that its value at time zero is equal to

$$\mathbb{E}(F_T(T) - K)_+)P(0, T).$$

We can evaluate this expectation to get

$$P(0, T) \left(F_T(0) N(h_1) - K N(h_2) \right), \tag{6.122}$$

with h_j as above.

We now have a formula for the price of a call option in a world with stochastic interest rates. Its principal input is the volatility of the forward price rather than the volatility of the spot price. Whilst these are essentially the same when we work with deterministic interest rates, there is no reason for this to be the case in general, and indeed for long-dated options the volatility coming from interest rates may be substantial. For further discussion of the philosophical aspects of these points, we refer the reader to Chapter 1 of [106].

When working with the spot, we had a neat method for deriving the Black–Scholes formula, Example 6.1, which involved dividing into two pieces and choosing a different numeraire for each. In particular the payoff could be written as

$$S_T I_{S_T > K} - K I_{S_T > K},$$

and we then took S_t as numeraire for the first piece and the riskless bond as numeraire for the second. The payoff can again be written as

$$F_T(T) I_{F_T(T) > K} - K I_{F_T(T) > K}.$$

We have to be slightly more careful now as we cannot take $F_T(T)$ as numeraire; it is not the price of a traded asset. However, remembering that the payoff occurs at time T, we can rewrite it as

$$F_T(T) P(T, T) I_{F_T(T) > K} - K P(T, T) I_{F_T(T) > K}.$$

We therefore take $F_T(t) P(t, T)$ as numeraire for the first piece and $P(t, T)$ as numeraire for the second. Note that it is OK to take $F_T(t) P(t, T)$ as numeraire since we showed above that it is the price of a forward contract struck at zero.

To evaluate the first piece, we need to find the process of $F_T(t)$ when $F_T(t) P(t, T)$ is numeraire. As the ratio with $P(t, T)$ must be a martingale, this means that $1/F(t, T)$ must be a martingale. A quick application of Ito's rule, just as in Example 6.1, shows that

$$dF_T(t) = \sigma^2 F_T(t) dt + \sigma F_T(t) dW_t. \tag{6.123}$$

We therefore have that

$$d \log F_T(t) = \frac{1}{2} \sigma^2 dt + \sigma dW_t, \tag{6.124}$$

and the first term of the Black formula easily follows.

We shall return to the issue of working with the forward when we study interest rate derivatives; there the forward is the natural underlying but it is not tradable, and so we must proceed as we have in this section.

6.16 Key points

We have covered a lot of ground in this chapter; in particular we have introduced the concept of an equivalent martingale measure and shown that it can be used for pricing.

- The set of call option prices at a single time horizon define a synthetic probability measure such that their prices are equal to their discounted expectations under this measure.
- The synthetic probability measure has the property that the mean growth rate for the stock is equal to that of the riskless bond.
- The synthetic probability measure in the Black–Scholes world is given by taking the growth rate of the stock to be r.
- A martingale is a random process such that its expectation is equal to its current value at all times.
- Two measures on a probability space are equivalent if they have the same sets of zero probability.
- An arbitrage exists under one probability measure if and only if it exists under an equivalent probability measure.
- There can be no arbitrage if every asset is a martingale.
- We can price in an arbitrage-free fashion if we set every derivative equal to its discounted future value at every time and world state.
- An equivalent martingale measure is found in the Black–Scholes world by setting the growth rate of the stock to be r.
- Absence of arbitrage is effectively equivalent to the existence of an equivalent martingale measure.
- A market is complete if every contingent claim can be replicated.
- A market is complete if and only if the equivalent martingale measure is unique.

6.17 Further reading

There is any number of books on martingales and risk-neutral pricing. We list a few that the author has found helpful.

An accessible book on martingales is Oksendal, [100]. A fully rigorous and standard textbook for the pure mathematician is Karatzas & Shreve, [78]. The sequel, [79], is a fully rigorous account of mathematical finance from a pure mathematical viewpoint.

Baxter & Rennie, [12], is accessible and is wholly devoted to the martingale pricing approach at a slightly more rigorous level than we have adopted here. A more rigorous and wide-ranging book on financial mathematics from the martingale point of view is Musiela & Rutowski, [96].

Björk, [17], derives risk-neutral valuation from the PDE approach and is accessible. It is very good at emphasizing the key ideas.

For the fundamental theorems on asset pricing see Harrison & Kreps, [57], and Harrison & Pliska, [58, 59] and also [39]. The result that the payoff of a general vanilla option could be written as an integral against the second derivative of call option prices is due to Breeden & Litzenberger, [21].

6.18 Exercises

Exercise 6.1 If S_t follows geometric Brownian motion, what is the process for the forward price of S_t at time T in the risk-neutral measure.

Exercise 6.2 Let W_t be a Brownian motion. Which of the following events are in the filtration $\mathcal{F}_s$?

(i) $W_s < 0$.
(ii) $W_r > W_s$ for $r > s$.
(iii) $W_{s-1} < 0$.
(iv) $W_{s+1} > 0$.
(v) $W_s > W_{s-1}$.
(vi) $W_r < 1$ for $r \leq s$.
(vii) W_r is increasing for $r < s$.

Exercise 6.3 Let W_t be a Brownian motion. Which of the following are stopping times?

(i) The first time such that W_t is at least 1.
(ii) The first time such that $W_{t-1} < 0$ and $W_t \geq 0$.
(iii) The first time such that $W_{t+1} < 0$ and $W_t \geq 0$.
(iv) The first time that the path W_t crosses the level 10.

Exercise 6.4 Let an asset follow a Brownian motion

$$dS = \mu dt + \sigma dW,$$

with μ and σ constant. The constant interest rate is r. What process does S follow in the risk-neutral measure? Develop a formula for the price of a call option and for the price of a digital call option. What is the analogue of the Black–Scholes equation for this asset?

Exercise 6.5 In general, how will increasing the dividend rate affect the price of a call option?

Exercise 6.6 Suppose a stock follows geometric Brownian motion in a Black–Scholes world. Develop an expression for the price of an option that pays $S^2 - K$ if $S > K$ and zero otherwise. What PDE will the option price satisfy?

Exercise 6.7 A non-dividend paying stock follows the process

$$dS = \mu S dt + \sigma' S dW_t.$$

Suppose we hedge a call option on S using the Black–Scholes Delta but with a value of volatility σ' not equal to σ. What will happen?

Exercise 6.8 Price a derivative paying $\log S_T$ on a non-dividend-paying stock following geometric Brownian motion.

Exercise 6.9 A *trigger call option* forces the holder to buy a stock S at a price K if the stock price is above H at the time of expiry. Develop an analytic formula for the price of this option in a Black–Scholes world.

7

The practical pricing of a European option

7.1 Introduction

We have developed several techniques for pricing an option: trees, PDEs, risk-neutral valuation and replication plus variants of each. The purpose of this chapter is to look at the practicalities involved in using each one. We therefore study the pricing of a derivative, C, on an underlying S_t which pays a function $f(S_T)$ at time T. To keep the analysis simple, we shall assume that f is piecewise smooth, that is it is an infinitely differentiable function except at a finite number of points, which is true of all the market instruments known to the author. Our purpose is as much to use the product C to illustrate issues which arise in the practical pricing of exotic options, as to discuss the pricing of C.

We recall some simple examples which we have studied already. A *forward contract* struck at K is defined by

$$f(S_T) = S_T - K. \tag{7.1}$$

A *call option* struck at K is defined by

$$f(S_T) = (S_T - K)_+ = \max(S_T - K, 0). \tag{7.2}$$

A *put option* struck at K is defined by

$$f(S_T) = (K - S_T)_+ = \max(K - S_T, 0). \tag{7.3}$$

A *digital call* option struck at K pays

$$f(S_T) = H(S_T - K) \tag{7.4}$$

at time T, where $H(s)$ is 1 for $s \geq 0$ and 0 otherwise. Similarly a *digital put* option is defined by

$$f(S_T) = H(K - S_T). \tag{7.5}$$

A *power call* option struck at K of order l pays

$$f(S_T) = \max((S_T - K)^l, 0). \tag{7.6}$$

Similarly for *power put* options.

A *straddle* pays

$$f(S_T) = |S_T - K|.$$

7.2 Analytic formulae

We have already developed analytic formulae for the Black–Scholes price of vanilla call and put options. We can similarly develop prices for digitals and power options. Note that a digital is really a power option of power zero. The easy way to get the digital price is to observe that

$$\frac{\partial}{\partial K} \max(S_T - K, 0) = -H(S_T - K), \tag{7.7}$$

where $H(x)$ is as above, and differentiation with respect to K commutes with the solution operator for the Black–Scholes equation, or with taking the risk-neutral expectation, so the price of a digital call is just the derivative of the call price with respect to K. (I have swept some technical points under the carpet here but it can be made rigorous.)

Alternatively, we can just evaluate the risk-neutral expectation directly. If C is the digital call then the price at time zero is equal to

$$e^{-rT} \mathbb{E}(C(S_T)).$$

As $C(S_T)$ is the indicator function of the set $S_T > K$, the expectation $\mathbb{E}(C(S_T))$ is just the risk-neutral probability that S_T is greater than K. We can work with the log: we need the probability that $\log S_T$ is greater than $\log K$. As

$$d \log S_t = \left(r - \frac{1}{2}\sigma^2\right) dt + \sigma dW_t, \tag{7.8}$$

we have that

$$\log S_T = \log S_0 + \left(r - \frac{1}{2}\sigma^2\right) T + \sigma \sqrt{T} N(0, 1), \tag{7.9}$$

where $N(0, 1)$ is a draw from a standard normal distribution. Thus we need the probability that

$$\log S_0 + \left(r - \frac{1}{2}\sigma^2\right) T + \sigma \sqrt{T} N(0, 1) > \log K. \tag{7.10}$$

This is the probability that

$$N(0, 1) > \frac{\log(K/S_0) - \left(r - \frac{1}{2}\sigma^2\right) T}{\sigma \sqrt{T}}. \qquad (7.11)$$

Using the fact that $N(-x)$, the probability that $N(0, 1)$ is less than $-x$, is equal to $1 - N(x)$, we have that this probability is equal to $N(d_2)$ where

$$d_2 = \frac{\left(r - \frac{1}{2}\sigma^2\right) T + \log(S_0/K)}{\sigma \sqrt{T}}. \qquad (7.12)$$

We therefore conclude that the price of the digital call is

$$e^{-rT} N(d_2).$$

As the price of a digital call plus a digital put is a zero-coupon bond, we conclude that the price of a digital put is

$$e^{-rT}(1 - N(d_2)) = e^{-rT} N(-d_2).$$

Note that we can obtain all the Greeks of the digitals by differentiating their prices. We have deduced the prices of the digital options via risk-neutral expectations. An alternative approach would be to take the Black–Scholes PDE, transform to the heat equation and solve with the transformed boundary condition. We leave this to the readers who like solving heat equations.

7.3 Trees

Suppose we decide that we wish to price our European option with pay-off $f(S_T)$ by using a tree. The key is to think of a risk-neutral tree; our objective is to compute the risk-neutral expectation, so we want a process that approximates the risk-neutral evolution well. The real-world tree was useful for justifying the risk-neutral expectation, but it is not particularly useful for actually computing prices.

We first consider a binary tree. We have that the log of the stock price evolves according to the process

$$d \log S_t = \left(r - \frac{1}{2}\sigma^2\right) dt + \sigma d W_t. \qquad (7.13)$$

We can discretize this by dividing the interval $[0, T]$ into many small time steps of length Δt. For concreteness suppose that we have N steps of size T/N.

At each step we discretize by

$$\log S_{j+1} = \log S_j + \left(r - \frac{1}{2}\sigma^2\right) \Delta t + \sigma \sqrt{\Delta t} X_j, \qquad (7.14)$$

7.3 Trees

where X_j takes the values 1 and -1 with equal probability. As X_j has mean 0 and variance 1, this process will converge to Brownian motion as N tends to infinity; this follows from the arguments in Chapter 3. If we take $e^{-rT}\mathbb{E}(f(S_N))$, we should therefore obtain a good approximation to the price of the derivative.

The important thing about (7.14) is that S_j has only $j+1$ possible values – an up-move followed by a down-move is the same as a down-move followed by an up-move. Following the procedure outlined in Chapter 3, we can now price the option. We take the pay-off of the option at each of the possible nodes in the final time step and store them. Then at the second last time step, the expectation at each possible node is just the average of the values after the up- and down-moves. We then just cascade back to get the expectation at time zero.

Of course, this is just an approximation to the price. We will want to know how good an approximation it is: one approach is simply to compute the value for lots of values of N and observe the trend. If the first four significant figures stop changing, then we can expect the price to be correct to four significant figures and so on. Note that for N steps we have $N(N-1)/2$ nodes. This is crucial in the tractability of the method in that it means we have to do N^2 calculations. If the drift of our stochastic process was state-dependent, or the volatility was time-dependent, then an up-move followed by a down-move would not be the same as a down-move followed by an up-move, and we would have 2^N nodes and therefore order 2^N computations to do. A non-recombining (or *bushy*) tree therefore quickly becomes impossible to do in reasonable time. For time-dependent volatility, we could however, use the observation that the same price is attained with a constant volatility equal to the root-mean-square of the volatility, and price with that volatility instead, or alternatively rescale the time-step sizes to make all steps have the same variance.

One problem with binary trees is that the price, as a function of steps, tends to display a zig-zag pattern. For example, suppose we wish to price a binary call option struck at K, which is a tiny amount above $S_0 \exp(-\frac{1}{2}\sigma^2 T)$, and suppose there are no interest rates. The price of the option is then just the fraction of the nodes in the final layer that are above the strike. If we have an even number of nodes in the final layer then this will be precisely $\frac{1}{2}$. However, if there is an odd number then there will be one at $S_0 \exp(-\frac{1}{2}\sigma^2 T)$ just below K, and if there are $2n+1$ nodes, the price will be $n/(2n+1)$. The price will therefore alternate between two levels which are converging together. See Figures 7.1 and 7.2.

This alternation property can make binary trees slow to converge. One way out is to look at the average of the even and odd prices for which the alternations cancel. Another alternative is to use a trinomial tree instead. Although we have repeatedly shown that the use of three branches at each node does not lead to

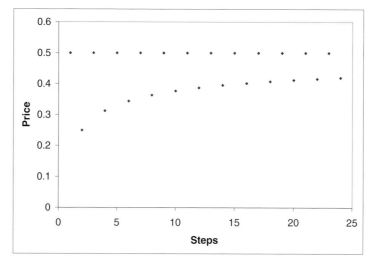

Fig. 7.1. A European digital call option struck at 100, with spot 100, no interest rates and one year to expiry, price plotted as a function of the number of steps on a tree.

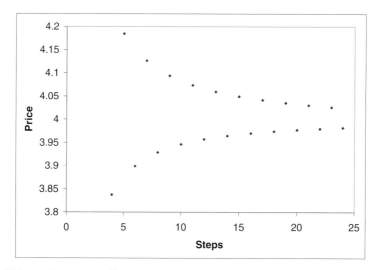

Fig. 7.2. A European call option struck at 100, with spot 100, no interest rates and one year expiry, price plotted as a function of the number of steps on a tree.

no-arbitrage pricing, the trinomial tree is nevertheless very useful. The reason is that we do not attempt to use it to justify the pricing methodology, but instead use it to approximate the evaluation of the risk-neutral expectation which has already been proven to give the unique arbitrage-free price.

We still use (7.14) but now let X_j take three values: $-\alpha, 0, \alpha$. The crucial point is that we want X_j to still have mean 0 and variance 1. If we let both α and $-\alpha$

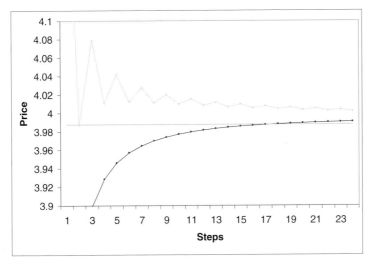

Fig. 7.3. Call with parameters as in Figure 7.2. The trinomial price (dark line) and successively averaged binomial price are plotted as a function of the number of steps. The correct price is the flat line.

have probability p then the mean is 0 and the variance is $2\alpha^2 p$. Thus if we take $\alpha = \sqrt{2}$ and $p = \frac{1}{4}$, the variance is 1. Note that there are other solutions which we could use instead if we wanted more control over the nodes' placements; this is an important advantage of trinomial trees.

The tree will recombine as before; the number of nodes at step j will be $2j + 1$ and so the total number of nodes will still be of order N^2. We can now apply the same approach as for the binary case. Compute the values of the payoff in the final layer. For the second last layer at each node, we take the expectation obtained by p times the up-node value plus p times the down-node value plus $1 - 2p$ times the zero-node value. We then just iterate back as before. The fact that the central line of the tree remains invariant across the number of steps means that the price of the option is more stable with respect to the number of steps. See Figures 7.3, 7.4 and 7.5.

As well as wanting to compute the price of an option, we shall generally also want to know its Greeks, as these are central to hedging the option. One method is to simply bump the relevant parameter slightly, recompute the price and then divide the difference by the amount of bumping of the parameter. Of course, one has to be careful that the simulation had converged both in terms of size of the bump, and in the number of steps. The number of steps for the Greek to converge may well be greater than for the original product.

A good approximation to the Delta is to take the value of the option after the first up- and down-moves and divide their difference by the distance between them. This

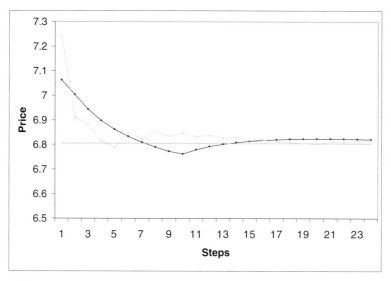

Fig. 7.4. Call with parameters as in Figure 7.4 but interest rates now 5%. The trinomial price (dark line) and successively averaged binomial price are plotted as a function of the number of steps. The correct price is the flat line.

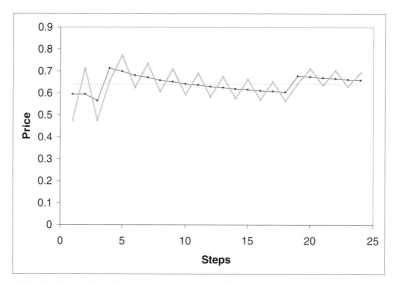

Fig. 7.5. Digital call with parameters as in Figure 7.4. The binomial price and successively-averaged binomial price (dark line) are plotted as a function of the number of steps. The correct price is the flat line.

is fairly accurate but one has to be aware that there is a slight error from the fact that the nodes are slightly forward in time. We discuss some other techniques which are equally valid for trees, in Sections 7.4 and 7.5.

The most powerful application of trees is to the pricing of options with early exercise features and we shall study this application in Chapter 12.

7.4 Numerical integration

We know that the price of the derivative C is given by

$$e^{-rT}\mathbb{E}(f(S_T)),$$

so all we need to do is evaluate this expectation. If $\Phi(S_T)$ is the density of S_T in the risk-neutral measure then

$$\mathbb{E}(f(S_T)) = \int f(S_T)\Phi(S_T)dS_T. \tag{7.15}$$

This is a simple Riemann integral and so can be evaluated numerically. Of course, to do so we need the value of $\Phi(S_T)$.

This is in fact quite easy to compute in the Black–Scholes model. The density of a random variable is just the derivative of the cumulative distribution function. Recall that the cumulative distribution $F(x)$ is defined by

$$F(x) = \mathbb{P}(S_T \leq x).$$

The price of a digital put option is therefore just e^{-rT} times $F(K)$, where K is strike. Thus to get the density we just differentiate the price of a digital put by K and multiply by e^{rT}. Putting $K = S_T$, we obtain the *Black–Scholes density*

$$\Phi(S_T)(x) = \frac{1}{\sqrt{2\pi}S_T\sigma\sqrt{T}} \exp\left(-\frac{1}{2}\left(\frac{\log\left(\frac{S_T}{S_0}\right) - (r - \frac{1}{2}\sigma^2)T}{\sigma\sqrt{T}}\right)^2\right). \tag{7.16}$$

Whilst this expression is not particularly nice, it is straightforward to evaluate, and we can now apply any method of numerical integration to evaluate (7.15).

In case the reader is not familiar with (or more probably has forgotten how to carry out) numerical integration, we sketch one simple method: the trapezium method. If we wish to integrate a function $g(x)$ over an interval $[a, b]$ then we divide the interval into N pieces of equal length. Thus we set

$$x_j = a + \frac{j}{N}(b-a), \quad (7.17)$$

for $j = 0, \ldots, N$. Across each interval $[x_j, x_{j+1}]$ we replace $g(x)$ by the unique linear (really affine) function, $L_j(x)$, such that $L_j(x_j) = g(x_j)$ and $L_j(x_{j+1}) = g(x_{j+1})$. In fact,

$$L_j(x) = g(x_j) + (x - x_j)\frac{g(x_{j+1}) - g(x_j)}{x_{j+1} - x_j}. \quad (7.18)$$

As $L_j(x)$ is linear, its integral, I_j, over $[x_j, x_{j+1}]$ is trivial to compute analytically. We therefore just compute the value of each I_j and sum. As N goes to infinity, the value of the sum will converge to the true value of the integral. The rate of convergence will depend on the amount of curvature of $g(x)$, i.e. the size of $g''(x)$, as this expresses how bad the linear approximation is.

Our integral is over an infinite domain rather than a finite interval, so the cut off has to be chosen at a suitably large value of b. As $\Phi(S_T)$ is rapidly decaying this is not unreasonable, or we can perform a change of variables to change the infinite interval into a finite one. We refer the reader to [104] for a discussion of many different methods of implementing numerical integration.

It is worth noting that we can perform the integration in a different way that avoids the need to know the value of $\Phi(S_T)$. We have that

$$S_T = S_0 \exp\left(\left(r - \frac{1}{2}\sigma^2\right)T + \sigma\sqrt{T}N(0,1)\right). \quad (7.19)$$

This means that we can equally well regard $N(0, 1)$ as our fundamental underlying variable to integrate against. This means that

$$\mathbb{E}(f(S_T)) = \int f\left(S_0 \exp\left(\left(r - \frac{1}{2}\sigma^2\right)T + \sigma\sqrt{T}x\right)\right) \frac{\exp\left(-\frac{1}{2}x^2\right)}{\sqrt{2\pi}} dx. \quad (7.20)$$

The equivalence of these methods can be seen by a change of variables in the integration.

It is unlikely that one would ever actually use numerical integration to evaluate the price of a European option in a Black–Scholes world; however, when using alternative models where no closed-form expression for the call price exists, but a closed-form expression for the density is known, it can be very useful.

As well as evaluating the price, the trader will want to know the value of the Greeks. There are a number of different ways of doing this. The simplest method is simply to bump the input parameter and divide the change in value by the size of the bump. This is handy when only the price function has been implemented.

7.4 Numerical integration

It's also good for testing that alternative implementations of the Greeks are correct. We would, however, prefer something more robust; the value of our option is of the form

$$e^{-rT} \int f(S_T) \Phi(S_T, S_0, r, d, \sigma, T) dS_T. \tag{7.21}$$

One approach is therefore to simply differentiate this formula and evaluate it. For example for the Vega we obtain

$$e^{-rT} \int f(S_T) \frac{\partial \Phi}{\partial \sigma}(S_T, S_0, r, d, \sigma, T) dS_T, \tag{7.22}$$

and for the Delta

$$e^{-rT} \int f(S_T) \frac{\partial \Phi}{\partial S_0}(S_T, S_0, r, d, \sigma, T) dS_T. \tag{7.23}$$

Of course, one would still have to analytically compute the derivatives of Φ.

For log-based evolutions, including the Black–Scholes setting, Φ takes a special form. By a log-based evolution, I mean any evolution for which the distribution of $\log S_T - \log S_0$ does not depend on the value of S_0. The probability density is then of the form $\psi(\log S_T - \log S_0) d(\log S_T)$, ignoring the other parameters. This means that we have a density function for S_T of the form

$$\Phi(S_T, S_0) = \frac{1}{S_T} g\left(\frac{S_T}{S_0}\right), \tag{7.24}$$

as

$$d(\log S_T) = \frac{dS_T}{S_T}.$$

This implies that

$$\frac{\partial \Phi}{\partial S_0} = -\frac{1}{S_0^2} g'\left(\frac{S_T}{S_0}\right) \tag{7.25}$$

and that

$$\frac{\partial}{\partial S_T}\left(\frac{S_T}{S_0} \Phi\right) = \frac{\partial}{\partial S_T}\left(\frac{1}{S_0} g\left(\frac{S_T}{S_0}\right)\right) = \frac{1}{S_0^2} g'\left(\frac{S_T}{S_0}\right). \tag{7.26}$$

This means that we can express the derivative with respect to S_0 in terms of Φ and its derivative with respect to S_T. This is particularly useful because we can then integrate by parts to shift the differentiation with respect to S_T onto f. In particular, we have

$$\frac{\partial \Phi}{\partial S_0} = -\frac{\partial}{\partial S_T}\left[\frac{S_T}{S_0} \Phi\right]. \tag{7.27}$$

Integrating by parts, we conclude that the Delta of our option is

$$e^{-rT} \int \frac{S_T}{S_0} f'(S_T) \Phi(S_T, S_0) dS_T. \tag{7.28}$$

We have to be slightly careful about interpreting (7.28) when f is not continuous. For example, if the option is a digital call then f has a jump singularity at the strike and the derivative of f is a delta function at the strike. (Just to make life confusing, we have the Delta of an option, and a delta function which have absolutely nothing in common except the name.) This means that we have to interpret (7.28) in distributional terms. (Distributions here means generalized functions, **not** probability distributions to make life even more confusing.) For example, for integration against a delta function, we just put $S_T = K$ in the rest of the integrand to obtain the value; thus for a digital call the Delta is

$$e^{-rT} \frac{K}{S_0} \Phi(K, S_0).$$

Alternatively, if the payoff, f, is not continuous, one can divide the integrand into areas where f is continuous and perform the integration by parts individually. One then obtains additional terms from the endpoints of the integral over each area, which match the terms from delta functions.

7.5 Monte Carlo

The Monte Carlo method is one of the most general techniques for the pricing of options. The basic idea is very simple but the implementation issues can often become very subtle. Monte Carlo is essentially the use of the law of large numbers to evaluate the expectation $\mathbb{E}(f(S_T))$.

Recall that the law of large numbers states that if Y_n is a sequence of identically distributed random variables which are independent, then we have

$$\lim_{N \to \infty} \frac{1}{N} \sum_{j=1}^{N} Y_j = \mathbb{E}(Y_1).$$

In other words, our intuitive notion of expectation, the long run average, agrees with the mathematical definition of expectation as an integral against a density function. The law of large numbers is therefore a numerical method for evaluating integrals! We just keep on drawing the random numbers, Y_j, and keep a running count of the average, this will eventually converge to any desired degree of accuracy, and this tells us the value of the expectation to that degree of accuracy.

7.5 Monte Carlo

We wish to find the value of $\mathbb{E}(f(S_T))$. We know that the solution to the SDE for S_T is

$$S_T = S_0 \exp\left(\left(r - \frac{1}{2}\sigma^2\right)T + \sigma\sqrt{T}N(0,1)\right). \tag{7.29}$$

Our procedure is therefore to draw a random $N(0, 1)$ variable, plug it into (7.29) and then take $f(S_T)$. We repeatedly do this, keeping note of the running sum and the average. The average will eventually converge to the expectation as desired. Unfortunately, the operative word here is *eventually*. The order of error is $\mathcal{O}(N^{-1/2}))$. Thus to get high levels of accuracy one needs a lot of samples, sometimes millions. As a Monte Carlo simulation is based on random numbers, the answer after n samples will be a random number depending upon precisely which random draws we have made. If the variance of a single trial is V, it can be shown, using the Central Limit theorem, that the result will be distributed approximately as

$$\mathbb{E}(X) + \sqrt{\frac{V}{n}}N(0,1).$$

The quantity $\sqrt{\frac{V}{n}}$ is called the *standard error*.

One practical point here is how to synthesize Gaussian random variables. There are a number of approaches. Most computer languages allow the generation of a uniform random variable between 0 and 1, or a random integer between 0 and some very large number. The second sort can be turned into the first by dividing by the very large number. The issue is therefore how to turn a uniform random variable on the interval $[0, 1]$ into an $N(0, 1)$ random variable. There are a number of approaches.

Let $N(x)$ denote the cumulative normal function; then its right inverse function, $I(y)$, is called the *inverse cumulative normal function*. That is I has the property that

$$N(I(y)) = y.$$

If r is a random draw from the uniform distribution of $[0, 1]$, then $I(y)$ is a draw from $N(0, 1)$,

$$\mathbb{P}(I(r) < x) = \mathbb{P}(N(I(r)) < N(x)) = \mathbb{P}(r < N(x)) = N(x), \tag{7.30}$$

with the final equality coming from the definition of the uniform distribution. But this shows that $I(r)$ has cumulative distribution function $N(x)$, which means that it is an $N(0, 1)$ random variable. This is very neat, except that it begs the question of how to compute the function I. We give one method in Appendix B. Note that this method would work for any sort of random variable for which one could compute the inverse cumulative distribution function.

A simple method which gives a reasonable, but not great, approximation is to simply add together 12 uniform variables and subtract 6. The results has correct mean, variance and third moment. This method is worthwhile for quick tests but not appropriate when real precision is required.

For the sort of option we are studying in this chapter, Monte Carlo simulation is not a particularly great method. Its convergence is not very fast and a straightforward numerical integration is generally quicker. However, in later chapters where we need to evolve many assets across many time steps it can become the optimal method. Indeed, when studying the evolution of interest rates, it is an indispensable tool.

Often Monte Carlo is the easiest method to implement. It therefore provides a good test of other methods. A quantitative analyst will generally want to implement any model in at least two different ways and not be happy until they agree. After all, how else can he be sure his implementation is correct? It is often also very important when studying alternative models of stock price evolution for which other methods can be hard to implement.

Variance reduction

As Monte Carlo simulation is slow to converge, a lot of research has gone into methods for increasing the speed of convergence. This essentially comes down to simulating the payoffs in such a way that their variance is reduced.

One simple method is *anti-thetic sampling*. With anti-thetic sampling one draws samples in pairs. If x is normally distributed then so is $-x$. For every random draw two paths are therefore simulated one with x and one with $-x$. This ensures that the mean of the draws is zero, and the symmetry of the normal distribution is achieved by construction.

A method which meshes well with anti-thetic sampling is *moment matching*. If we have decided how many samples we wish to take then we can rescale our random numbers to ensure that their moments are correct. This requires two passes. We first draw all the random numbers, computing all the moments we wish to match. We then reset the random number generator to generate the same random numbers, and draw them again but this time rescaling them in such a way as to make the final moments equal to those of the desired distribution. If we have used anti-thetic sampling on a normal distribution then all the odd moments have already been made zero by construction, so it is only the even moments which have to be matched. For example, if the variance of our sample is V and we rescale every random number by $V^{-1/2}$, we obtain a sample with variance 1.

Another fairly easy method is *importance sampling*. If we know our payoff function f is zero outside an interval $[a, b]$, then any draw which makes S_T lie outside

$[a, b]$ is wasted. We therefore only sample uniform distributions which cause S_T to lie in the interval $[a, b]$ and multiply the result by the probability that S_T lies in $[a, b]$. How would we carry out this procedure? We have a deterministic function map from the uniforms to the positive reals which turns a uniform random variable into S_T. We can invert this map to find the interval $[x_1, x_2]$ which is mapped onto $[a, b]$. The probability of S_T lying in $[a, b]$ is therefore $x_2 - x_1$.

To compute the expectation we therefore draw random variables from the uniform distribution on $[0, 1]$, multiply by $(x_2 - x_1)$ and add x_1 to force them to lie in $[x_1, x_2]$. This random variable is then turned into S_T in the usual way and the pay-off computed. We then average over many draws as usual and multiply the result by $(x_2 - x_1)$. Clearly the effectiveness of this method is dependent on the size of $x_2 - x_1$. If one wishes the price of an at-the-money option, it is unlikely to be much help but for a far out-of-the-money option, it may help considerably.

One of the most powerful methods for improving Monte Carlo simulation is that of low-discrepancy numbers. The important aspect of random numbers that makes Monte Carlo simulation work is the fact that they eventually cover the unit interval in an even manner. However, in the short term, they may cluster around certain values which is why the simulation takes a long time to converge. Therefore, instead of using random numbers, why not use a deterministic sequence of numbers which does a very good job of covering the interval? Such a sequence is called a *low-discrepancy sequence*. Note that the idea of using a deterministic sequence, is not very strange, since any computer random number generator is actually one. It is therefore just a question of taking the deterministic sequence which makes simulations converge the fastest. The great advantage of low-discrepancy sequences is that their rate of convergence is $\mathcal{O}(N^{-1})$ rather than $\mathcal{O}(N^{-1/2})$, which greatly increases the competitiveness of Monte Carlo simulations. However, the techniques for generating low-discrepancy sequences are outside our scope and we refer the reader to [104] or [68]. Using low-discrepancy sequences to carry out Monte Carlo, is sometimes called *quasi-Monte Carlo*.

The Greeks

As usual, as well as pricing the option, we want to compute its Greeks. The simplest approach is to simply bump the parameter by ϵ and run the simulation again. The Greek is then the difference divided by ϵ. However this has its problems. The final answer in a Monte Carlo simulation is sensitive to which random numbers have been used and there is no guarantee that the error will be in the same direction in both simulations. An additional error has therefore been introduced that will magnify when divided by the small number ϵ rendering the computation meaningless! One simple way to avoid this problem is to use the same random number stream for both simulations. Remember there is no such thing as a truly random number on a computer. Any biases then introduced are equal in the two simulations and not magnified by ϵ.

This technique works reasonably well for many options. However, it breaks down when the payoff function is not benign. To see why, note that if we are using the same random number streams for the two simulations, then we are really taking the difference approximation to the derivative on a path-by-path basis. Thus if we are computing the Delta in a Black–Scholes world, we are computing the mean of

$$X_j = \epsilon^{-1} e^{-rT} \left(-f\left(S_0 e^{(r-\frac{1}{2}\sigma^2)T + \sigma\sqrt{T}W_j}\right) + f\left((S_0 + \epsilon) e^{(r-\frac{1}{2}\sigma^2)T + \sigma\sqrt{T}W_j}\right) \right) \quad (7.31)$$

for a sequence of normal variables W_j. If f were a Heaviside function, and the option a digital call option, then the value of X_j would be zero except for the very small number of paths where adding ϵ to S_0 made the option go from below the strike to above the strike, and for those paths X_j would be very big: it would be approximately ϵ^{-1}.

We therefore need a more subtle method when f is not continuous. Remember that Monte Carlo is really just a method of numerical integration, so we can adapt the techniques from Section 7.4. We observed there that differentiating with respect to a parameter, including S_0, yielded an integral of the form

$$e^{-rT} \int f(S_T) \Psi(S_0, r, d, T, S_T, \ldots) dS_T,$$

where Ψ is the derivative of the density Φ by the parameter, and ... denotes any other parameters. This integral is not immediately amenable to Monte Carlo simulation as there is no obvious density in it. However, we can reintroduce Φ, and evaluate

$$e^{-rT} \int f(S_T) \frac{\Psi}{\Phi}(S_T) \Phi(S_T) dS_T,$$

instead. We therefore take a random draw from the terminal distribution of S_T implied by Φ, plug it into $e^{-rT} f(S_T) \frac{\Psi}{\Phi}(S_T)$, and average over a large number of draws. This method is sometimes called the *likelihood ratio* method, as we are essentially reweighting the draws by the ratio Ψ/Φ, which is of course just the derivative of $\log \Phi$, with respect to the relevant parameter. Note that as we are using the same density for pricing as for computing the Greeks, we actually only need to run the Monte Carlo once; we can compute the price and Greeks simultaneously.

Another approach when computing Deltas for log-type evolutions is to use (7.28) and evaluate

$$e^{-rT} \int \frac{S_T}{S_0} f'(S_T) \Phi(S_T, S_0) dS_T, \quad (7.32)$$

by Monte Carlo. This is sometimes called the *pathwise* method. The main difficulty with this method is how to interpret $f'(S_T)$ when f is discontinuous. Jump discontinuities will give rise to delta functions in the derivative. We can therefore write $f = g + h$, with g continuous and h piecewise constant. Then g' is well-behaved, and its contribution to the Delta can now be evaluated by Monte Carlo. The derivative of h is a sum of Delta functions so the integral can be computed analytically as a finite sum and we are done.

For an f without jumps, if one runs a simulation of the pathwise method and compares to the finite difference method, one finds that the convergence is almost identical. The reason is that if one lets ϵ tend to zero in (7.31), one obtains the Monte Carlo simulation for (7.32). Thus the main advantages of the pathwise method are the ability to explicitly remove the delta functions, and the removal of the small bias introduced by using finite differencing to evaluate the derivative of f.

7.6 PDE methods

We have shown that the price of an option satisfies the Black–Scholes equation so we can use this fact to price them. This comes down to solving the problem

$$\frac{\partial C}{\partial t} + \frac{1}{2}\sigma^2 S^2 \frac{\partial^2 C}{\partial S^2} + rS\frac{\partial C}{\partial S} - rC = 0, \tag{7.33}$$

$$C(T, S) = f(S). \tag{7.34}$$

We saw in Section 5.8 that the Black–Scholes equation can be reduced to the heat equation. This means that any technique used for solving the heat equation can be used to solve the Black–Scholes equation. In particular, one can attempt to solve analytically or to use numerical methods. However, evaluating the expectation directly tends to be easier than analytically solving the Black–Scholes equation. However, one can also apply one's favourite numerical method. In general, one attains better numerical stability if one works in log space but then there is really very little difference, except in terminology, between a trinomial tree and a finite difference method. We refer the reader who is interested in PDE approaches to Wilmott's *Derivatives*, [118].

7.7 Replication

In fact, if one wishes to price a European option in a market with liquid calls and puts, most of this chapter is useless. The reason is that the most important thing is to price the contract in such a way as to make it compatible with the liquid instruments, and the effective method of doing that is by replication. What does compatible mean? The pricing method must price all the market-observable liquid instruments to agree with their market prices.

For example, suppose we wish to price a digital call option struck at K. We have a Black–Scholes formula but we have to choose a volatility. As market-traded options typically display a volatility smile the choice is not so obvious. An obvious choice is to use the volatility of a call option struck at K. This would however be a mistake and leave us open to arbitrage. An alternative method of pricing is to approximate the digital by vanilla options. In particular, we can create a portfolio consisting of being long $1/2\epsilon$ call options struck at $K - \epsilon$, and short $1/2\epsilon$ call options struck at $K + \epsilon$. This will approximate the digital very well and as $\epsilon \to 0+$ will become the digital (see Section 6.3). The price this gives will be different from the Black–Scholes formula with the implied volatility of the call at K plugged in.

To see this, we write σ, the volatility, as a function of strike. Our approximating portfolio is then worth

$$\frac{-C(K+\epsilon, \sigma(K+\epsilon)) + C(K-\epsilon, \sigma(K-\epsilon))}{2\epsilon}.$$

The price of the digital should be the limit as ϵ goes to zero. From Taylor's theorem, we have

$$C(K+\epsilon, \sigma(K+\epsilon)) = C(K-\epsilon, \sigma(K-\epsilon)) + 2\epsilon \frac{\partial C}{\partial K}(K-\epsilon, \sigma(K-\epsilon))$$
$$+ 2\epsilon \frac{\partial \sigma}{\partial K}\frac{\partial C}{\partial \sigma}(K-\epsilon, \sigma(K-\epsilon)) + \mathcal{O}(\epsilon^2). \quad (7.35)$$

Thus we conclude on letting ϵ tend to zero that the price of the digital call is

$$-\frac{\partial C}{\partial K}(K, \sigma(K)) - \frac{\partial C}{\partial \sigma}(K, \sigma(K))\frac{\partial \sigma}{\partial K}(K).$$

If we had just plugged the Black–Scholes implied volatility into the digital call option formula, we would have only got the first term and not the second. The error is therefore the Vega of the option times the slope of smile. The moral here is that one should be very careful in interpreting implied volatilities. Indeed, to quote Rebonato, [106],

> The implied volatility is the wrong number to put in the wrong formula to get the right price.

The point is that the implied volatility has been defined in such a way as to make this tautologically true for call and puts, but not true for any other option.

We therefore conclude that the way to price any European option is to approximate it as well as possible by vanilla options. In fact, if the payoff function f is piecewise linear then it can synthesized precisely using call and puts. To see this divide the positive real axis into a number of intervals, $[x_j, x_{j+1}]$, such that on the interior of each interval f is linear, with the possibility of a jump at x_j. We can first remove all the jumps arbitrarily accurately by approximating digital options with calls as we saw above. The remaining function, f_1, will now be linear on each interval and continuous everywhere. Taking $x_0 = 0$, we can synthesize its payoff on $[x_0, x_1]$ by a number of zero-coupon bonds with expiry T with principal $f(0)$ and a set of forwards struck at 0 in volume equal to the gradient of f at 0. At x_1, if the gradient of f changes by α we simply add α call options struck at x_1 to the portfolio where the number α is possibly negative. At each x_j we then just add a number of call options struck at x_j, where the number of call options is just the change in gradient of f.

Having constructed this portfolio, we can value the derivative C just by taking the value of the portfolio – if all the options in the model are traded then we just observe their prices in the market and no further mathematics is required. If they are not traded then we can interpolate their volatilities from those that are. We have the added bonus that the Greeks of C are just the sum of the Greeks for the approximating portfolio.

In practice, the way things work in the market is often the reverse of this in that the trader wants exposure to a certain part of the smile and he therefore makes prices on certain combinations of vanilla options that express his views on how the smile will change. These views could be that the smile will flatten, or tighten, or that the smile will tilt. The prices of these combination instruments are then used to infer the shape of the smile.

The techniques of this section are an example of *static* replication where an option is hedged by a portfolio which is set up once and not changed as opposed to the *dynamic* replication of the Black-Scholes theory. In conclusion, static replication is the best method of option pricing when it is available as it automatically takes all smile effects into account. It also has the advantages that the Greeks are easy to compute and that it does not depend upon continuous rehedging.

7.8 Key points

In this chapter, we have looked at a number of different approaches to pricing an option with payoff depending on the value of the underlying at a single time horizon with a view to applying the techniques to exotic options.

- There are many approaches to evaluating the price of European option including:

    (i) analytic formulas;
    (ii) PDE solving;
    (iii) numeric integration;

(iv) Monte Carlo;
(v) replication.

- Replication is the safest method of pricing as it automatically takes smile effects into account.
- Using the implied volatility to price options other than vanillas can lead to pricing errors.
- Monte Carlo pricing relies on the fact that the price of an option is its discounted expected payoff in the risk-neutral measure.
- The law of large numbers underlies Monte Carlo pricing techniques.
- Formulas for Greeks can be deduced from the integral for the price via differentiation under the integral sign and integration by parts.

7.9 Further reading

For lists of option-pricing formulas, see [55].

For discussion of various techniques and code for carrying out numerical integration (also known as quadrature) see *Numerical Recipes in C++* [104]. Another good source on implementing numerical methods is [32].

For further discussion of PDE methods see Wilmott, Howison & Dewynne, [119] or Wilmott, [118].

A good book on many different aspects of Monte Carlo is *Monte Carlo: Methodologies and Applications for Pricing and Risk Management*, edited by Bruno Dupire, [45] which is a collection of accessible research papers on various financial applications of Monte Carlo techniques. A recent book on the applications of Monte Carlo methods to financial problems is [68].

Another book which is strong on the implementation of various models is *Equity Derivatives and Market Risk Models*, [27].

The Monte Carlo techniques for computing Greeks discussed here were introduced by Broadie & Glasserman, [24].

For a discussion of the issues involved in implementing tree models see [33].

7.10 Exercises

Exercise 7.1 Suppose we discretize Brownian motion by taking a trinomial tree. What conditions (if any) on the probabilities and the branches will get the third and fourth moments of one step to agree with those of Brownian motion across the same time step?

Exercise 7.2 Show that the trapezium rule can be simplified by observing that the integral across each step is equal to the size of the step multiplied by the average of the values at the beginning and end of the step.

Exercise 7.3 How do we construct a random draw from the Cauchy distribution, if we start with a draw from a uniform distribution? The Cauchy distribution has density function
$$\frac{1}{\pi}\frac{1}{1+x^2}.$$

Exercise 7.4 Prove that anti-thetic sampling makes all the odd moments of a sample of normals equal to zero.

Exercise 7.5 An option pays $|S_T - K|$. Decompose it into vanilla options.

Exercise 7.6 An option pays $90 - S_T$, for $S_T < 95$. It pays -5 otherwise. Decompose it into vanilla options.

Exercise 7.7 Suppose two smiles have the same implied volatility at 100. One smile is downwards sloping and the other one is upwards sloping. How will the prices of digital calls struck at 100 compare?

Exercise 7.8 Prove that if a European derivative is replicated by a portfolio of vanilla options then it has the same Greeks as that portfolio.

Exercise 7.9 Suppose f is twice-differentiable and $f''(x)$ is non-zero. Show that
$$\lim_{h \to 0+} \frac{f(x+h) - f(x-h)}{2h}$$
converges to $f'(x)$ faster than
$$\lim_{h \to 0+} \frac{f(x+h) - f(x)}{h}.$$

8

Continuous barrier options

8.1 Introduction

One of the simplest and most commonly traded exotic options is the continuous barrier option. This is an option with the ordinary call or put pay-off but the pay-off is contingent on a second event. This second event is typically whether some level has been crossed or not during the life of the option.

For example, a *down-and-out call option* struck at K and with barrier at B will pay

$$(S_T - K)_+,$$

unless at any time, $t < T$, the value of spot passes below B. The option is said to *knock out* when the spot passes below B, and it is said to be a *knock-out option*. Similarly, the *down-and-in call* will pay

$$(S_T - K)_+,$$

provided that at some point during the life of the option spot passes below B. The option is said to *knock in* when the barrier is crossed, this option is called a *knock-in option*.

Clearly, precisely one of these two options will pay $(S - K)_+$ at time T and the other will pay zero. We therefore have the simple relationship that

$$\boxed{\text{knock-out} + \text{knock-in} = \text{knockless}}$$

We illustrate this in Figure 8.1. The relationship means that we need not study the pricing of knock-in options since their values are immediately deducible from those of knock-out options. As both in and out options have non-negative and possibly positive payoffs, we have that their values are always positive (before knock-out) and always less than the value of the vanilla option. This is not surprising in that they both carry fewer rights than a vanilla option and so must have smaller value.

8.1 Introduction

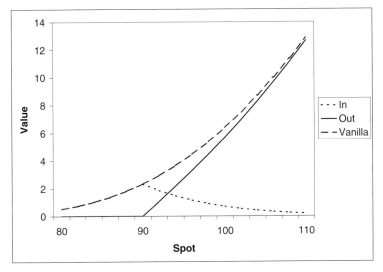

Fig. 8.1. The value of a down-and-out call, a down-and-in call and a vanilla call with barrier at 90 and strike at 100.

We can have any combination of up/down, in/out and call/put that we like. The most interesting combinations are those for which the barrier is in-the-money. For example, an up-and-out call option with barrier at 120 and strike at 100, has an interesting payoff profile. Below 100 the value is zero, from 100 to 120 the value increases in a straight line to 20, and then above 120 it drops immediately to zero.

At preceding times, the value of the option will increase until 120 is getting close, and then as 120 looms up, the value will plummet to zero as the probability of knocking-out dominates over the payoff. See figures 8.2 and 8.3.

Why buy a barrier option? For the purpose of hedging they are not particularly useful (except as a component of the hedging of some complicated exotic.) However, they are cheaper than vanilla options, so if a speculator has very strong views on how he believes an asset price will move then he can make more money by purchasing an option which expresses those views precisely. For example, if the speculator believes that the asset will greatly increase in value and will not go below 90 in the mean time, he could purchase a down-and-out call with barrier at 90 saving a little on the option's premium. There are also sometimes regulatory restrictions on the number of options a company or fund can hold or issue. The use of barrier options allows the options to automatically disappear when they are too far out-of-the-money.

A technical issue with continuous barrier options is the question of how to agree what it means for the asset price to cross the barrier. The asset price is only observable when a trade is actually made, and then there is the issue of recording

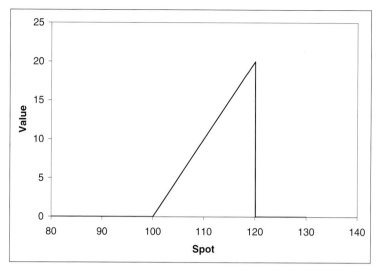

Fig. 8.2. The value of an up-and-out call option struck at 100 with barrier at 120 at expiry.

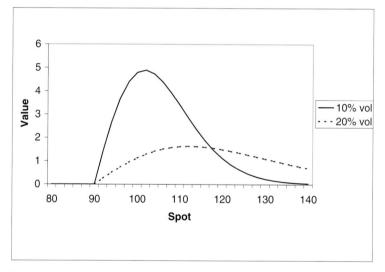

Fig. 8.3. The value of a down-and-out put option struck at 110 with barrier at 90 and one year to expiry. Note that volatility is bad near the barrier and good away from the barrier.

and checking all trades made to see whether they actually crossed the barrier. For this reason, truly continuous barrier options are rare. Instead, the option price is generally sampled on a daily basis. The option is therefore really a discrete barrier option; however the continuous barrier is a very good approximation to the daily sampled barrier.

In this chapter, we examine various methods of pricing continuous barrier options in a Black–Scholes world. As a part of the study of risk-neutral valuation, we develop a better understanding of Girsanov's theorem.

8.2 The PDE pricing of continuous barrier options

We focus on the case of a down-and-out option for concreteness. An up-and-out option can be handled similarly, and as we observed above, the knock-in option prices can be simply deduced.

Let the option be struck at K and have barrier at B. Suppose that the expiry is T and that the asset follows a geometric Brownian motion in a perfect Black–Scholes world. Away from the barrier, we can apply the arguments of either Chapter 5 or Chapter 6 to deduce that C must satisfy the Black–Scholes equation. The new aspect is that as well as the terminal condition that the value must equal the payoff (which may be truncated by the barrier), we also require the option to be of zero value on the barrier.

We therefore now have a boundary value problem. Our option satisfies the following equations:

$$\frac{\partial C}{\partial t}(S, t) + rS\frac{\partial C}{\partial S} + \frac{1}{2}\sigma^2 S^2 \frac{\partial^2 C}{\partial S^2}(S, t) = rC, \tag{8.1}$$

$$C(S, T) = f(S), \tag{8.2}$$

$$C(B, t) = 0. \tag{8.3}$$

We now simply have to solve this equation. This is a little tricky. Suppose we copy the approach of Section 5.8. If we do the first three changes of variables

$$Z = \log S, \tag{8.4}$$

$$\tau = T - t, \tag{8.5}$$

$$C = e^{-r\tau} D, \tag{8.6}$$

then we obtain

$$\frac{\partial D}{\partial \tau} - \left(r - \frac{1}{2}\sigma^2\right)\frac{\partial D}{\partial Z} - \frac{1}{2}\sigma^2\frac{\partial^2 D}{\partial Z^2} = 0. \tag{8.7}$$

What have the boundary conditions become? The barrier condition is now

$$D(\log B, \tau) = 0, \tag{8.8}$$

and the final condition is just

$$D(Z, 0) = f(e^Z). \tag{8.9}$$

If we now continue the previous argument by putting
$$y = Z + \left(r - \frac{1}{2}\sigma^2\right)\tau,$$
we have a problem; the barrier has become a function of time. Indeed, we obtain
$$D\left(\log B + \left(r - \frac{1}{2}\sigma^2\right)\tau, \tau\right) = 0, \tag{8.10}$$
when writing D as a function of y. Whilst we have reduced the problem to a diffusion equation, we have done so at the cost of making the barrier level non-constant.

We therefore look for a different approach to eliminating the first-order term which does not involve changing Z. Our approach is to multiply D by a function which we will choose in such a way to eliminate the extra terms. In particular, we try
$$D(Z, \tau) = e^{aZ + b\tau} E(Z, \tau). \tag{8.11}$$
Differentiating and substituting into (8.7), we obtain
$$\frac{\partial E}{\partial \tau} + bE - \left(r - \frac{1}{2}\sigma^2\right)\left(aE + \frac{\partial E}{\partial Z}\right) - \frac{1}{2}\sigma^2\left(a^2 E + 2a\frac{\partial E}{\partial Z} + \frac{\partial^2 E}{\partial z^2}\right) = 0 \tag{8.12}$$
If we collect terms, the coefficient of E is
$$b - \left(r - \frac{1}{2}\sigma^2\right)a - \frac{1}{2}\sigma^2 a^2,$$
and the coefficient of $\frac{\partial E}{\partial Z}$ is
$$-\left(r - \frac{1}{2}\sigma^2\right) - \sigma^2 a.$$

We are now in a position to solve the problem. Picking a to make the coefficient of $\frac{\partial E}{\partial Z}$ equal to zero and then picking b to make the coefficient of E equal to zero, we have reduced the problem to a simple heat equation and the boundary condition is just
$$E(\log B, t) = 0, \tag{8.13}$$
and the initial condition at time zero is easily computed.

There is a standard approach to such problems. It relies on the principle of reflection. As we are interested only in the value of the solution for $Z \geq \log B$, we can modify the initial condition below $\log B$ without affecting the problem. We therefore solve the problem by making the initial condition odd when reflected in $\log B$. Thus if the initial condition was
$$E(Z, 0) = g(z) \quad \text{for} \quad z > \log B, \tag{8.14}$$

we extend it via

$$E(Z, 0) = -g(\log B + (\log B - z)) \text{ for } z < \log B. \quad (8.15)$$

Why does this do what we want? At any point along $\log B$, there is an equal amount of positive heat coming from above $\log B$ and negative heat coming from below; these cancel each other to give zero. Alternatively, if we think financially we have defined a new option payoff which is negative below the barrier in such a way that the value of the amount we may owe below the barrier is equal to the value of the amount we may make above the barrier. Indeed, in Chapter 10, we look at an approach to barrier pricing based on this point of view.

How do we prove the solution has the correct property? The function $E(Z, \tau)$ solves the diffusion equation if and only if $E(2 \log B - Z, \tau)$ does, as we get a double negative sign when differentiating with respect to Z. Since the boundary conditions are odd on reflection in $\log B$, we conclude that $E(Z, \tau)$ and $-E(2 \log B - Z, \tau)$ solves the diffusion equation with the same boundary conditions. By uniqueness of solutions, it follows that

$$E(Z, \tau) = -E(2 \log B - Z, \tau), \quad (8.16)$$

for all Z and τ. Putting Z equal $\log B$, we obtain

$$E(\log B, \tau) = -E(\log B, \tau) \quad (8.17)$$

which immediately implies that $E(\log B, \tau)$ is equal to zero as desired.

To derive the solution, we now just need to solve on the entire space using the reflected final condition, and carry out multiple changes of variables to get back to the original problem. We do not carry this out here, but instead we will use an alternative method based on risk-neutral expectation to derive the solution. However, the underlying techniques used are similar, relying both on reflection and a less mysterious multiplication by a function of the form $e^{aZ+b\tau}$, and the reader should try to hold both points of view firmly in mind.

8.3 Expectation pricing of continuous barrier options

We can equally well price by risk-neutral expectation. We focus on a down-and-out option for concreteness as before. The price of an option will then be given by its discounted risk-neutral expectation. To compute this expectation we will need to know the probability that the barrier will be breached and the distribution of the final value of spot given that the barrier has not been breached. To compute these what we really need is the joint distribution of the minimum and the terminal value for a Brownian motion with drift. In order to compute the joint distribution we need to develop a better understanding of Brownian motion, stopping times and Girsanov's theorem.

Our programme for this rest of this chapter is therefore to develop the necessary techniques to carry this out. We start with the reflection principle, move on to computing with Girsanov's theorem, compute the joint distributions of minima and terminal values and then return to the pricing.

8.4 The reflection principle

We present an argument that allows us to deduce the joint law of the minimum and the terminal value for a driftless Brownian motion. Our argument is not wholly rigorous and we refer the reader who wishes to see the technical details to [78].

Let W_t be a Brownian motion. Let m_T denote the minimum value of W_t over the interval $[0, T]$. We want to compute the probability of the event, E, defined by

$$m_T \leq y, \quad W_T \geq x$$

for $x \geq y$ and $y < 0$. If the event occurs then for some t_0 we have that W_{t_0} is equal to y as Brownian paths are continuous, and there is certainly some value of t for which W_t is less than or equal to y. The Brownian motion therefore descends at least as far as y and then comes back up to level x. Suppose that instead of continuing the Brownian motion after time t_0, we restart it and replace it by its value reflected in the level y. We thus define a second random process via

$$W'_t = \begin{cases} W_t & \text{for } t < t_0, \\ 2y - W_t & \text{for } t \geq t_0. \end{cases} \tag{8.18}$$

The event $W_T \geq x$ becomes $W'_T \leq 2y - x$. The crucial point here is that the event $W'_T \leq 2y - x$ can only occur if $m_T \leq y$ also occurs, as otherwise W_t has been above y at all times, and therefore above $2y - x$ which is less than y at all times. Thus the event E for W_T is equivalent to the much simpler event

$$W'_T \leq 2y - x.$$

Of course, we need to know the distribution of W'_T for this to be of any use. In fact, W'_t is also a Brownian motion. Let τ be the first time that W_t equals y. Then for $s \geq 0$ we have

$$W'_{\tau+s} - W'_\tau = W_\tau - W_{\tau+s}. \tag{8.19}$$

The crucial issue now is therefore the distribution of $W_{\tau+s} - W_\tau$. If τ were a constant then there would be no issue, it would follow from the properties of Brownian motion that the distribution is just a normal of mean 0 and variance s. But τ is not

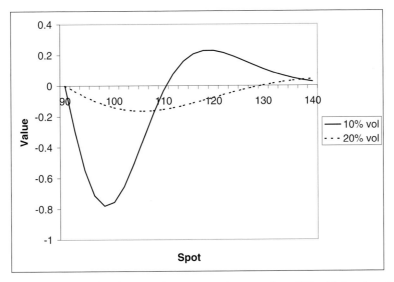

Fig. 8.4. The Vega of a down-and-out put option struck at 110 with barrier at 90 and one year to expiry.

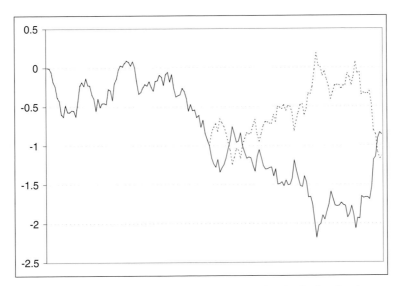

Fig. 8.5. A Brownian motion and its reflection in the level -1.

constant; it does however have the property of being a stopping time. Recall that a stopping time is a random time such that the event that the time is before a time t is determined by information available at time t. The time to reach a given level is clearly such a time, as $\tau \leq t$ is the statement that the minimum value of W_s for $s \leq t$ is less than or equal to y.

The key point is that as the event that the stopping time has occurred only relies on information already available, the distribution of $W_{\tau+s} - W_\tau$ will be totally unaffected, since it is, by definition, independent of the behaviour of W_r for $r \leq \tau$.

In conclusion, we have

$$\mathbb{P}(W_T \geq x, m_T \leq y) = \mathbb{P}(W_T \leq 2y - x) \tag{8.20}$$

for $y \leq 0, x \leq y$.

The property that Brownian motion starts anew at stopping times is sometimes called *the strong Markov property*.

8.5 Girsanov's theorem revisited

We have derived a formula for the joint law of minimum and terminal value for a driftless Brownian motion. Unfortunately this is a not a great deal of help, as a stock in a non-zero interest rate environment will have drift equal to the risk-free rate and the log of the stock will have drift depending on the risk-free rate and the volatility. Note that working with discounted prices will not help here, as the barrier will depend upon the actual price, not the discounted price.

The standard tool for changing the drift of a Brownian motion is, of course, Girsanov's theorem. Previously, we have treated Girsanov's theorem as a black box and avoided looking at how the measure is changed. Here we will need to explicitly understand the measure change in order to be able to compute its effect on the joint law.

A measure change really consists of reweighting the probability of paths. We therefore construct them by multiplying probabilities by a random variable. Let A be an event in the filtration $\mathcal{F}_T$. Let 1_A be the random variable which is 1 if A occurs and 0 otherwise. We can define a measure via

$$\tilde{\mathbb{P}}(A) = \mathbb{E}(1_A X) \tag{8.21}$$

for some random variable X in the same filtration. What properties should X have? Probabilities should be non-negative so X should be non-negative. The probability of the global event should be 1. Taking A to be the entire sample space is the same as taking 1_A to be identically 1, so we conclude that we must have

$$\mathbb{E}(X) = 1. \tag{8.22}$$

Recall that an equivalent change of measure involves having the same sets of probability 0 and 1. We will therefore need the probability of X being zero (in the original measure) to be zero, as otherwise the set where it is zero will go from being of positive probability to zero probability. We therefore assume that X is positive everywhere in the following.

8.5 Girsanov's theorem revisited

One trickiness is that we will actually need to use different random variables for each filtration, $\mathcal{F}_t$, as we cannot have a single simple random variable which is in the filtration $\mathcal{F}_\infty$. Let the random variable for $\mathcal{F}_t$ be X_t. For an event, A, in $\mathcal{F}_s$ for $s < t$, we will then have two different candidates for $\tilde{\mathbb{P}}(A)$, namely

$$\mathbb{E}(1_A X_s) \quad \text{and} \quad \mathbb{E}(1_A X_t).$$

Clearly, we will want these two values to agree. Agreement is equivalent to the condition that

$$\mathbb{E}(1_A(X_s - X_t)) = 0 \tag{8.23}$$

for any event A in $\mathcal{F}_s$. As A is arbitrary, this equation really says that no matter what value X_s takes and no matter how it got there, the expected value of X_t at time s should be equal to X_s. That is, we must require X_s to satisfy the martingale condition

$$\mathbb{E}(X_t | \mathcal{F}_s) = X_s. \tag{8.24}$$

We then have

$$\mathbb{E}(X_t 1_A) = \mathbb{E}(X_s 1_A) + \mathbb{E}((X_t - X_s) 1_A). \tag{8.25}$$

As $1_A \in \mathcal{F}_s$ and $X_t - X_s$ is independent of $\mathcal{F}_s$, we can rewrite the last term as

$$\mathbb{E}(X_t - X_s)\mathbb{E}(1_A),$$

which will be equal to zero as the first factor is zero.

We conclude that our measure change must be given by a collection of positive random variables X_t which form a martingale with respect to the filtration generated by the Brownian motion W_t. There is one such process that we have repeatedly studied in this book: geometric Brownian motion. We therefore take $X_0 = 1$ and

$$dX_t = \nu X_t dW_t, \tag{8.26}$$

or equivalently,

$$X_t = e^{-\frac{1}{2}\nu^2 t + \nu W_t}. \tag{8.27}$$

We want to show that this really does give the right measure change. First, we check that W_t is distributed correctly in the new measure. Recall that W_t is distributed as $\sqrt{t} N(0, 1)$. We therefore have that

$$\mathbb{P}(W_t < x) = \mathbb{P}\left(N(0, 1) < xt^{-\frac{1}{2}}\right) = \frac{1}{\sqrt{2\pi}} \int_{-\infty}^{xt^{-\frac{1}{2}}} e^{-\frac{1}{2}s^2} ds. \tag{8.28}$$

Changing variables, we obtain

$$\mathbb{P}(W_t < x) = \frac{1}{\sqrt{2\pi t}} \int_{-\infty}^{x} e^{-\frac{s^2}{2t}} ds. \tag{8.29}$$

This is equivalent to saying that the density of W_t is

$$\frac{1}{\sqrt{2\pi t}} e^{-\frac{s^2}{2t}}.$$

We therefore have

$$\tilde{\mathbb{P}}(W_t < x) = \mathbb{E}(1_{W_t < x} X_t) = \frac{1}{\sqrt{2\pi t}} \int_{-\infty}^{x} e^{-\frac{s^2}{2t}} e^{-\frac{1}{2}v^2 t + vs} ds. \tag{8.30}$$

Collecting terms, we have

$$\tilde{\mathbb{P}}(W_t < x) = \frac{1}{\sqrt{2\pi t}} \int_{-\infty}^{x} e^{-\frac{(s-vt)^2}{2t}} ds. \tag{8.31}$$

A simple change of variables, $r = s - vt$, shows that this is equal to

$$\mathbb{P}(W_t < x - vt). \tag{8.32}$$

We have shown that the probability that $W_t < x$ in the new measure is equal to the probability that $W_t + vt$ is less than x in the old measure. In other words, the process W_t is distributed as a Brownian motion with drift v in the new measure.

We are not yet finished. We have only shown that the distribution of W_t is correct; we must also show that the marginal increments are correct.

We compute

$$\tilde{\mathbb{P}}(W_t - W_s < x) = \mathbb{E}\left(1_{W_t - W_s < x} e^{-\frac{1}{2}v^2 t + vW_t}\right),$$
$$= \mathbb{E}\left(1_{W_t - W_s < x} e^{-\frac{1}{2}v^2(t-s) + v(W_t - W_s)} e^{-\frac{1}{2}v^2 s + vW_s}\right). \tag{8.33}$$

As $W_t - W_s$ is independent of W_s, we can factorise the last term into a separate term of expectation 1. This leaves us with

$$\tilde{\mathbb{P}}(W_t - W_s < x) = \mathbb{E}\left(1_{W_t - W_s < x} e^{-\frac{1}{2}v^2(t-s) + v(W_t - W_s)}\right). \tag{8.34}$$

As the distribution of $W_t - W_s$ is identical to that of W_{t-s}, we are now back in the situation where $s = 0$ which we already covered above. We conclude that $W_t - W_s$ is distributed as a Brownian motion with drift v.

In conclusion, we have proven

Theorem 8.1 *Let W_t be a Brownian motion; then we can define a new measure by*

$$\tilde{\mathbb{P}}(A) = \mathbb{E}\left(1_A e^{-\frac{1}{2}v^2 t + v W_t}\right).$$

Under this new measure W_t is a Brownian motion with drift v.

The term X_t which we used to change the probability weightings, is sometimes called the *Radon–Nikodym derivative*. Note the analogue with ordinary integration, a change of variables leads to an extra term which is the derivative of the variable change. Note also that we can easily change back. We simply change the drift by $-v$ instead of by v. We change the drift by using the Brownian motion in the new measure rather than the old. Thus if W_t was our original Brownian motion, then W_t has drift v in the new measure, so the Brownian motion in the new measure is

$$\tilde{W}_t = W_t - vt. \tag{8.35}$$

The Radon–Nikodym derivative for changing back is therefore

$$e^{-\frac{1}{2}v^2 t - v\tilde{W}_t} = e^{-\frac{1}{2}v^2 t + v^2 t - v W_t} = \frac{1}{e^{-\frac{1}{2}v^2 t + v W_t}}, \tag{8.36}$$

that is, the reciprocal of the original derivative.

If we want to compute expectations under $\tilde{\mathbb{P}}$ then we reweight by X_t also. This is clear when computing the expectation of a piecewise constant function as then we just have a linear sum of probabilities. The general case follows from approximating by piecewise constant functions.

8.6 The joint distribution of minimum and terminal value for a Brownian motion with drift

In Section 8.4, we derived the joint law of the minimum and terminal value for a Brownian motion without drift. In this section, we combine that result with our results on Girsanov's theorem to derive the joint law for a Brownian motion with drift.

Let W_t be a Brownian motion. Let $Y_t = \sigma W_t$, and m_t^Y be the minimum of Y_t up to time t. We then have for $y < 0$ and $x > y$ that

$$\mathbb{P}(Y_t \geq x, m_t^Y \leq y) = \mathbb{P}(Y_t \leq 2y - x). \tag{8.37}$$

This follows from the result for Brownian motion, (8.20), as the volatility term makes no real difference.

We wish to prove an analogous result for a Brownian motion with drift. Let

$$Z_t = v\,dt + \sigma\,dW_t \tag{8.38}$$

and m_t^Z denotes its minimum up to time t. Our main result is

Theorem 8.2 *If $y < 0$ and $x \geq y$, then*

$$\mathbb{P}(Z_t \geq x, m_t^Z \leq y) = e^{2\nu y \sigma^{-2}} \mathbb{P}(Z_t \leq 2y - x + 2\nu t)$$
$$= e^{2\nu y \sigma^{-2}} N\left(\frac{2y - x + \nu t}{\sigma \sqrt{t}}\right).$$

Proof The volatility σ scales through everything and it is straightforward to reduce to the case where $\sigma = 1$. We therefore assume that σ is equal to 1.

We use a change of measure to remove the drift of Z_t. The change of measure is given by

$$e^{-\frac{1}{2}\nu^2 t - \nu W_t},$$

and to change back we take

$$e^{-\frac{1}{2}\nu^2 t + \nu Z_t}.$$

We denote expectation under the original measure by $\mathbb{E}$ and under the new measure by $\tilde{\mathbb{E}}$. We have, by the results of the previous section, that

$$\mathbb{P}(A) = \tilde{\mathbb{E}}\left(1_A e^{-\frac{1}{2}\nu^2 t + \nu Z_t}\right), \tag{8.39}$$

for any event $A \in \mathcal{F}_t$.

In particular, it holds for

$$A = \{Z_t \geq x, m_t^Z \leq y\}. \tag{8.40}$$

We therefore wish to compute

$$\tilde{\mathbb{E}}\left(1_{\{Z_t \geq x, m_t^Z \leq y\}} e^{-\frac{1}{2}\nu^2 t + \nu Z_t}\right),$$

with Z_t a Brownian motion under this measure.

We use the reflection principle. As $x \geq y$, we have that if the Brownian motion, Z_t, touches the level y anywhere then the terminal distribution of $2y - Z_t$ is equal to that of Z_t. As our indicator function is zero unless the level y is breached, our expectation must be equal to

$$\tilde{\mathbb{E}}\left(1_{\{2y - Z_t \geq x, m_t^Z \leq y\}} e^{-\frac{1}{2}\nu^2 t + \nu(2y - Z_t)}\right).$$

However, $2y - Z_t \geq x$ is equivalent to $Z_t \leq 2y - x$ and $2y - x$ is less than x, which means that the condition on the minimum is now redundant. The expectation is therefore equal to

$$\tilde{\mathbb{E}}\left(1_{\{Z_t \leq 2y - x\}} e^{-\frac{1}{2}\nu^2 t + \nu(2y - Z_t)}\right) = e^{2y\nu} \tilde{\mathbb{E}}\left(1_{\{Z_t \leq 2y - x\}} e^{-\frac{1}{2}\nu^2 t - \nu Z_t}\right)$$

We wish to eliminate the exponential term in the expectation. We can regard it as the Radon–Nikodym derivative of a Girsanov transformation which changes the drift of Z_t by $-\nu$. We use $'$ to denote the corresponding measure.

8.6 Joint distribution

So

$$\tilde{\mathbb{E}}\left(1_{\{Z_t \leq 2y-x\}}e^{-\frac{1}{2}v^2 t - vZ_t}\right) = \mathbb{E}'\left(1_{\{Z_t \leq 2y-x\}}\right) = \mathbb{P}'(Z_t \leq 2y - x).$$

Under the new measure Z_t has drift $-v$ so the final term is equal to the probability that a Brownian motion with drift $-v$ is less than $2y - x$. This is equal to the probability that a Brownian motion with drift v is less than $2y - x + 2vt$ and we are done. □

We can easily deduce the law of the minimum of a Brownian motion with drift. We have that

$$\mathbb{P}(m_t^Z \leq y) = \mathbb{P}(m_t^Z \leq y, Z_t \leq y) + \mathbb{P}(m_t^Z \leq y, Z_t \geq y). \quad (8.41)$$

The event that the minimum is less than y and the terminal value is less than y is the same as the event that the terminal value is less than y. We therefore have

$$\mathbb{P}(m_t^Z \leq y) = \mathbb{P}(Z_t \leq y) + \mathbb{P}(m_t^Z \leq y, Z_t \geq y). \quad (8.42)$$

We conclude

Corollary 8.1

$$\mathbb{P}(m_t^Z \leq y) = N\left(\frac{y - vt}{\sigma\sqrt{t}}\right) + e^{2vy\sigma^{-2}} N\left(\frac{y + vt}{\sigma\sqrt{t}}\right),$$

or equivalently

$$\mathbb{P}(m_t^Z \geq y) = N\left(\frac{vt - y}{\sigma\sqrt{t}}\right) - e^{2vy\sigma^{-2}} N\left(\frac{y + vt}{\sigma\sqrt{t}}\right).$$

As

$$\mathbb{P}(Z_t \geq x, m_t^Z \leq y) + \mathbb{P}(Z_t \leq x, m_t^Z \leq y) = \mathbb{P}(m_t^Z \leq y),$$

we also have

Corollary 8.2 *For $y \leq 0$ and $x \geq y$,*

$$\mathbb{P}(Z_t \leq x, m_t^Z \leq y) = N\left(\frac{y - vt}{\sigma\sqrt{t}}\right) + e^{2vy\sigma^{-2}} N\left(\frac{y + vt}{\sigma\sqrt{t}}\right) \\ - e^{2vy\sigma^{-2}} N\left(\frac{2y - x + vt}{\sigma\sqrt{t}}\right).$$

Similarly, we can prove

Corollary 8.3 *For $y \leq 0$ and $x \geq y$,*

$$\mathbb{P}(Z_t \geq x, m_t^Z \geq y) = N\left(\frac{vt - x}{\sigma\sqrt{t}}\right) - e^{2vy\sigma^{-2}} N\left(\frac{2y - x + vt}{\sigma\sqrt{t}}\right).$$

We have concentrated on studying the distribution of the minimum of a Brownian motion which is relevant when studying down-and-out options. If we wish price up-and-out options, we will need similar theorems for the maximum. Let M_t^Z denote the maximum over the interval $[0, t]$. Fortunately, the fact that the negative of a Brownian motion with drift is a Brownian motion with drift means that the law for the maximum is easily deducible from the law for the minimum. We can write

$$M_t^Z = \max(\sigma W_t + vt) = \max(-(\sigma W_t + vt)) = -\min(-\sigma W_t - vt). \quad (8.43)$$

As $-W_t$ is a Brownian motion, we have, letting Z_t' denote a process with drift $-v$ and volatility σ, that

$$\mathbb{P}(M_t^Z \leq y) = \mathbb{P}(m_t^{Z'} \geq -y),$$
$$= N\left(\frac{y - vt}{\sigma\sqrt{t}}\right) - e^{2vy\sigma^{-2}} N\left(\frac{-y - vt}{\sigma\sqrt{t}}\right).$$

A similar argument shows

$$\mathbb{P}(Z_t \leq x, M_t^Z \geq y) = e^{2vy\sigma^{-2}} N\left(\frac{x - 2y - vt}{\sigma\sqrt{t}}\right), \quad (8.44)$$

which immediately implies

$$\mathbb{P}(Z_t \leq x, M_t^Z \leq y) = N\left(\frac{x - vt}{\sigma\sqrt{t}}\right) - e^{2vy\sigma^{-2}} N\left(\frac{x - 2y - vt}{\sigma\sqrt{t}}\right) \quad (8.45)$$

for $x \leq y$ and $y > 0$.

8.7 Pricing the continuous barrier by risk-neutral expectation

We pull together our results in order to derive formulas for the price of barrier options. Suppose our call option is struck at K and is down-and-out with barrier at H. We first suppose that $H < K$. The payoff is then

$$(S_T - K)1_{\{S_T \geq K, m_T \geq H\}}.$$

As in Section 6.13, we divide the payoff into two pieces:

$$S_T 1_{\{S_T \geq K, m_T \geq H\}} - K 1_{\{S_T \geq K, m_T \geq H\}}.$$

To price the first piece, we take S as numeraire. Then, the value is

$$S_0 \mathbb{E}\left(1_{\{S_T \geq K, m_T \geq H\}}\right) = S_0 \mathbb{P}(S_T \geq K, m_T \geq H) \quad (8.46)$$
$$= S_0 \mathbb{P}\left(\log(S_T) \geq \log(K), m_T^{\log S} \geq \log(H)\right). \quad (8.47)$$

8.7 Pricing continuous barriers by expectation

In this measure, $\log(S_T) - \log(S_0)$ is a Brownian motion with drift $r + \frac{\sigma^2}{2}$. We can therefore apply Corollary 8.3, with

$$x = \log(K) - \log(S_0) = -\log(S_0/K),$$

$$y = \log(H) - \log(S_0) = \log(H/S_0)$$

and

$$\nu = r + \frac{1}{2}\sigma^2$$

to obtain

$$S_0 N\left(\frac{\log\left(\frac{S_0}{K}\right) + \left(r + \frac{1}{2}\sigma^2\right)T}{\sigma\sqrt{T}}\right)$$

$$- S_0 \left(\frac{H}{S_0}\right)^{1+2r\sigma^{-2}} N\left(\frac{\log\left(\frac{H^2}{S_0 K}\right) + \left(r + \frac{1}{2}\sigma^2\right)T}{\sigma\sqrt{T}}\right).$$

To price the second piece we take the continuous compounding money market account as numeraire. The value of the second part is therefore $-Ke^{-rT}$ times the probability that the pay-off occurs. In this measure, as usual, we have that $\log(S)$ has drift

$$\nu = r - \frac{1}{2}\sigma^2.$$

Applying Corollary 8.3 again, we obtain

$$-Ke^{-rT} N\left(\frac{\log\left(\frac{S_0}{K}\right) + \left(r - \frac{1}{2}\sigma^2\right)T}{\sigma\sqrt{T}}\right)$$

$$+Ke^{-rT} \left(\frac{H}{S_0}\right)^{-1+2r\sigma^{-2}} N\left(\frac{\log\left(\frac{H^2}{S_0 K}\right) + \left(r - \frac{1}{2}\sigma^2\right)T}{\sigma\sqrt{T}}\right).$$

Pulling all this together, we have

Theorem 8.3 *In a Black–Scholes world with interest rate r, and volatility σ, the value of a down-and-out call option struck at K with barrier $H < K$ is equal to*

$$S_0 N(d_1) - Ke^{-rT} N(d_2) - \left(\frac{H}{S_0}\right)^{1+2r\sigma^{-2}} S_0 N(h_1) + \left(\frac{H}{S_0}\right)^{-1+2r\sigma^{-2}} Ke^{-rT} N(h_2)$$

where
$$d_j = \frac{\log\left(\frac{S_0}{K}\right) + \left(r + (-1)^{j-1}\frac{1}{2}\sigma^2\right)t}{\sigma\sqrt{T}}$$

and
$$h_j = \frac{\log\left(\frac{H^2}{S_0 K}\right) + \left(r + (-1)^{j-1}\frac{1}{2}\sigma^2\right)t}{\sigma\sqrt{T}}.$$

Note that we can regard this value as the price of a vanilla call option minus a correction term. The correction term arises from the decrease in value caused by the barrier. This correction, whilst more complicated than the original option, has a similar form. We will see in Section 10.6 that it can be interpreted as the price of an option whose payoff is the reflection of the original payoff in the barrier.

Thus far, we have only looked at the case where the barrier is less than the strike. When the barrier is above the strike, the condition of being in-the-money at expiry is redundant, because if the minimum is above the barrier then the terminal value is certainly above the barrier, and hence the strike. The option's payoff is therefore

$$1_{m_T^S \geq H}(S_T - K).$$

As before, we can tackle this by dividing into pieces with coefficients S and K, and then using the stock and money market account as numeraires respectively. The value is therefore

$$S_0 \mathbb{P}_S\left(1_{m_T^S > H}\right) - e^{-rT} K \mathbb{P}_B\left(1_{m_T^S > H}\right),$$

where $\mathbb{P}_S$ is the probability measure with stock as numeraire and $\mathbb{P}_B$ is the measure with the money market as numeraire. Using Corollary 8.1, we have for the first term, taking $\nu = r + \frac{1}{2}\sigma^2$, and x, y as above,

$$S_0 N\left(\frac{\log\left(\frac{S_0}{H}\right) + \left(r + \frac{1}{2}\sigma^2\right)T}{\sigma\sqrt{T}}\right) - S_0\left(\frac{H}{S_0}\right)^{1+2r\sigma^{-2}} N\left(\frac{\log\left(\frac{H}{S_0}\right) + \left(r + \frac{1}{2}\sigma^2\right)T}{\sigma\sqrt{T}}\right).$$

For the second term, we have $\nu = r - \frac{1}{2}\sigma^2$, and obtain

$$-Ke^{-rT} N\left(\frac{\log\left(\frac{S_0}{H}\right) + \left(r - \frac{1}{2}\sigma^2\right)T}{\sigma\sqrt{T}}\right)$$

$$+Ke^{-rT}\left(\frac{H}{S_0}\right)^{-1+2r\sigma^{-2}} N\left(\frac{\log\left(\frac{H}{S_0}\right) + \left(r - \frac{1}{2}\sigma^2\right)T}{\sigma\sqrt{T}}\right).$$

In conclusion, we have

Theorem 8.4 *In a Black–Scholes world with interest rate r, and volatility σ, the value of a down-and-out call option struck at K with barrier $H \geq K$ is equal to*

$$S_0 N(d_1) - K e^{-rT} N(d_2) - \left(\frac{H}{S_0}\right)^{1+2r\sigma^{-2}} S_0 N(h_1) + \left(\frac{H}{S_0}\right)^{-1+2r\sigma^{-2}} K e^{-rT} N(h_2)$$

where

$$d_j = \frac{\log\left(\frac{S_0}{H}\right) + \left(r + (-1)^{j-1}\tfrac{1}{2}\sigma^2\right) T}{\sigma \sqrt{T}}$$

and

$$h_j = \frac{\log\left(\frac{H}{S_0}\right) + \left(r + (-1)^{j-1}\tfrac{1}{2}\sigma^2\right) T}{\sigma \sqrt{T}}.$$

We have only looked at the case of a down-and-out call option. The same techniques can be applied with little difficulty to each of the other out options, i.e. down-and-out put options, up-and-out call options and up-and-out put options. We refer the reader to Haug, [55], or Wilmott, [118], for the formulas.

8.8 American digital options

There are many variants of the American digital option. One particular variant can be regarded as a barrier option. We study an option which pays 1 at expiry if at any point during the life of the option a given barrier has been breached. The option is American in the sense that it is equivalent to the holder early exercising a digital. As there is never any advantage to not exercising, the contract's American features are innocuous; the maximum payoff is 1 so there is never any advantage to not exercising in the money and exercise will occur at the instant the barrier is breached. Note that the option pays 1 at expiry rather than at time of exercise which is slightly different from a standard American option.

An American digital put struck at K therefore has payoff equal to

$$1_{m_T^S \leq K}.$$

Taking the money-market account as numeraire the value is then just

$$e^{-rT} \mathbb{P}(m_T^S \leq K),$$

with the probability taken in the appropriate risk-neutral measure. This is similar to the term which arose when pricing the strike part of a down-and-out call option with barrier above the strike. We conclude that the price is

$$e^{-rT} - e^{-rT} N\left(\frac{\log\left(\frac{S_0}{K}\right) + (r - \frac{1}{2}\sigma^2)T}{\sigma\sqrt{T}}\right)$$

$$+ e^{-rT}\left(\frac{K}{S_0}\right)^{-1+2r\sigma^{-2}} N\left(\frac{\log\left(\frac{K}{S_0}\right) + (r - \frac{1}{2}\sigma^2)T}{\sigma\sqrt{T}}\right).$$

A digital call option could be handled similarly.

8.9 Key points

In this chapter, have examined the pricing of a continuous barrier options using both PDE methods and risk-neutral valuation.

- An out option only pays off if a given barrier level is not breached.
- An in option only pays off if a given barrier level is breached.
- An in option plus an out option is equivalent to a vanilla so in option prices can always be deduced from the vanilla and out option prices.
- A knock-out option can have negative Vega.
- Knock-out options satisfy the Black–Scholes equation with an additional boundary condition at the barrier.
- A key component of all approaches to pricing barrier options is to use reflection in the barrier.
- To price a down-and-out option by risk-neutral evaluation, we need the joint law of the minimum and terminal value for a Brownian motion with drift.
- We can use Girsanov's theorem to compute probabilities of events for Brownian motions with drift.
- To change measure, we multiply expectations by a random process which is a positive martingale called the Radon–Nikodym derivative.
- The measure change for changing the drift of a Brownian motion uses geometric Brownian motion as Radon–Nikodym derivative.
- The formula for a down-and-out call is most easily deduced by dividing the pay-off into two pieces and using a different numeraire for each piece.
- An American digital option is really just a barrier option.

8.10 Further reading

For lists of formulas for pricing barrier options see [55] or [118].

Björk, [17], gives an alternative approach to deriving prices of barrier options. His approach is related to that of Carr, Ellis & Gupta, [30], and put-call symmetry. We discuss replication methods and put-call symmetry in Chapter 10.

8.11 Exercises

Exercise 8.1 If the price of a knock-in option plus the corresponding knock-out option is not equal to the price of the corresponding vanilla, construct an arbitrage portfolio.

Exercise 8.2 A stock follows geometric Brownian motion with time-dependent volatility. How will the time-dependence affect the price of a down-and-out call? Distinguish the two cases where interest rates are zero and interest rates are positive. Suppose the knock-out is determined by the forward rate for the same expiry instead of the spot price, what happens?

Exercise 8.3 The first passage time to a given level is the first time at which a Brownian motion reaches that level. Use the distribution of the maximum of a Brownian motion to derive the density of the first passage time.

Exercise 8.4 Explain why increasing volatility can decrease the price of a barrier option.

Exercise 8.5 How will increasing volatility affect the price of an American digital option?

Exercise 8.6 Sketch the price of a down-and-in put with barrier in-the-money as a function of spot.

Exercise 8.7 Sketch the price of a down-and-out put with barrier out-of-the-money as a function of spot.

Exercise 8.8 Check that the price of a down-and-out put satisfies the Black–Scholes equation.

Exercise 8.9 Develop a pricing formula for an American digital put option.

Exercise 8.10 What we can say about the relative prices of American digital put and European digital puts in general?

Exercise 8.11 Suppose an asset follows Brownian motion and there are no interest rates. What can we say about the relative prices of out-of-the-money American and European digital calls?

9

Multi-look exotic options

9.1 Introduction

In this chapter, we look at the pricing of a derivative which depends upon the value of the underlying not just at one time but rather at several times. We concentrate on two examples, the Asian option and the discrete barrier option. To motivate the Asian option, consider a company that has regular cashflows in a foreign currency. At the end of each accounting year, it has been exposed to a certain amount of currency risk arising from fluctuations in the exchange rate. However, the company does not care about the cashflow for individual months, instead it wants to smooth the average cashflow for the year. The company therefore purchases a call option on the average of the exchange rates rather than the final one. This option is much cheaper than buying an option for each month as there is only one decision involved, and is cheaper even than an option on the final exchange rate for the year as the averaging effect reduces the overall volatility.

The discrete barrier option is a vanilla option that either knocks in or knocks out on any one of a certain set of pre-specified dates. That is, unless spot is within a certain range on one of the look-at dates, the option ceases to exist. A knock-in option will only exist if a barrier is crossed on one of the dates while a knock-out option will only exist if it is not. As discrete barrier options carry fewer rights than vanilla options they will always be cheaper. For that reason they tend to be popular; they can however be a false economy as their cheapness reflects their reduced optionality. We shall concentrate on knock-out options here as we have the relation

$$\boxed{\text{knock-in} + \text{knock-out} = \text{knock-less} = \text{vanilla}}$$

The Asian option and the discrete barrier option can be placed in the same framework. They depend upon the value of spot on a discrete set of times $t_1, \ldots t_n$, and at time t_n they pay the sum

$$f(S_{t_1}, \ldots, S_{t_n}),$$

for a fixed function, f. We shall call such an option a *multi-look option*. They are often called *path-dependent exotic options*.

For the Asian call option struck at K, we have

$$f_A(S_{t_1},\ldots,S_{t_n}) = \max\left(\frac{1}{n}\sum_{j=1}^{n} S_{t_j} - K, 0\right). \tag{9.1}$$

For a down-and-out call option with barrier B and struck at K, we have

$$f_D(S_{t_1},\ldots,S_{t_n}) = \prod_{j=1}^{n-1} H(S_{t_j} - B)\max(S_{t_n} - K, 0), \tag{9.2}$$

where H is, as usual, the Heaviside function.

In Chapter 8, we studied continuous barrier options; the discrete barrier option with frequent barrier dates is very similar in price to a continuous barrier option, and each can be regarded as an approximation to the other. Indeed, the difficulty of defining continuous sampling means that in practice most continuous barrier options are really discrete barrier options with daily sampling.

9.2 Risk-neutral pricing and Monte Carlo simulation for path-dependent options

We can apply the techniques of risk-neutral pricing to value these options. If we pass to the risk-neutral measure then we immediately have that for any multi-look option, C, the price is given by

$$C(0) = e^{-rT}\mathbb{E}(f(S_{t_1},\ldots S_{t_n})), \tag{9.3}$$

where the expectation is taken in the risk-neutral measure. Thus, if we are working in a Black–Scholes world with constant volatility then the expectation is associated to the process for the spot which, as usual, is given by

$$dS = rS\,dt + \sigma S\,dW. \tag{9.4}$$

To compute the price, we need to compute the joint density function, Φ, of $(S_{t_1},\ldots,S_{t_n})$ in the risk-neutral measure and then compute the integral

$$\int f(S_{t_1},\ldots,S_{t_n})\Phi(S_{t_1},\ldots,S_{t_n})dS_{t_1}\ldots dS_{t_n}. \tag{9.5}$$

Unfortunately, analytically evaluating the expectation is not so trivial, and, in general, not possible. However, the risk-neutral evolution of the spot is easy to write down. We simply use the solution of the stochastic differential equation for geometric Brownian motion. If the initial value of S is S_0, then

$$S_{t_1} = S_0 e^{(r-\frac{1}{2}\sigma^2)t_1 + \sigma\sqrt{t_1}N(0,1)}, \tag{9.6}$$

and

$$S_{t_{j+1}} = S_{t_j} e^{(r-\frac{1}{2}\sigma^2)(t_{j+1}-t_j)+\sigma\sqrt{t_{j+1}-t_j}N(0,1)}. \tag{9.7}$$

We can therefore easily simulate a path $(S_{t_1}, S_{t_2}, \ldots, S_{t_n})$. This means that to evaluate (9.5) is straightforward by Monte Carlo simulation. One simply takes a vector of n independent $N(0, 1)$ draws, and uses (9.6) to simulate all the relevant spot prices, and plugs them into f to get one realized price. We then average over many paths. (Note that we have really carried out a change of variables here in order to make the Gaussian draws the variables to simulate instead of the spot prices.)

One important issue with this sort of problem is dimensionality. As we are integrating over n realizations of spot, we are really studying an n-dimensional problem. This has a number of implications. The first is that we have to be sure that we draw n truly independent $N(0, 1)$ variables each time. If we synthesize our $N(0, 1)$ variables from uniform random variables using the inverse cumulative normal function, this means that the vector of underlying uniforms must be drawn from the uniform density on the hypercube $[0, 1]^n$. If a single draw from the random generator was truly random and all draws were truly independent, then this would be easy. Just take n draws and we are done. However, a computer cannot produce random numbers: it simply produces numbers from a deterministic sequence that look reasonably random. The problem then becomes that whilst many draws of a single number are reasonably random, if one thinks in terms of a vector of uniforms they may not be so random. In particular, there may be non-obvious relationships between succeeding draws from the number generator which may cause all vectors of n draws to lie inside a fixed lower-dimensional space, and render the Monte Carlo integration meaningless. The moral here is that one needs to be careful in the choice of random number generator, and ensure that the one used is certified to work for the dimensionality required, rather than relying on the one provided by the computer language's implementation. It is important to realize that for many languages including C, the definition of the random number generator is left up to the writer of the compiler rather than being specified in the language standard. This means that if you use the inbuilt generator you are placing yourself at the mercy of your compiler writer.

The second issue arising from dimensionality is that it slows down convergence; many more variables are contributing and it takes many more draws to fill out a hypercube uniformly than to fill out the unit interval. In fact, whilst convergence is slower in higher dimensions, it is still of order $\mathcal{O}(n^{-\frac{1}{2}})$ but the constant in front of the $n^{-\frac{1}{2}}$ may be higher.

This is the great advantage of Monte Carlo: other methods always get much worse in higher dimensions. In dimension d, if we divide a unit hypercube into little cubes of side length k, then the number of cubes required is $(1/k)^d$. The amount of time required to carry out numerical integration by approximating over little cubes will therefore increase much more rapidly.

A one-year option may well involve weekly look-at dates and thus be 52-dimensional so high dimensionality is a very real issue. Monte Carlo simulation gives an effective method of carrying out the integration but it is still not a very fast one. There are various methods of speeding up the convergence. We examined some of them in Section 7.5, and these will work as well for multi-time-step evolutions as for single-step evolutions.

One nice feature of both the Asian option and discrete barrier option is that at the penultimate (i.e. second last) look-at date, the option becomes a vanilla option or is zero. For the Asian call option with strike K, after the penultimate date, the option pay-off can be written as

$$\frac{1}{n} \max \left(S_{t_n} - \left(nK - \sum_{j<n} S_{t_j} \right), 0 \right),$$

i.e. it has become a vanilla call with strike $nK - \sum_{j<n} S_{t_j}$. This means that there is no need to simulate the final step, and the payoff at time t_n can be replaced with the Black–Scholes value of the option at time t_{n-1}, for the relevant strike.

For the discrete barrier option, at the last barrier date either the option has knocked out, in which case the option is valueless, or it cannot any longer knock out in which case it has become a vanilla call option, and we can substitute the Black–Scholes value for the final pay-off.

In the remainder of this chapter, we look at some alternative methods for pricing these options and at some of the practicalities.

9.3 Weak path dependence

The discrete barrier option is different from the Asian option in that the path dependence is fairly mild in that there are only two possible states before expiry: either the option has knocked out and is valueless, or the option has not knocked out. Precisely where it knocked out or how much it managed to avoid knocking out by are totally irrelevant. For the Asian option, we need to know more about where the spot was, which makes it the harder problem.

A consequence of this is that we can apply a backwards method such as trees and PDEs to the discrete barrier option. From the PDE point of view, the option must satisfy the Black–Scholes equation. Let the boundary level be B; we study a down-and-out option stuck at K for concreteness. We must simply identify the boundary conditions. The first is just that the value of the option at maturity is the payoff of the option. The additional boundary conditions are that at each barrier date, t_j, the value of the option is zero along the set $[0, B]$. This means that as a function of (S, t) the boundary conditions are

$$C(S, t_n) = (S - K)_+, \tag{9.8}$$

$$C(S, t_j) = 0, \quad \text{for } S \leq B, \ j = 1, \ldots, n-1. \tag{9.9}$$

Thus to solve the PDE, we solve backwards from t_n to t_{n-1}; this yields a function f_{n-1} at time t_{n-1}. Outside the barrier, the option is then worth f_{n-1} and zero otherwise. Let $g_{n-1} = H(S - B) f_{n-1}$ be the value. We then solve back to time t_{n-2} with g_{n-1} as final condition at time t_{n-1} and so on. Eventually we obtain the value at time 0.

We could similarly apply tree methods. Solve back to time t_{n-1}, set the values at nodes outside the barrier to zero, solve back to time t_{n-2}, and so on. For both the tree and numerical PDE method, the subtleties are in the details of the implementation. Whilst both methods are guaranteed to converge eventually, we must be careful to implement them in such a way that the results obtained are stable. In the case of a PDE finite difference method, this means adapting the grid points to the locations of the barriers, and for a tree it means adapting the nodes to lie on the barrier.

In Section 10.3, we adapt these methods to give a method of pricing discrete barriers by replication.

9.4 Path generation and dimensionality reduction

We have described a very simple and intuitive way of constructing the requisite points of our Brownian path. We just take a random draw to describe the first step, then a second to describe the next step and so on. Such a method of path construction is said to be *incremental*. If we think in terms of drawing an entire path at once rather than stepping along, then there is no necessity to determine the points in order. If we are not to generate the points in order, then we need to think about the relationships between the points. Clearly, the location of the second point, if already known, will affect the distribution of the location of the first point.

Suppose W_t is a Brownian motion starting at 0, and we are interested in its values at $\{t_1, \ldots, t_n\}$. There will be a certain amount of correlation between W_{t_j} and W_{t_k}. If $j \leq k$, we can write

$$\mathbb{E}(W_{t_j} W_{t_k}) = \mathbb{E}(W_{t_j}(W_{t_k} - W_{t_j})) + \mathbb{E}(W_{t_j}^2). \tag{9.10}$$

The first of these terms is zero by the independence of increments for Brownian motion, and the second is t_j, by the definition of Brownian motion. We thus have

that
$$\mathbb{E}(W_{t_j} W_{t_k}) = \min(t_j, t_k), \qquad (9.11)$$
for any j and k.

The objective in path generation is therefore to generate a vector of normal variables $X_1, \ldots, X_n$ such that the covariance of X_j and X_k is the minimum of t_j and t_k. Suppose we draw a set of independent $N(0, 1)$ variables, $Z_1, \ldots, Z_n$, and try to write the random variables X_j as a linear combination of Z_j. We know this is possible as this is essentially what happens in incremental path generation where
$$X_1 = \sqrt{t_1} Z_1, \qquad (9.12)$$
and
$$X_{j+1} = X_j + \sqrt{t_{j+1} - t_j} Z_j. \qquad (9.13)$$
We want to find other possible methods. Thus suppose we write
$$X_j = \sum_{k=1}^{n} a_{jk} Z_k, \qquad (9.14)$$
or equivalently
$$\mathbf{X} = A\mathbf{Z}, \qquad (9.15)$$
where $\mathbf{X}$ and $\mathbf{Z}$ are now vectors, and A is an $n \times n$ matrix. The covariance of X_i and X_j is then
$$\sum_{k} a_{ik} a_{jk}. \qquad (9.16)$$
In other words, the covariance matrix is AA^T, where A^T is the transpose of A. Our problem is therefore to find the solutions of the matrix equation
$$AA^T = C \qquad (9.17)$$
where $C_{ij} = \min(t_i, t_j)$. Any covariance matrix has certain properties. It is always symmetric and positive semi-definite. It will in general be strictly positive definite. A matrix A satisfying (9.17) is said to be a *pseudo-square root* of C.

There will always be many pseudo-square roots, and there are a number of algorithms for generating them. In fact, there is always a unique symmetric positive-definite square root, but this is rarely the best choice for performing Monte Carlo simulations. There are generally two popular choices, one of which is easy to compute, whilst the second has other desirable properties (see Appendix C).

If we restrict the class of A we wish to use to the lower triangular matrices, then we get a unique solution which is easy to compute. The process of finding this lower triangular matrix is called *Cholesky decomposition*. If we solve for the elements of A row by row, and column by column in each row, then we find that we have precisely one undetermined element at each step which is easy to compute. Note that the fact that the choice of element is forced at each stage means that the Cholesky decomposition is unique. If A is positive semi-definite rather than positive-definite then the decomposition will not be unique but will still exist.

In particular, we find that

$$a_{11}^2 = c_{11}, \qquad (9.18)$$

implying that $a_{11} = \sqrt{c_{11}}$. From the second row, we obtain

$$a_{21}a_{11} = c_{21}, \quad a_{21}^2 + a_{22}^2 = c_{22}, \qquad (9.19)$$

which imply that

$$a_{21} = \frac{c_{21}}{a_{11}}, \quad a_{22} = \sqrt{c_{22} - a_{21}^2}. \qquad (9.20)$$

We can continue similarly for the remaining rows.

In fact for Brownian motion, Cholesky decomposition is nothing new for us: inspecting our incremental path generation, we see that it is equivalent to multiplying by a lower triangular matrix and thus must be the Cholesky decomposition by uniqueness. In particular, if we take time points which are 1 apart then incremental path generation is equivalent to multiplying the random draws by a matrix which is all 1s on and below the diagonal.

For example,

$$\begin{pmatrix} 1 & 1 & 1 \\ 1 & 2 & 2 \\ 1 & 2 & 3 \end{pmatrix} = \begin{pmatrix} 1 & 0 & 0 \\ 1 & 1 & 0 \\ 1 & 1 & 1 \end{pmatrix} \begin{pmatrix} 1 & 1 & 1 \\ 0 & 1 & 1 \\ 0 & 0 & 1 \end{pmatrix}. \qquad (9.21)$$

What other square roots can we find? A simple approach is to relabel the random variables, X_j. If we let

$$Y_k = X_{\sigma(k)}, \qquad (9.22)$$

with σ a permutation of $1, \ldots, n$ then the covariance matrix, C', of the variables Y_k can be obtained by reordering C. If we now take the Cholesky decomposition, B, of C', and use that to generate the variables Y_k, then permuting back it must also generate the variables X_j. In order words, if we permute the rows of B back, then we obtain a pseudo-square root of C.

Another approach is to use spectral theory. If we recall the spectral theory of symmetric matrices from elementary linear algebra, then we have that there exists a basis of eigenvectors, $\mathbf{e}_1, \ldots, \mathbf{e}_n$ such that

$$C\mathbf{e}_j = \lambda_j \mathbf{e}_j, \qquad (9.23)$$

9.4 Path generation and dimensionality reduction

for some $\lambda_j \geq 0$, and the vectors $\mathbf{e}_j$ are orthonormal. If we let P be the matrix with the jth column equal to $\mathbf{e}_j$, then as the vectors $\mathbf{e}_j$ are orthonormal we have that

$$PP^\mathrm{T} = P^\mathrm{T}P = \mathrm{I}, \tag{9.24}$$

and we can write

$$C = PDP^\mathrm{T}, \tag{9.25}$$

where D is a diagonal matrix with the numbers λ_j on the diagonal. This gives us two possible pseudo-square roots. Let $D^{1/2}$ be the diagonal matrix with diagonal elements $\lambda_j^{1/2}$. Clearly, we have that $D^{1/2}$ is the square root of D. It therefore follows that

$$PD^{1/2}(PD^{1/2})^\mathrm{T} = PDP^\mathrm{T} = C. \tag{9.26}$$

This means that $PD^{1/2}$ is a pseudo-square root for C. Similarly, $PD^{1/2}P^\mathrm{T}$ is also a pseudo-square root as $P^\mathrm{T}P = \mathrm{I}$.

If we now order the eigenvalues λ_j so that $\lambda_j \geq \lambda_{j+1}$. We can regard each eigenvector as being a different component of the Brownian motion, then the first eigenvector expresses the largest component. Each successive eigenvalue represents the weighting of higher frequency vibrations. If we take the pseudo-square-root to be of the form $PD^{1/2}$, the form of our map from the independent draws, Z_j, to the correlated variates will be of the form

$$Z \mapsto \sum_{j=1}^{n} \lambda_j^{1/2} Z_j \mathbf{e}_j. \tag{9.27}$$

One notices two things about the decomposition into the principal components. The first is that the first eigenvalue is much bigger than the others and the contribution of the latter eigenvalues rapidly decays. The second is that the higher components all consist of lots of up and down movements. These two things mean that we can expect most of the value of an Asian option to come from the first few components. See Figure 9.1. The upshot of this is that instead of using $PD^{1/2}$ as our pseudo-square root, we can use a truncated map,

$$Z \mapsto \alpha \sum_{j=1}^{k} \lambda_j^{1/2} Z_j \mathbf{e}_j, \tag{9.28}$$

with k being small, to reduce the dimensionality of our problem. We have introduced the scaling factor α in order retain the total overall variance, and it would be chosen to make

$$\alpha^2 \sum_{j=1}^{k} \lambda_k = \sum_{j=1}^{n} \lambda_n. \tag{9.29}$$

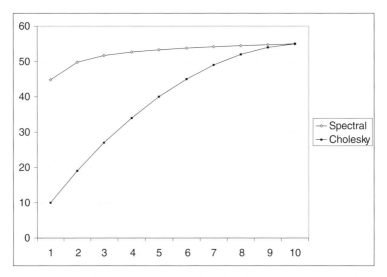

Fig. 9.1. The sum of the squares of the first j columns for the spectral pseudo-square root and the Cholesky decomposition of the covariance matrix of Brownian motion for times 1 to 10.

Having reduced the dimensionality we can expect faster convergence but at the price of some loss of accuracy.

We remark that another advantage of such a method of path generation is that when using low-discrepancy numbers one finds that the lower dimensions achieve greater uniformity of coverage. For optimal convergence, we therefore need to have the maximum possible weight placed upon the low dimensions, and this analysis into components does precisely that. This allows more rapid convergence even if we do not truncate the number of components.

Spectral decomposition can be time consuming. However, almost as rapid convergence can be obtained by using a pseudo-square root obtained from reordering if we reorder in an optimal fashion. In particular, if we wish to simulate a path from Brownian motion, then we can proceed by placing the last point first, as that is the point of greatest variation. We then fill in each succeeding point by placing it in the largest empty gap. This procedure is called the *Brownian bridge* and results in a substantial increase in the rate of convergence for low-discrepancy-based simulations at little computational cost. It is equivalent to taking the covariance matrix for Brownian motion, rearranging the order of the rows and columns, and then performing a Cholesky decomposition. Whilst the computational burden in computing the spectral square root is not great when the covariance matrices are small, it can become prohibitive when large numbers of time steps are involved and then the Brownian bridge is much more appropriate. The fact that each point in the path is

9.5 Moment matching

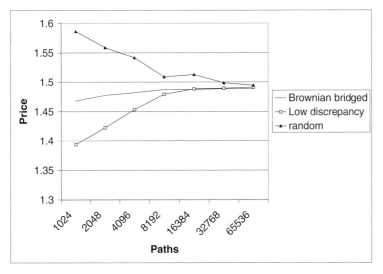

Fig. 9.2. The convergence of a Monte Carlo simulation for the pricing of a twenty-step Asian option using random numbers, low-discrepancy numbers and a Brownian bridge combined with low-discrepancy numbers

determined by the location of its neighbours and a random number means that it is not necessary to carry out a matrix multiplication for each path, which also speeds things up. See Figure 9.2 for a comparison of convergence speeds. (See [68] for further discussion.)

9.5 Moment matching

Another approach to pricing Asians is to try and simulate the average variable directly. As an average of random variables, it is itself a random variable so we can just simulate that random variable directly. We can then either use Monte Carlo, or if the distribution of the random variable is benign, an analytic formula can be developed.

The main problem with this approach is that the average of a set of log-normal variables does not have an easily computable distribution. We can therefore try to approximate it by a similar distribution. One standard method of carrying out such an approximation is *moment matching*. Recall that the moments of a distribution are the expectations

$$\mathbb{E}(X^k), \quad k = 1, \ldots, \infty.$$

Under certain technical conditions, a distribution is determined by its moments. This means that we can approximate a distribution by matching its first few moments. Typically, one works with a family of distributions depending on some parameters. The parameters are then chosen to match the moments of the target distribution.

Of course, this all begs the question of how to compute the moments. If the underlying is log-normally distributed with volatility σ and interest-rate r, and we wish to average over the times $T_1, \ldots, T_n$, then the spot at time T_j is distributed as

$$X_j = S_0 e^{(r-\frac{1}{2}\sigma^2)T_j + \sigma\sqrt{T_j}N(0,1)}. \tag{9.30}$$

We want the moments of

$$X = \frac{1}{n}\sum_{j=1}^{n} X_j.$$

The first moment is just the expectation and easy to compute, we have

$$\mathbb{E}(X) = \frac{1}{n}\sum_{j=1}^{n}\mathbb{E}(X_j) = \frac{1}{n}\sum_{j=1}^{n} S_0 e^{rT_j}. \tag{9.31}$$

For the second moment, we have

$$\mathbb{E}(X^2) = \frac{1}{n}\sum_{j,k=1}^{n} \mathbb{E}(X_j X_k). \tag{9.32}$$

If $j \leq k$, we can write the individual expectations as the expectation of

$$e^{(r-\frac{1}{2}\sigma^2)(T_k-T_j)+\sigma\sqrt{T_k-T_j}N(0,1)} X_j^2.$$

The two terms in the product are independent since the increments of a Brownian motion are independent. The expectation of the product is therefore the product of the expectations. Recalling that

$$\mathbb{E}\left(e^{\alpha N(0,1)}\right) = e^{\frac{1}{2}\alpha^2}, \tag{9.33}$$

we have that the expectation is

$$S_0^2 e^{r(T_k-T_j)+2rT_j+\sigma^2 T_j}.$$

Summing over all the possible j,k terms we can compute the second moment. A similar argument works for the higher moments. We can thus compute as many moments of X as we choose.

Having done so, we then match the moments with our favourite family of distributions. One popular choice for matching the first two moments is the log-normal distribution, [116]. If

$$Y = \mu e^{-\frac{1}{2}\nu^2 + \nu N(0,1)}, \tag{9.34}$$

then the mean is μ and, arguing as above, the second moment is $\mu^2 e^{\nu^2}$. We can thus easily solve for μ and ν to match the first two moments. As Y is log-normally distributed a variation of the Black–Scholes formula is easily developed for $\mathbb{E}((Y-K)_+)$ which yields a price for the option.

We may feel that matching the first two moments of X is not sufficiently accurate (and it probably is not). To match three moments, we therefore need a third parameter. One simple choice is to displace the log-normal distribution by adding a constant parameter a. Our new random variable is therefore $W = a + Y$. As a is constant the moments of W are easily computed in terms of moments of Y. We also have

$$\mathbb{E}((W-K)_+) = \mathbb{E}\left((Y-(K-a))_+\right), \qquad (9.35)$$

so the pricing formula for W is just the pricing formula for Y with K displaced.

One can match any number of moments one likes by picking sufficiently complicated families of distributions. Of course, at some point the amount of time spent computing and matching moments will become prohibitive, and then it will be easier to use other techniques. The main disadvantage of moment matching is that there is no natural concept of convergence so one is not sure how good the approximation is. On the other hand, it can be faster than Monte Carlo techniques which makes it usable for trading in a way that Monte Carlo is not.

9.6 Trees, PDEs and Asian options

Between sampling dates, the price of an Asian option will, like any other option, satisfy the Black–Scholes equation. To see this, just compute the drift of the option price divided by the riskless bond. This drift must be zero as the ratio is a martingale.

The trickiness is that the option's value will depend not just on the current value of spot but also on the value of spot on the previous sampling dates. Suppose that the sampling dates are $t_1 < t_2 < \cdots < t_n$, and $t_j < t < t_{j+1}$. When the first j sampling dates have passed, we can rewrite the payoff of the Asian call option struck at K as

$$\max\left(\frac{1}{n}\sum_{l=j+1}^{n} S_{t_l} - \left(K - \frac{1}{n}\sum_{k=1}^{j} S_{t_k}\right), 0\right).$$

The second sum here is determined at time t whereas the first is not. If we want to apply backwards methods here then we need to know the value of the second sum at time t but in a backwards method it is not known. This is why Monte Carlo is the natural approach to pricing Asian options.

A solution to this problem is simply to solve for all possible values of the second sum! That is, we develop a one-dimensional family of solutions indexed by an auxiliary variable α_j which expresses the running average.

We therefore proceed as follows. At time t_n the payoff is

$$\max\left(\frac{1}{n}S_{t_n} - (K - \alpha_n), 0\right),$$

and we solve back to time t_{n-1} with this payoff. In fact, we can just use the Black–Scholes formula as the payoff is equivalent to a call option struck at $n(K - \alpha_n)$, with expiry $t_n - t_{n-1}$ and notional $1/n$.

This gives us a one-dimensional family of solutions at time t_{n-1} indexed by α_n. Our definition of α_n implies that part of its value comes from S_{n-1}, so as we pass over t_{n-1} we have to take this dependence into account. We now develop a second family of solutions in the domain $[t_{n-2}, t_{n-1}]$, indexed by a new variable

$$\alpha_{n-1} = \frac{1}{n-1} \sum_{j=1}^{n-2} S_{t_j}.$$

We need to join this new family naturally onto the preceding one. As α_n is the average of the first $n-1$ terms and α_{n-1} is the average of the first $n-2$ terms, we have the condition that

$$C(S, t_{n-1}-, \alpha_{n-1}) = C\left(S, t_{n-1}+, \frac{n-1}{n}\alpha_{n-1} + \frac{S}{n}\right), \tag{9.36}$$

where C is the value of the option. We can now numerically solve back to the previous time step, t_{n-2}, using trees or PDEs, and repeat the algorithm all the way back to time zero. In Section 10.4, we adapt this technique to construct a trading strategy which replicates Asian options.

The key to this argument was that although the value of the Asian option was path dependent, the effect of the path dependency was such that one only needed a one-dimensional set of data to express its effect. This allowed the backwards pricing by simultaneously solving for all possible values of this piece of data.

One could therefore price any option for which the path dependency can be expressed by a one-dimensional auxiliary quantity in a similar fashion. More generally, one could use two (or more) auxiliary variables if the dependency could not be reduced to one dimension.

9.7 Practical issues in pricing multi-look options

We have developed various methods for pricing Asian options and discrete barrier options in a perfect Black–Scholes world. Unfortunately, we will never want to price them in a perfect Black–Scholes world. The reason is that we want to develop prices which are compatible with the prices of the vanilla options which we use to hedge them. Even if we ignore smile effects, which is often done, we find that the

9.7 Practical issues in pricing multi-look options

volatility is not a constant function. In particular, if we observe the prices of at-the-money call options, then for each of the look-at dates we get a different value for volatility. Let the value of volatility for the interval $I_j = [0, t_j]$ be σ_j.

We need a volatility function, $\sigma(t)$, such that the effective volatility over each interval I_j is σ_j. Recall from Section 6.11 that it is the root-mean-square volatility that matters in Black–Scholes pricing. We therefore need $\sigma(t)$ to be such that

$$\int_0^{t_j} \sigma(s)^2 ds = \sigma_j^2 t_j, \qquad (9.37)$$

for all j. This can be achieved by letting $\sigma(t)$ be piecewise constant, and on the interval $[t_j, t_{j+1}]$ giving it the value

$$\sqrt{\frac{1}{t_{j+1} - t_j}(\sigma_{j+1}^2 t_{j+1} - \sigma_j^2 t_j)},$$

and up to time t_1, the value σ_1. This breaks down if for any j,

$$\sigma_j^2 t_j > \sigma_{j+1}^2 t_{j+1}.$$

However, at that point (9.37) is definitely insolvable which means that either we abandon the Black–Scholes model, or conclude that there is an arbitrage to be had by selling options with t_j expiry and buying ones with expiry t_{j+1}.

Pricing a multi-look option with time-dependent volatility is, in fact, little harder than with constant volatility. One simply has to use a different value of σ over each time step $[t_j, t_{j+1}]$ but otherwise everything is the same. The ideas are all the same, it is just that the formulas are slightly more delicate. Note also that the only way $\sigma(t)$ manifests itself is as the integral of its square over an interval $[t_j, t_{j+1}]$. It therefore does not matter whether we use a smoothly time-varying volatility function or a piecewise constant one as long as (9.37) is satisfied.

We similarly need to address issues arising from the non-constancy of the continuously compounding interest rate r. In the market, we can observe discount factors, P_j, for each of the times t_j and infer from them a different r_j given by $P_j = e^{-r_j t_j}$. However, as with volatility, this problem can be removed by using a piecewise constant r, and indeed it will be necessary to do so. To carry out a Monte Carlo simulation we therefore infer values of σ and r which are constant across each interval $[t_j, t_{j+1}]$ and use them stepwise.

Having placed our multi-look option in the same framework as the traded vanilla options, we can now hedge the former with the latter. The price will be sensitive to the values of the volatilities over the different time steps. We will wish to hedge these sensitivities; we can do so by hedging with vanilla call options. We use a vanilla call option for each maturity in such a way that the Asian option minus the sum of the call options has zero sensitivity to the volatility over each time segment. That is, the derivative of the price with respect to each volatility is zero. We then Delta hedge to make our portfolio instantaneously neutral to price changes in the underlying.

Note that as well as placing the multi-look option in the same framework as the vanillas for hedging, matching the volatilities gives us the market's best estimate of the volatility of the underlying over the life of the option.

9.8 Greeks of multi-look options

For hedging purposes we will need to compute the Greeks of the multi-look option. In this section, we look at how the Monte Carlo methods we discussed in Section 7.5 go over to the multi-period case. As before, to run two Monte Carlo simulations with different variates with slightly different parameter values is a recipe for disaster – the errors in the convergence are magnified and one simply obtains noise.

As before, bumping the starting parameter for example spot by a small amount and recomputing using the same variates is reasonably effective in a lot of cases but not all. As in the one-dimensional case, the breakdown occurs precisely when the pathwise method has problems.

9.8.1 Pathwise method

We rederive the pathwise method in this setting. We derive it in a more general setting in Black–Scholes as it can be useful in the contexts of the other models we discuss in Chapters 15, 16 and 17.

Thus suppose our derivative pays $f(S_1, S_2, \ldots, S_n)$ at time T where S_j is the value of spot at time t_j. The value of the derivative is then, of course, equal to

$$e^{-rT}\mathbb{E}(f(S_1, S_2, \ldots, S_n)), \tag{9.38}$$

with the expectation taken in the risk-neutral measure. We must differentiate (9.38) and the various methods really come down to different ways of writing the expectation.

In the Black–Scholes world, we have

$$S_{j+1} = S_j e^{r(t_{j+1}-t_j)+\sigma\sqrt{t_{j+1}-t_j}Z_j}, \tag{9.39}$$

with Z_j independent $N(0, 1)$ draws. It follows that we can write

$$S_j = S_0 X_j, \tag{9.40}$$

where X_j is a random variable independent of S_0.

The derivative price is therefore equal to

$$e^{-rT}\mathbb{E}(f(S_0X_1, S_0X_2, \ldots, S_0X_n)).$$

Note that the random variables are not independent of each other. If we now differentiate with respect to S_0, we obtain

$$e^{-rT}\mathbb{E}\left(\sum_{j=1}^{n} X_j \frac{\partial f}{\partial S_j}(S_0X_1, S_0X_2, \ldots, S_0X_n)\right).$$

As $X_j = S_j/S_0$, we have the expression

$$\frac{\partial D}{\partial S_0} = e^{-rT}\mathbb{E}\left(\sum_{j=1}^{n} \frac{S_j}{S_0} \frac{\partial f}{\partial S_j}(S_1, \ldots, S_n)\right). \qquad (9.41)$$

This is the multi-look pathwise method. Note that the derivation only depended on (9.40).

For the second derivative, we can repeat the argument to get

$$\frac{\partial^2 D}{\partial S_0^2} = e^{-rT}\mathbb{E}\left(\sum_{i,j=1}^{n} \frac{S_j S_i}{S_0^2} \frac{\partial^2 f}{\partial S_j \partial S_i}(S_1, \ldots, S_n)\right). \qquad (9.42)$$

As in the one-dimensional case, we are essentially proceeding by perturbing S_0 and we need the derivatives of f to exist; when these derivatives do not exist will be precisely when the finite difference method is not effective. Note, however, that just as in the one-dimensional case, one can make the pathwise method work by interpreting the derivatives in a distributional sense.

9.8.2 Likelihood ratio method

If we write our expectation in an alternate way and then differentiate, we obtain a different expression. Let

$$\Phi(S_1, \ldots, S_n; \alpha),$$

be the joint density of the spot variables, with α any parameter such as S_0 or volatility, or indeed any other parameter we might want the Greek for. We can write the derivative with respect to α_j as

$$e^{-rT}\int f(S_1, S_2, \ldots, S_n) \frac{\partial \Phi(S_1, \ldots, S_n; \alpha)}{\partial \alpha_j} dS_1 \ldots dS_n.$$

Dividing and multiplying by Φ, we obtain

$$e^{-rT}\int f(S) \frac{\partial \log \Phi}{\partial \alpha_j}(S; \alpha) \Phi(S; \alpha) dS,$$

where $S = (S_1, \ldots, S_n)$. Rewriting as an expectation, have

$$e^{-rT} \mathbb{E} \left(f(S) \frac{\partial \log \Phi}{\partial \alpha_j} (S; \alpha) \right). \tag{9.43}$$

The main advantages of the likelihood ratio method are that it is quite general, and that the weighting factor is independent of the choice of f. This means that we can build an engine for pricing derivatives that computes the Greeks simultaneously, with the only input required being the pay-off f. For the pathwise method, we would have to compute the derivatives analytically, and when they exist only in a distributional sense we would have to adjust the integration appropriately.

The disadvantage of the likelihood ratio method is that it can be fiddly to compute the density and its derivatives. It can also be slow to converge close to expiry.

We can easily extend the method to cope with higher derivatives. If we take a second derivative with respect to α_j, we obtain

$$e^{-rT} \int f(S) \left(\frac{\partial^2}{\partial \alpha_j^2} (\log \Phi) + \left(\frac{\partial \log \Phi}{\partial \alpha_j} \right)^2 \right) \Phi dS,$$

and thus the second derivative can be written as

$$e^{-rT} \mathbb{E} \left(f(S) \left(\frac{\partial^2}{\partial \alpha_j^2} (\log \Phi) + \left(\frac{\partial \log \Phi}{\partial \alpha_j} \right)^2 \right) \right).$$

9.8.3 Central method

We have derived the pathwise and likelihood ratio method, by making choices for the changes of variables and then differentiating. What other choices can we make? Suppose we make our random variables a Brownian motion starting at $\log S_0$, and observe the value of the Brownian motion at the times $t_1, \ldots, t_n$. Let W_j denote the Brownian motion at time t_j, and let ϕ denote their joint density function. Then the derivative is equal to

$$e^{-rT} \int f(e^{\mu t_1 + \sigma W_1}, \ldots, e^{\mu t_n + \sigma W_n}) \phi(W_1, \ldots, W_n; \log S_0) dW_1 \ldots dW_n,$$

with $\mu = r - d$. We can now argue as we did as for the likelihood ratio method to obtain

$$e^{-rT} \int f(e^{\mu t_1 + \sigma W_1}, \ldots, e^{\mu t_n + \sigma W_n}) \frac{\partial}{\partial S_0} (\log \phi(W)) \phi(W) dW_1 \ldots dW_n,$$

where $W = (W_1, W_2, \ldots, W_n)$. This expression is very similar to the one we obtained for the likelihood ratio method. The main advantage lies in the fact that the density ϕ is much simpler than Φ. One also finds greater stability in certain circumstances.

9.9 Key points

We have looked at two path-dependent exotic options in detail as they exemplify most of the issues involved in general.

- A path-dependent or multi-look exotic option depends on the value of spot at many times.
- A discrete barrier option pays off according to whether a certain barrier is breached on a finite set of dates.
- An arithmetic Asian option pays off according to the value of the average of spot across a number of days.
- Path-dependent exotic options can be easily priced by Monte Carlo.
- Stock price paths can be generated by many different techniques.
- It is much easier to price a barrier option than a general exotic option by backwards methods as it can only be in two possible states.
- Asian options can be priced by backwards method by solving for all possible values of an auxiliary variable.
- A rapid method for pricing Asian options is to approximate the distribution of the average via moment matching.
- When pricing exotic options it is not enough to use a constant volatility and interest rate as the vanilla options used for hedging will not be priced correctly.
- The vanilla option prices gives us the market's best guess of the volatility over their lives.

9.10 Further reading

Chapter 9 of [27] has a good discussion of Monte Carlo techniques for security pricing including the estimation of Greeks.

Discussion of the rate of convergence as a function of dimension for various methods of numerical integration can be found in [45], along with lots of helpful discussion of the application of Monte Carlo to various option pricing problems.

The pathwise and likelihood ratio methods for computing Greeks by Monte Carlo were introduced by Broadie & Glasserman, [24]. An alternative approach relying on Malliavin calculus was introduced by Fournie, Lasry, Lebuchoux, Lions & Touzi, [46]. Their approach appears to yield similar results to the central method discussed here.

A lot of work has been done on the problem of rapidly pricing Asian options. A good survey of the state of the art in 1997 is given in [34]. Of particular interest is the method of Curran, [37, 38], which depends upon conditioning on the geometric mean and is highly effective.

For further discussion of the PDE method see [118], [119]. Early work on the topic was done by Bergman, [15]. See also the book by Ingersoll, [66].

Benhamou & Duguet have shown that the PDE method for pricing arithmetic Asian options can be greatly speeded up by using homogeneity ideas, see [14]. Their ideas extend results from the continuously-sampled case due to Rogers & Shi, [111].

In the continuously sampled case, Geman & Yor have shown that it is possible to develop a formula for the Laplace transform of the solution, [54].

9.11 Exercises

Exercise 9.1 How will the price of a discrete barrier option compare to the price of a vanilla option and the price of a continuous barrier option?

Exercise 9.2 How will the price of an Asian option compare to the price of a vanilla option with the same strike and final maturity?

Exercise 9.3 A max option pays the maximum value of the stock on a number of dates $T_1, \ldots, T_n$ at time T_n. Describe how to price this option using PDE techniques and Monte Carlo.

Exercise 9.4 Express the moments of a displaced log-normal distribution in terms of the moments of the log-normal one.

Exercise 9.5 Show that the auxiliary variable PDE method can be used to price a discrete barrier option.

Exercise 9.6 Discuss how a geometric mean Asian option would be priced by the auxiliary variable method and by Monte Carlo. The geometric mean Asian pays the positive part of the geometric mean minus the strike. (The geometric mean is the exponential of the average of the logs.) Develop an analytic price. How will the price of a geometric Asian option compare to the price of an ordinary Asian?

Exercise 9.7 Suppose an option pays the maximum value of spot minus the minimum value of spot across a number of dates. Discuss how to price this option using PDEs and Monte Carlo.

Exercise 9.8 Show that if the prices of discount bonds are a decreasing function of time to maturity then the implied continuously-compounding rates across periods $[t_j, t_{j+1}]$ are always non-negative.

10

Static replication

10.1 Introduction

The Black–Scholes no-arbitrage argument is an example of dynamic replication; by continuously trading in the underlying and riskless bonds, we can precisely replicate a vanilla option's payoff. The cost of setting up such a replicating portfolio is then the unique arbitrage-free price of the option.

If we do not allow continuous trading in the underlying, but instead restrict ourselves to portfolios in the underlying and zero-coupon bonds set up today, the only derivative we can precisely replicate is the forward contract which does not get us very far. However, if our principal interest is the pricing of exotic options, we can allow ourselves to hedge with vanilla options including forwards, calls, puts and digitals. Given this ability, we can do much better. We do not really need digitals as we can always replicate them arbitrarily well with calls or puts; however it will be convenient to include them.

We have already seen in Chapter 7 that any derivative paying off a function of the underlying at a single fixed-time horizon can be approximated arbitrarily well by calls and puts with the same payoff time. This is an example of *strong* static replication. The only real assumption here is the existence of a liquid market in calls and puts today. We require no assumptions on the process of the underlying, nor on the values of calls and puts at future times.

Unfortunately, very few options can be strong-statically-replicated. In this chapter, we therefore concentrate on various versions of *weak* static replication which means that we are allowed a (reasonably small) finite number of trades, and in particular, we are allowed to sell vanilla options before their expiry.

If we are to replicate using a trading strategy involving selling options before their expiry, we have to make assumptions about how their values change.

Definition 10.1 We shall say a model admits a deterministic future smile if the value of a call option struck at K and expiry T is a known function of the current time, t, and spot, S.

Thus we are assuming that we have a function

$$F(S, K, t, T)$$

which tells us the future price of any call option. Note that we then get the put prices for free by put-call parity. Whilst this assumption appears quite strong, it is certainly implied by the Black–Scholes model in which we have

$$F(S, K, t, T) = \text{BS}(S, K, T - t, \sigma). \tag{10.1}$$

The deterministic future smile assumption also holds for the jump-diffusion model and the Variance Gamma model studied in Chapters 15 and 17. However, it does not hold for stochastic volatility models. It is effectively equivalent to using a model in which the only state variable is spot. Once we add in a second stochastic quantity such as volatility the assumption breaks down.

The importance of this chapter has several facets. The first is the demonstration that the assumption of the future smile dynamics enforces unique prices for a large class of exotic options. The second facet is that as well as being of theoretical interest, the approaches presented here are effective numerical methods which are often the optimal method of pricing. The third facet is that we can gain additional hedging methods. If we can create a weak static replication then our risk is solely that the function F may change; we are no longer exposed to the behaviour of the underlying.

Of course, F will change. However, these methods allow us to assess how changes affect the prices of options, and to see to what extent we are exposed to model risk. By model risk, I mean the risk coming from the fact that our model for price movements is imperfect.

10.2 Continuous barrier options

The down-and-out put option

Consider a down-and-out put option struck at K with barrier at B, and expiry at T. If spot ever passes below B the option ceases to exist. We want to replicate this option with vanilla options.

The final payoff is $(K - S)_+$ if S is greater than or equal to B and zero otherwise. It will also be zero if spot has ever passed below the barrier. We first replicate the final payoff. This is easy: we just use the techniques of Chapter 7. We take a put

10.2 Continuous barrier options

option struck at K with expiry T. The pay-off of the put is wrong by

$$(K - S)H(B - S),$$

where H is the Heaviside function. If we now short a digital put struck at B of notional $K - B$, the error in our final pay-off is

$$(B - S)_+,$$

which is the pay-off of a put option struck at B. Shorting this put, we have replicated the final pay-off.

Our replicating portfolio now dominates the down-and-out put option. If the spot finishes below B they agree and if spot never goes below B they agree. However, if spot touches B and then goes back up the replicating portfolio pays off and the down-and-out put does not.

As the down-and-out put option ceases to exist when the spot touches B, suppose we dissolve our replicating portfolio immediately this happens. We then receive a sum of cash. It is this cash that reflects the amount by which we are over-replicating. The deterministic future smile assumption guarantees that this sum is precomputable and is purely a function of the *hitting time* (i.e. the time it reaches B.)

If we make the assumption that the movements of spot are continuous, then we can be sure that we can actually dissolve the portfolio at B, whereas if we allow jumps, the spot could jump from above B to below it in one go. For the rest of this section, we therefore assume that the spot moves continuously.

We now want to construct a portfolio which kills the value of the difference portfolio along the barrier and has zero value at the final pay-off time. Our assumption of continuity and the fact we dissolve the portfolio on touching the barrier means that we do not have any constraints on the portfolio's value below the barrier.

Rather than trying to pin down the portfolio's value at every time on the barrier, we divide time into N steps, t_j, with $t_N = T$. We now successively kill the difference portfolio's value at the points (B, t_j) with j counting downwards. If we add a put option struck at B, and expiring at time t_N to our portfolio, then the final pay-off above B is not affected, but the put option will have non-zero value along the barrier B, so there will be some notional which makes it kill the value, precisely at (B, t_{N-1}). In fact, the notional will just be the negative of the ratio of the value of the difference portfolio to the value of the put at that point.

To kill the value at (B, t_{N-2}), we now do the same thing again but use an option struck at B expiring at time t_{N-1} to ensure that the portfolio value at (B, t_{N-1}) is not affected. We can now work all the way back to $j = 0$, and we have a portfolio which replicates the final payoff above B, and is zero at all the points (B, t_j). This last fact means that we can expect its value to be small for any (B, t).

The difference in price between the replicating portfolio and the down-and-out put must be less than the maximum value of the replicating portfolio on the line (B, t) (actually the discounted maximum value) or we have an arbitrage. As N goes to infinity this difference will go to zero, and the replicating portfolio value will converge to the price of the down-and-out put.

In fact, the convergence is reasonably rapid and an accurate price can be quickly obtained. Note that since the portfolio replicates the option without any trading in an area around the initial values of spot and time, the time and spot Greeks of the replicating portfolio must equal those of the down-and-out put. We therefore get the Delta, Gamma and the Theta of the down-and-out put for free. Note that this argument does not extend to the other Greeks as changing other parameters in the model will affect the composition of the replicating portfolio.

There was little special about a down-and-out put option in this argument. We could equally do an up-and-out option or a call option. The same basic procedure would apply. First, replicate the final pay-off, then kill the value on the barrier by selling options which pay off behind the barrier.

Note that we do not actually have to replicate the final payoff everywhere, we only really need to be accurate above the barrier. However, a more accurate replication of the final pay-off vastly decreases the number of steps required to get a good price. This happens because one ends up needing to sell large notional put options to kill the value at the boundary because of the large value of the pay-off behind the boundary.

We have only discussed out options; however, using the fact that 'in + out = vanilla', we can immediately replicate in options by purchasing the vanilla, and shorting the replicating portfolio for the out option.

Double barrier options

We can apply similar techniques to the replication of double barrier options. For concreteness consider an American double digital option. It pays 1 at expiry if spot remains in the interval (B_1, B_2) at all times and zero otherwise.

We can replicate by a similar argument. We approximate the final profile using digitals (or call spreads). In particular, the final pay-off is accurately replicated by a digital call struck at B_1 minus (i.e. short) a digital call struck at B_2.

As before, we divide time into N steps and construct a portfolio which replicates backwards. Things are slightly trickier now in that we have to kill the values at both barriers simultaneously. To kill the values at (B_1, t_{n-1}) and (B_2, t_{n-1}), we use a call option struck at B_2 and a put option struck at B_1. Unfortunately, the put option struck at B_1 will have non-zero value at (B_2, t_{n-1}) as well as substantial value at (B_1, t_{n-1}). Similarly for the call option struck at B_2.

This is only a slight obstacle in that we obtain a 2 × 2 system for the notionals. However, it will always be invertible. Let $P(S)$ denote the value of the put at S, and $C(S)$ the value of the call. We get

$$\begin{pmatrix} P(B_1) & P(B_2) \\ C(B_1) & C(B_2) \end{pmatrix} \begin{pmatrix} N_1 \\ N_2 \end{pmatrix} = - \begin{pmatrix} V_1 \\ V_2 \end{pmatrix}, \tag{10.2}$$

where N_j are the notionals and V_j are the values to kill. As $B_1 < B_2$, we must have $P(B_1) > P(B_2)$ and $C(B_1) < C(B_2)$, the determinant must be non-zero, and the system is solvable.

As in the single barrier case, we can now repeat the argument. To kill the values at (B_1, t_{n-2}) and (B_2, t_{n-2}) we use a put option struck at B_1 with expiry at t_{n-1}, and a call option struck at B_2 with the same expiry. We again obtain a 2 × 2 system which will always be solvable. We now just iterate back to time 0.

10.3 Discrete barriers

The method of the last section worked very well for the continuous barrier option; however, it relied on the assumption that the underlying was continuous. In this section, we present a method for pricing discrete barrier options which makes no assumptions beyond the fact that the future smile is deterministic. As a continuous barrier option can be regarded as a discrete barrier option in which the barrier times are very close together, it also gives us a method for replicating continuous barrier options when the underlying is discontinuous.

For concreteness, we once again study a down-and-out put option. Let the option be struck at T_N with strike K, and suppose the barrier times are $T_1, \ldots, T_{N-1}$ with $T_{N-1} < T_N$. Let the barrier be at B. Our option pays off $(K - S)_+$ at time T_N, provided it is above B at the times T_j for $j < N$. Note that there is nothing forcing B to be above K.

Were we to try to repeat the argument of the previous section, we would hit the problem that we cannot dissolve the portfolio as soon as the spot passes the barrier B; the spot could pass below B and then come up again before the next barrier time. If we wait to dissolve until the next barrier time then the spot will not be at B but possibly far below. The replicating portfolio will therefore need to be zero not on the set (B, t), but instead on the union of the sets (S, T_j) for $j = 1, \ldots, N-1$ and $S < B$. We shall call these sets the *barrier sets*.

Our objective is therefore to construct a portfolio which replicates the final pay-off, and has zero value on the barrier sets. The idea here is that if the option knocks out then the replicating portfolio would be immediately liquidated at zero cost. To construct this portfolio, we backwards induct. First, we choose a portfolio of vanilla options with expiry T which approximates the final pay-off as accurately as we desire. Of course, if we were modelling a knock-out call or put, this would just be the call or put without the knock-out condition. Call this initial portfolio P_0. For any portfolio P, we denote by $P(S, t)$ its value at the point (S, t).

As we know the price of any unexpired vanilla option for any value of spot and time, we can value P_0 along the last barrier $[0, B] \times \{T_{N-1}\}$. We can kill the value of P_0 at the point (B, T_{N-1}) by shorting a digital put option with notional equal to $P_0(B, T_{N-1})$, struck at B. Call our new portfolio P_1^0. This portfolio then has correct final payoff profile and has zero value at (B, T_{N-1}) but may have non-zero value along $[0, B) \times \{T_{N-1}\}$. We partition $[0, B)$ into $[0, x_1], [x_1, x_2], \ldots [x_{k-1}, B)$. We then approximate the value of P_1^0 along $[0, B) \times \{T_{N-1}\}$ by assuming it is affine on each of these subintervals (i.e. its value is a straight line on each subinterval.) We now remove this value by moving successively inward. The portfolio P_1^0 has zero value at (B, T_{N-1}) so if we short put options struck at B with expiry at T_{N-1} and notional $P_1^1(x_{k-1}, T_{N-1})$, we obtain a portfolio P_1^1 which has zero value at both (B, T_{N-1}) and (x_{k-1}, T_{N-1}). If our partition is suitably small, the linear approximation will be close in value to the original portfolio, and the value will be small on the interval $[x_{n-1}, B] \times \{T_{N-1}\}$. We now iterate along the barrier at each stage shorting put options struck at x_{n-j} with expiry T_{N-1} and notional $P(x_{n-j-1}, T_{N-1})$ to obtain a sequence of portfolios, P_1^j. Note that the put options only affect value for spot below their strike so will not affect the value of the portfolio at the points already fixed. The portfolio P_1^{k-1} will then have close to zero value along the barrier as desired. Let P_1 denote the portfolio P_1^{k-1}.

We can repeat this procedure along the barrier $[0, B] \times \{T_{N-1}\}$ using the portfolio P_1 instead of P_0 to obtain a portfolio P_2. Repeating, we obtain a sequence of portfolios, P_l. If we regard an option as having zero value after expiry then the portfolio P_N has the property of having close to zero value along all the barriers $[0, B] \times \{T_j\}$ and approximates the final payoff profile. We are justified in considering the options to have zero value after their expiry because, if at any of the expiry times, the option are in the money, our hedging procedure is to liquidate the portfolio as this is precisely when the original option knocks out. The value of $P_n(0, S_0)$ is then the approximate value of the knock-out option today.

By increasing the fineness of the partition size, we can replicate the price arbitrarily well, and thus price the option as accurately as we desire. As with our strategy in the continuous case, we get the Delta, Gamma and Theta for free, just by evaluating them for the replicating portfolio.

A similar procedure would be effective for up barriers; simply replace puts by calls and induct upwards instead of downwards. To do double barriers, we simply do each barrier independently at each knock-out time as there is no interaction between the two pieces at a given knock-out time; the portfolio is, of course, affected by both barriers at previous times. Note also that our procedure does not require the barrier level to be constant.

The procedures for both discrete and continuous barriers involve setting up a portfolio initially which we then dissolve when the option knocks out. At that time, our replicating portfolio has been constructed to have zero value. This is where the deterministic future smile condition is crucial: for the portfolio to have zero value, all the unexpired options must have precisely the value predicted.

10.4 Path-dependent exotic options

One interpretation of replication methods is that they are really approximating a solution to the Black–Scholes equation by using a basis for the space of solutions. Whilst this interpretation is a little narrow in that replication methods do not require the existence of a PDE describing option prices, it does suggest that problems which can be tackled by PDE methods in the Black–Scholes world can be tackled by replication for any deterministic future smile model.

In this section, we adapt the auxiliary variable method presented in Section 9.6 to the pricing of Asian options by replication. In fact, there is nothing special about Asian options; any option which can be priced using the auxiliary variable PDE method can be priced by replication.

The method we present in this section is weaker than the one for continuous barrier options in that it involves trading in options at each reset time of the Asian option, unlike our barrier hedging strategies which only involved trading at setup and one other time – knock-out.

Suppose our Asian has reset times $t_1, \ldots, t_n$ and for concreteness is a call option struck at K. Our auxiliary variable, A_j, is defined to be equal to

$$\frac{1}{j} \sum_{k=1}^{j} S_{t_k}.$$

We have

$$A_j = \frac{1}{j}(S_{t_j} + (j-1)A_{j-1})), \qquad (10.3)$$

where $A_0 = 0$. The final payoff of the option is

$$(A_n - K)_+. \qquad (10.4)$$

The fact that the final payoff depends only on A_n and that the value of A_j can always be computed in terms of S_{t_j} and $A_{t_{j-1}}$ means that the value of the Asian option at any time, t, is a function purely of S_t and the value for A_j, where j is such that $t_j \leq t < t_{j+1}$.

At the final time, $T = t_n$, the value of the option is $(A_n - K)_+$. We can rewrite this as

$$\left(\frac{n-1}{n}A_{n-1} + \frac{1}{n}S_{t_n} - K\right)_+ = \frac{1}{n}\left(S_{t_n} - (nK - (n-1)A_{n-1})\right)_+. \quad (10.5)$$

This means that at time t_{n-1} a call option struck at $(nK - (n-1)A_{n-1})$ of notional $1/n$ precisely replicates the final payoff. Note that the replicating portfolio here depends upon A_{n-1} but not $S_{t_{n-1}}$.

We have assumed the existence of a pricing function so we immediately have that the value of this replicating call option is determined as a function of $S_{t_{n-1}}$ for each value of A_{n-1}. By no-arbitrage, the value of the Asian call option must be equal to the value of this replicating option and we therefore know its value as a function of $S_{t_{n-1}}$ and A_{n-1}.

At time t_{n-2}, we wish to construct portfolios of options expiring at time t_{n-1} whose payoffs precisely replicate the value of the Asian call at time t_{n-1}.

Recall that the value of A_{n-1} is equal to

$$\frac{n-2}{n-1}A_{n-2} + \frac{1}{n-1}S_{t_{n-1}}.$$

This means that the set of points in the $(S_{t_{n-1}}, A_{n-1})$-plane reachable from a point $(S_{t_{n-2}}, A_{n-2})$ is a line, one that depends purely upon the value of A_{n-2} and not $S_{t_{n-2}}$. In particular, the reachable line of points is

$$\left(S_{t_{n-1}}, \frac{n-2}{n-1}A_{n-2} + \frac{1}{n-1}S_{t_{n-1}}\right).$$

Thus given a value of A_{n-2}, the value of $S_{t_{n-1}}$ determines a point in the $(S_{t_{n-1}}, A_{n-1})$-plane and a price for the Asian option at time t_{n-1}. Call this price $f_{A_{n-2}}(S_{t_{n-1}})$. This means that for each value of A_{n-2}, we can replicate the value of the Asian call at time t_{n-1} by a European option paying $f_{A_{n-2}}(S_{t_{n-1}})$ at time t_{n-1}.

This European option can be approximated arbitrarily well by using a portfolio of vanilla call and put options. The given pricing function can then be used to assess the value of this portfolio for any value of $S_{t_{n-2}}$, at time t_{n-2}. Thus by valuing the portfolio associated to each value of A_{n-2} for every value of $S_{t_{n-2}}$, we develop the price of the Asian call option as a function of $S_{t_{n-2}}$ and A_{n-2} at time t_{n-2}.

We can repeat this method to get the value of the call option across each plane (S_{t_j}, A_j) for $j = 1, \ldots, n-1$.

At time t_1, we have, of course, that $A_1 = S_1$ and only the values along that line are relevant. Thus in setting up the initial replicating portfolio at time zero, we replicate the values along this line in the (S_1, A_1) at time t_1.

The value of this initial replicating portfolio is now the value of the Asian call option at time zero. Note that as with barrier options, we get the value of the Delta,

Gamma and Theta just by evaluating the relevant quantity for the initial replicating portfolio.

Having used a replicating argument to price the derivative, what is the actual trading strategy? We set up the initial replicating portfolio and hold it until the first reset time, when we dissolve the portfolio by exercising all the options which are in the money, as all the options are at their expiry. The sum of money received is by construction precisely the cost of setting up a new portfolio which depends upon the value of A_1 and which replicates out to the second reset time. We then exercise again and use the money to buy a new portfolio out to the third reset time and so on. At all stages, the new portfolio set up will depend on the value of the auxiliary variable, and will be equal in value to the exercised value of the previous portfolio.

The existence of a deterministic pricing function is crucial because any indeterminacy in the set-up cost of the later replicating portfolios will destroy our argument.

We have described the method as if we could do everything continuously; in practice we would work with a grid of values in spot and A at each time t_j.

As we mentioned above, there is nothing special about Asian options, and any multi-look option that can be priced using PDE auxiliary variables techniques can also be replicated in this fashion. One interesting consequence of this is an alternative method for replicating a discrete barrier option. If we define $A_0 = 1$, and let A_j equal A_0 if the barrier is not breached at time t_j and 0 otherwise, then a knock-out call's final payoff is

$$A_n(S_T - K)_+.$$

This means that we can replicate the discrete barrier by a trading strategy that involves at each stage buying European options which approximate the profile of the option's value across the next knock-out time. This is conceptually not as nice as the first method we presented, in that it requires trading at multiple times. However, it has the practical advantage that the total number of options in the portfolio at any given time will not be as large. As we need to repeatedly evaluate the price of the portfolio, this can cause this pricing method to be substantially faster. In particular, the time will grow linearly with the number of barrier dates rather than quadratically.

10.5 The up-and-in put with barrier at strike

Suppose we wish to replicate an up-and-in continuous-barrier put option with strike and barrier at K. Suppose also that there are no interest rates nor dividend rates. Spot is initially below K (otherwise the option would just be a vanilla option). We suppose that spot is continuous. Let the expiry of the option be T.

Our hedge is as follows: we initially hold one call option struck at K with expiry T. If spot never touches K both our hedge and our target option finish valueless and we have replicated. If spot does touch K then at that instant, the target option becomes a vanilla put. As spot is equal to K at that instant, a forward with strike equal to K will have zero value. By put-call parity, this means that the value of the put and the call will be equal at the touching time. In addition, we can transform the call into the put by shorting a forward contract, of zero value, struck at K.

Our hedging strategy is therefore to buy the call struck at K today and to short a forward contract struck at K when the barrier is crossed. This replicates the target option's payoff precisely.

We conclude that the up-and-in put and the call option have the same value. Note that we have used the fact that interest rates are zero in a non-trivial way to guarantee that the forward price is equal to the current price. Our strategy also heavily relies on the fact that spot is continuous.

10.6 Put-call symmetry

In this section, we look at an approach to replicating barrier options, one that relies on reflecting the pay-off in the barrier in an appropriate fashion.

10.6.1 A simple model

To illustrate the ideas of this section, consider a simple model in which there are no interest rates nor dividend rates, and the stock (not its log) follows a Brownian motion. So in the risk-neutral measure we have

$$dS = \sigma dW. \tag{10.6}$$

Suppose the current value of spot is B. Consider a call option struck at K and a put option struck at $B-(K-B)=2B-K$, with the same expiry. These two options will have the same value. The easy way to see this is that reflecting the Brownian motion in the level B takes the pay-off of one option to the payoff of the other. As the reflection of a Brownian motion is still a Brownian motion, we conclude that the risk-neutral expectations and therefore the prices are equal.

Suppose now that the current value of spot S_0 is above B, and we set up a portfolio consisting of a call option struck at $K > B$ and short a put option struck at $2B - K$. Above B, the final payoff is that of the call option. Along B at any time the argument above shows that the portfolio is of zero value. If we dissolve our portfolio when spot touches B, which we can do at zero cost, then we have precisely replicated a down-and-out call.

10.6 Put-call symmetry

The key point in this argument is that the call option has a symmetric partner obtained by reflection in the barrier which has the same value at any point on the barrier. This partner was particularly easy to find because the Brownian motion could be reflected in the barrier. This approach to replication is intimately related to the reflection arguments we used in Chapter 8.

10.6.2 The zero interest-rates log-normal case

Suppose we try the same argument when the process is log-normal instead of normal. As the log is the natural variable, we try reflecting the log in the log of the barrier rather than the spot in the barrier. As the pay-off is exponential in the log we cannot expect perfect symmetry; nevertheless we can still do quite well. We have to rescale the pay-off to compensate.

Reflecting the log of the strike, K, in the log of the barrier, B, is equivalent to a geometric reflection of the strike in B. The strike for our put option will therefore be

$$\frac{B^2}{K}.$$

If we value this option, we discover that its value is not the same as that of the original call option; however the adjustment to the notional required to obtain the same price is independent of time. This is sufficient for our barrier replication argument.

In general, we can prove

Theorem 10.1 *Suppose*

$$dS = \sigma S \, dW, \tag{10.7}$$

and interest and dividend rates are zero. If a European option, C, pays $f(S_T)$ at time T, then the European option D with the pay-off at time T equal to

$$\frac{S_T}{B} f\left(\frac{B^2}{S_T}\right),$$

has the same value as C when $S_t = B$.

To prove this we simply write down the risk-neutral valuation price and change variables. Let $\tau = T - t$. We have, using the expression (7.16) for the log-normal density

$$C(B, t) = \int_0^\infty f(S_T) \frac{1}{S_T \sqrt{2\pi \sigma^2 \tau}} e^{-\frac{(\log(S_T/B) + \frac{1}{2}\sigma^2 \tau)^2}{2\sigma^2 \tau}} \, dS_T. \tag{10.8}$$

Letting $\tilde{S} = B^2/S_T$, we get

$$\left|\frac{d\tilde{S}}{\tilde{S}}\right| = \left|\frac{dS_T}{S_T}\right|. \tag{10.9}$$

It follows that

$$C(B,t) = \int_0^\infty f\left(\frac{B^2}{\tilde{S}}\right) \frac{1}{\tilde{S}\sqrt{2\pi\sigma^2\tau}} e^{-\frac{(\log(B/\tilde{S}) + \frac{1}{2}\sigma^2\tau)^2}{2\sigma^2\tau}} d\tilde{S}. \tag{10.10}$$

Given $\log(B/\tilde{S}) = -\log(\tilde{S}/B)$, by expanding the exponent, it immediately follows that

$$C(B,t) = \int_0^\infty \frac{\tilde{S}}{B} f\left(\frac{B^2}{\tilde{S}}\right) \frac{1}{\tilde{S}\sqrt{2\pi\sigma^2\tau}} e^{-\frac{(\log(\tilde{S}/B) + \frac{1}{2}\sigma^2\tau)^2}{2\sigma^2\tau}} d\tilde{S}, \tag{10.11}$$

which is the price of an option with pay-off

$$\frac{S_T}{B} f\left(\frac{B^2}{S_T}\right).$$

If C is a call option then $f(S_T) = (S_T - K)_+$, and the reflected option's pay-off is

$$\frac{S_T}{B}\left(\frac{B^2}{S_T} - K\right)_+ = \frac{K}{B}\left(\frac{B^2}{K} - S_T\right)_+. \tag{10.12}$$

Thus we have proven that on the barrier B, at any time, a call option of strike K and notional 1 has the same value as a put option of strike B^2/K and notional K/B.

A down-and-out call is therefore replicated by the portfolio consisting of being long the vanilla call and short the vanilla put with modified notional. As usual, we dissolve the portfolio when the barrier is touched. This gives us an immediate formula for the option.

An interesting fact about this replicating portfolio is that the composition of the portfolio does not depend on σ. This means that as a hedge to the down-and-out call, it is robust under changes in volatility. In fact, all that really matters is that when spot is on the barrier B the smile should be symmetric – a call option of strike K/B must trade with the same implied volatility as a put (or call) option of strike B/K.

This symmetry condition is certainly satisfied by a Black–Scholes model with time-dependent volatility. One consequence of this is that the price of a down-and-out call option in a zero interest rate world is not affected by the time-dependence of volatility. All that matters is the final root-mean-square volatility, just as for a vanilla option.

10.6 Put-call symmetry

Of course, whilst these results are nice, a zero interest rate world is not realistic. However, the important thing in the above arguments is that there is zero drift; this means that the techniques immediately apply to barrier options on futures or forwards. We would use the Black formula rather than the Black–Scholes formula but the argument would otherwise be identical.

10.6.3 The non-zero interest rates log-normal case

Once we bring interest rates back in, put-call symmetry is not as simple. The reflected option is, in fact, no longer a put.

In general, we can prove

Theorem 10.2 *Suppose*

$$dS = (r - d)Sdt + \sigma S dW. \tag{10.13}$$

If a European option, C, pays $f(S_T)$ at time T, then the European option D with the payoff equal to

$$\left(\frac{S_T}{B}\right)^p f(B^2/S_T),$$

where $p = 1 - \frac{2(r-d)}{\sigma^2}$, has the same value as C when $S_t = B$.

To prove this theorem, we repeat the proof of Theorem 10.1. The interest rates add an extra term which gives rise to the extra factor. We have, in this case,

$$C(B, t) = \int_0^\infty f(S_T) \frac{1}{S_T\sqrt{2\pi\sigma^2\tau}} e^{-\frac{[\log(S_T/B) - (r - d - \frac{1}{2}\sigma^2)\tau]^2}{2\sigma^2\tau}} dS_T. \tag{10.14}$$

We then work through the same argument to get the payoff of the reflected option.

For a call option, the reflected option's payoff is

$$\left(\frac{B}{K}\right)\left(\frac{S_T}{K}\right)^{-\frac{2(r-d)}{\sigma^2}} \left(\frac{B^2}{K} - S_T\right)_+.$$

Clearly, we no longer have a put option. However, it is still a European option, which we can replicate arbitrarily well by a portfolio of vanilla put options. Our first option is a put struck at

$$B^2/K$$

of notional

$$\left(\frac{B}{K}\right)^{1 - \frac{2(r-d)}{\sigma^2}}.$$

This replicates the reflecting option to first order at B^2/K, and most of the value will be in this option. We can replicate further by approximating the difference in the usual manner, as discussed in Chapter 7.

As well as the fact that we need more options in the non-zero interest rates case, we also have the fact that the replicating portfolio depends upon r, d and σ. This means that the replication is no longer robust under changes in volatility. A consequence of this is that the barrier option's value will now be affected by time-dependence in the volatility. If volatility is time-dependent but deterministic, the pay-off needed by the reflecting option will depend upon time; the replication argument breaks down as we will no longer have zero value along the barrier at all times.

The fact that the time-dependence of volatility affects the knock-out option's value is not surprising. If $r - d$ is non-zero, the spot price will have non-zero drift in the risk-neutral world. If the volatility occurs before the spot has had time to drift away from the barrier, the option is clearly more likely to knock out than if it occurs after the spot has already drifted a long way.

Although we have seen that this method does not work in the presence of time-dependent volatility, it is still useful in that we can construct a first approximating portfolio using the methods of this section, and then complete the replication using the methods of Section 10.2.

10.7 Conclusion and further reading

In this chapter, we have looked at a number of methods for replicating path-dependent exotic options under the assumption that the future smile is known as a function of spot and time. The precise assumptions needed to replicate the options varied according to the exact method used. In this section, we try to categorize these methods and assumptions.

There are really two sorts of assumptions we can make:

(i) about the behaviour of the underlying;
(ii) about the behaviour of vanilla options.

Whilst generally in option pricing it is the first sort of assumption that is important, for static replication it is the second sort that really matters. This distinction is a little arbitrary in that the behaviour of vanilla option prices is intimately related to that of the underlying! (After all that's the basis of much of this book.)

Assumptions on the underlying are:

A1 there exists a liquid market in the underlying (or forwards) at all times;
A2 the underlying follows a Markovian process;
A3 the underlying follows a continuous process;

10.7 Conclusion and further reading

A4 the underlying follows a diffusive process;

A5 the underlying follows a log-normal process.

Possible assumptions on the vanilla options markets include:

B1 there exists a liquid market in calls and puts of all strikes and maturities today;

B2 there exists a liquid market in calls and puts of all strikes and maturities at all times;

B3 the prices of calls and puts satisfy the 'put-call symmetry' condition at all times;

B4 the price of calls and puts are a known deterministic function of calendar time, spot, strike and maturity.

Subject to these assumptions, there are a lot of different forms of replications, we could achieve for a given option.

We also give a classification of replication methods.

C1 strong static: the option pay-off can be perfectly replicated by a finite portfolio of calls, puts and the underlying set-up today with no further trading.

C2 mezzo static: the option pay-off can be perfectly replicated by a finite portfolio of calls, and puts setup today together with a finite number of trades in the underlying.

C3 weak static: the option pay-off can be perfectly replicated by setting up a finite portfolio of calls and puts today which may be sold before their own expiries.

C4 feeble static: the option pay-off can perfectly replicated by trading a finite number of calls and puts at a finite number of times.

C5 dynamic: the option pay-off can be perfectly replicated by continuous trading in the underlying.

We shall also use the term 'almost' to indicate that the pay-off can be replicated arbitrarily well rather than perfectly with a finite portfolio. If the underlying satisfies A1–5 then C5 holds; this is the fundamental result of Black and Scholes, [19]. This still holds under A1–A4; see Dupire [44].

Under assumption B1 then C1 holds for a straddle with no assumptions on the underlying. We also have, under the assumption B1, that digital European options can be almost strong-statically-replicated, by approximating using a call-spread. Unfortunately, strong static replication holds for very few options. This is the replication method we discussed in Chapter 7.

If we make the assumptions A1, A3 as well as B1 then we mezzo-statically-replicated the up-and-in put option. We saw this in Section 10.5. This was proven in [28, 62].

Under B1, B2, B3, A3 we saw in Section 10.6 that a class of barrier options can be weak-statically-replicated. This was proved in [30].

If we assume B1, B2 and B4 then we are in the situation of Section 10.3. The method we present for hedging discrete barrier options under these assumptions is then an almost weak static replication. Note that we can also hedge continuous barrier options arbitrarily well by approximating with a discrete barrier option with an arbitrarily large number of sampling dates.

The method of Section 10.4 for the replication of a general path-dependent exotic option makes the same assumptions, but it is almost feeble static in that it requires trading in options at multiple times.

Note that whilst these methods make no assumptions on the underlying, it is difficult to imagine a situation where B1, B2 and B4 hold but A2 does not.

If we make the additional assumption of continuity of sample paths, A3, then the simpler method of Section 10.2 can be used to almost-weak-replicate continuous barrier options. This method relies on dissolution of the portfolio at the instant the barrier is crossed, and therefore only requires the replicating portfolio to be of zero value on the barrier rather than behind it. Under these assumptions, it is also possible to replicate American options, [70]. This is further discussed in Chapter 12. In fact, one does not really need continuity of sample paths: the crucial property is that the spot cannot jump across the barrier for knock-out options or into the exercise domain for American options. These techniques could therefore be applied in markets where only down jumps occur – for example equity indices – to the pricing of up-and-out barrier options and American call options.

We have studied the problem of perfect replication under the assumption that the future smile is known. An alternative approach is to assume that the future smile is unknowable and therefore allow oneself only to trade in options today and hold them to expiry. Whilst one will not obtain perfect replication, one can obtain bounds on prices which can be strong. Such approaches are developed in [62] and [28].

10.8 Key points

In this chapter, we have studied the application of replication techniques to the pricing and hedging of exotic options.

- Replication is a powerful technique for taking account of our views on the evolution of vanilla option prices when pricing exotic options.
- Strong static replication allows us to price options purely by using the prices of vanilla options observable today without any modelling assumptions.
- Weak static replication allows us to price exotic options providing the future smiles of the vanilla options are known.
- The auxiliary variable technique for pricing exotic options using PDEs in a Black–Scholes world can be adapted to work with replication techniques in any model that implies a deterministic future smile.

- Replication can be made much simpler when it is assumed that the stock price process is continuous.
- Options which can be replicated include barrier options and Asian options.
- Very few options can be strong-statically-replicated which means that assumptions on the behaviour of future smiles strongly affect the price of exotic options.
- Weak static replication is generally not applicable in stochastic volatility models.

10.9 Exercises

Exercise 10.1 A range-accrual option pays £1 at expiry for each day the underlying has spent between two given levels. Show that the range-accrual can be almost-strong-statically-replicated given deterministic interest rates.

Exercise 10.2 Which of the techniques discussed in this chapter require the spot price to be continuous?

Exercise 10.3 For the weak static replication of a discrete barrier option, approximately how many price evaluations will be required if N options are used per barrier time and there are n barrier times? How many will be required for the auxiliary variable replication technique?

Exercise 10.4 Show that under the Black–Scholes model the pricing function satisfies
$$F(S, K, t, T) = F(S, K, 0, T-t) = \frac{S}{S_0} F\left(S_0, K\frac{S_0}{S}, 0, T-t\right).$$
What relation is satisfied in the Black–Scholes model with time-dependent volatility?

Exercise 10.5 It is possible to pick the function F in such a way that today's non-arbitrageable smile is matched, and arbitrages are implied in the future whilst F satisfies
$$F(S, K, t, T) = \frac{S}{S_0} F\left(S_0, K\frac{S_0}{S}, 0, T-t\right).$$
Find such an F.

11

Multiple sources of risk

11.1 Introduction

We have so far concentrated on the case of an option on a single asset driven by a single Brownian motion. In this chapter, we look at derivatives which are dependent on more than one stochastic quantity. Whilst some of the options we study may appear a little contrived, the mathematics we introduce will be essential in the development of interest-rate models and for the construction of more complicated models for asset price movements. We will see that, by careful use of the numeraire, the pricing of some derivatives can be reduced to a one-dimensional problem; however for others multi-dimensionality is unavoidable. It is these latter derivatives that bear the most similarity to interest-rate derivatives.

We start with some examples. A US investor is exposed to some Japanese equities risk. He therefore wishes to buy an option on the Nikkei. His banker sells him a put option which pays \$1 for each point the Nikkei is below 15000 a year from now. The Nikkei is made up of yen-priced equities, yet our pay-off is in dollars. The mismatch here between the currency of pay-off and the currency of the underlying means that the hedger is exposed to the exchange rate, and the option is not just a vanilla put option. Such options which pay off in the 'wrong' currency are known as *quanto* options.

An investor is not sure which of two stocks he will wish to hold in a year. He therefore buys the first one, and purchases an option which allows him to exchange it for the second one. If we denote the stocks, $S_t^{(1)}$, $S_t^{(2)}$, then this option's pay-off is

$$\max\left(S_t^{(2)} - S_t^{(1)}, 0\right).$$

Such an option is generally called a *Margrabe option*, after the first person to analyze its pricing.

More generally, we could allow the possibility to exchange for any one of a number of stocks $S_t^{(j)}$, and get the payoff

$$\max\left(\max\left(S_t^{(j)} - S_t^{(1)}\right), 0\right).$$

Another common product is an option on a basket. Thus we could take any ordinary derivative, and then make its payoff dependent on the average

$$\frac{1}{n}\sum_{k=1}^{n} S_t^{(k)};$$

or more generally the pay-off could be dependent a weighted average of the underlyings.

For all the products we examine, an important issue will be the correlation between the assets involved. For example, if two stocks are perfectly correlated then an option to exchange will simply be worth the value of exchanging them today. Whereas if they are perfectly negatively correlated, that is if one goes up when the other goes down and vice versa, then it will be worth a great deal.

In order to understand multi-asset options, we will need to understand multi-dimensional Brownian motions, how to understand correlations between Brownian motions and how to extend the Ito calculus to higher dimensions.

11.2 Higher-dimensional Brownian motions

Recall that a one-dimensional Brownian motion, X_t, is defined so that the distribution of $X_t - X_s$ is always a normal distribution of mean 0 and variance $t - s$, for $t > s$. This holds regardless of the value of X_s and the path of X_r for $r < s$.

We can make a similar definition in higher dimensions; we simply have to replace the one-dimensional normal by a higher-dimensional normal. The higher-dimensional normal with variance 1 and mean 0 is really just a vector of independent one-dimensional random variables. Thus to construct a k-dimensional normal, we take independent one-dimensional normals $X^{(j)}$, and set

$$X = \left(X^{(1)}, X^{(2)}, \ldots, X^{(k)}\right).$$

This means that X has a probability density function

$$p(x) = \left(\frac{1}{2\pi}\right)^{n/2} e^{-\frac{x_1^2}{2}} e^{-\frac{x_2^2}{2}} \cdots e^{-\frac{x_n^2}{2}}. \tag{11.1}$$

We therefore define a k-dimensional Brownian motion to be a k-dimensional random process such that the covariance matrix of $X_t - X_s$ is $(s - t)I$, where I denotes the identity matrix, and the behaviour of $X_t - X_s$ is independent of the behaviour of X_r for $r \leq s$. A k-dimensional Brownian motion path is a map from $\mathbb{R}$ to $\mathbb{R}^k$. The measure on the space of paths will therefore be a probability density on the space of k-dimensional paths. Conceptually there is very little difference from the one-dimensional case.

One interesting and important aspect of higher-dimensional Brownian motions is our ability to use them to construct other similar processes. First, trivially, any subset of the coordinates defines a Brownian motion as all the properties are immediately satisfied. More importantly, we can construct a Brownian motion from a linear combination of two coordinates. Let ρ be between -1 and 1. Let

$$Y_t = \rho X_t^{(1)} + \sqrt{1-\rho^2} X_t^{(2)}. \tag{11.2}$$

We then have

$$Y_t - Y_s = \rho\big(X_t^{(1)} - X_s^{(1)}\big) + \sqrt{1-\rho^2}\big(X_t^{(2)} - X_s^{(2)}\big). \tag{11.3}$$

We know that $X_t^{(j)} - X_s^{(j)}$ is a normal distribution of mean zero and variance $t-s$. Recall that a sum of two independent normal distributions is a normal distribution with mean equal to the sum of the means, and variance equal to the sum of the variances. This means that $Y_t - Y_s$ is a normal with mean zero, and with variance equal to

$$\rho^2(t-s) + (1-\rho^2)(t-s) = t-s.$$

Thus $Y_t - Y_s$ has mean zero and variance $t-s$. It is also clearly independent of the value of Y_s, as each of its constituent components are independent of the behaviour up to time s. We therefore have that Y_t defines a Brownian motion.

If we now compare Y_t with $X^{(1)}$, the fact that $X_t^{(1)}$ was used to construct Y_t means that their movements are correlated. In particular, we have that

$$\mathbb{E}\big((Y_t - Y_s)(X_t^{(1)} - X_s^{(1)})\big) = \rho \mathbb{E}\big((X_t^{(1)} - X_s^{(1)})^2\big) \\ + \sqrt{1-\rho^2}\mathbb{E}\big((X_t^{(2)} - X_s^{(2)})(X_t^{(1)} - X_s^{(1)})\big). \tag{11.4}$$

Since $X^{(1)}$ and $X^{(2)}$ are independent, the second expectation is zero and so

$$\mathbb{E}\big((Y_t - Y_s)(X_t^{(1)} - X_s^{(1)})\big) = \rho(t-s). \tag{11.5}$$

As $Y_t - Y_s$ and $X_t^{(1)} - X_s^{(1)}$ both have variance $t-s$, this means that the correlation coefficient is ρ. Thus we have constructed a Brownian motion whose increments are correlated to those of $X^{(1)}$ with correlation ρ.

More generally, we could construct a Brownian motion from any vector

$$\alpha = (\alpha_1, \ldots, \alpha_k)$$

with $\sum \alpha_i^2 = 1$, by taking $\sum_{j=1}^{k} \alpha_j X^{(j)}$.

The existence of such correlated Brownian motions will be crucial in pricing multi-asset options. In general, we may want a whole vector of Brownian motions with a specified correlation matrix. To construct such a vector, we can proceed in a similar fashion as to constructing a vector of correlated normal variates as we discussed in Section 9.4. In particular, given a positive-definite correlation matrix, R, we take any pseudo-square root, A, and set

$$Y^{(i)} = \sum_{j=1}^{n} a_{ij} W^{(j)}. \tag{11.6}$$

11.3 The higher-dimensional Ito calculus

When we are pricing derivatives, we will need to understand the process followed by a function of random variables which are each following Ito processes, that is we need a multi-dimensional Ito rule.

Thus suppose we have correlated Brownian motions $W_t^{(j)}$. Associated to each Brownian motion, we have an Ito process $X^{(j)}$,

$$dX^{(j)} = \mu_j(X^{(j)}, t)dt + \sigma_j(X^{(j)}, t)dW_t^{(j)}. \tag{11.7}$$

We want to understand what process is followed by a smooth function of all the $X^{(j)}$. Let $f(t, x_1, \ldots, x_n)$ be a smooth function from $\mathbb{R}^{n+1}$ to $\mathbb{R}$.

Following our arguments in the one-dimensional case, we can attempt to approximate the derivative via Taylor's theorem. Taylor's theorem tells us that

$$f(t + \Delta t, x_1 + \Delta x_1, \ldots, x_n + \Delta x_n) = f(x, t) + \frac{\partial f}{\partial t}(x, t)\Delta t + \sum_{j=1}^{n} \frac{\partial f}{\partial x_j} \Delta x_j$$

$$+ \frac{1}{2} \sum_{i,j=1}^{n} \frac{\partial^2 f}{\partial x_i \partial x_j} \Delta x_i \Delta x_j, \tag{11.8}$$

plus an error of order three in Δx_l, order two in Δt and order one in $\Delta x_l \Delta t$.

If we wish to imitate our arguments in the one-dimensional case, we have to understand how the terms $\Delta x_i \Delta x_j$ behave when $x_l = X_t^{(l)}$. Recall that our definition of the process for $X^{(j)}$, implies that

$$X_{t+\Delta t}^{(j)} = X_t^{(j)} + \mu_j(X^{(j)}, t)\Delta t + \sigma_j(X^{(j)}, t)\left(W_{t+\Delta t}^{(j)} - W_t^{(j)}\right) + \text{small error} \tag{11.9}$$

where, of course,

$$W_{t+\Delta t}^{(j)} - W_t^{(j)} = \sqrt{\Delta t} Z_j, \tag{11.10}$$

and the Z_j are ordinary $N(0, 1)$ draws. However, our definition of correlated Brownian motions means that the variates Z_j are correlated to the same extent as the Brownian motions. Thus if we let ρ_{jk} be the correlation between $W^{(j)}$ and $W^{(k)}$, we can write

$$Z_j = \rho_{jk} Z_k + \sqrt{1 - \rho_{jk}^2} e_{jk}, \tag{11.11}$$

where e_{jk} is $N(0, 1)$ and independent of Z_k. This means that

$$\left(W_{t+\Delta t}^{(j)} - W_t^{(j)}\right)\left(W_{t+\Delta t}^{(k)} - W_t^{(k)}\right) = \Delta t \rho_{jk} Z_k^2 + \Delta t \sqrt{1 - \rho_{jk}^2} e_{jk} Z_k. \tag{11.12}$$

The second term has mean zero and variance of order Δt^2 so we can discard it as small, whereas the first term has mean $\rho_{jk}\Delta t$ and variance of order Δt^2 and therefore contributes. This gives us a new rule for the multi-dimensional Ito calculus:

$$dW_t^{(j)} dW_t^{(k)} = \rho_{jk} dt. \tag{11.13}$$

To summarize, we have

Theorem 11.1 (Multi-dimensional Ito lemma) *Let $W_t^{(j)}$ be correlated Brownian motions with correlation coefficient ρ_{jk} between the Brownian motions $W_t^{(j)}$ and $W_t^{(k)}$. Let X_j be an Ito process with respect to $W_t^{(j)}$. Let f be a smooth function; we then have that*

$$df(t, X_1, X_2, \ldots, X_n) = \frac{\partial f}{\partial t}(t, X_1, \ldots, X_n) dt + \sum_{j=1}^n \frac{\partial f}{\partial x_j}(t, X_1, \ldots, X_n) dX_j$$

$$+ \frac{1}{2} \sum_{j,k=1}^n \frac{\partial^2 f}{\partial x_j \partial x_k}(t, X_1, \ldots, X_n) dX_j dX_k, \tag{11.14}$$

with

$$dW_t^{(j)} dW_t^{(k)} = \rho_{jk} dt. \tag{11.15}$$

When collecting terms, the final double sum will be absorbed into the dt term.

We still need to think a little about what a process of the form

$$dY_t = \mu dt + \sum_{j=1}^n \sigma_j dW_t^{(j)}, \tag{11.16}$$

means. Here we have a sum of Brownian motions driving the asset instead of just one. In a small time step Δt, we can simulate the Ito process by

$$Y_{t+\Delta t} - Y_t = \mu \Delta t + \sum_{j=1}^n \sigma_j \left(W_{t+\Delta t}^{(j)} - W_t^{(j)}\right), \tag{11.17}$$

where we have constructed the $W^{(j)}$ as correlated Brownian motions.

11.3 The higher-dimensional Ito calculus

Alternatively, if we are not interested in the individual $W^{(j)}$ and simply wish to simulate Y_j, we can do so by observing that a linear sum of correlated Brownian motions can be expressed in terms of a single Brownian motion.

For example,

$$\sigma_1\left(W^{(1)}_{t+\Delta t} - W^{(1)}_t\right) + \sigma_2\left(W^{(2)}_{t+\Delta t} - W^{(2)}_t\right) = \sqrt{\Delta t}\left(\sigma_1 Z_1 + \sigma_2 \rho_{12} Z_1 + \sigma_2\sqrt{1-\rho_{12}^2} Z_3\right), \quad (11.18)$$

with Z_1 and Z_3 independent $N(0, 1)$ variables. We can rewrite this as

$$\sqrt{\Delta t}(\sigma_1 + \sigma_2 \rho_{12})Z_1 + \sqrt{\Delta t}\sigma_2\sqrt{1-\rho_{12}^2} Z_3.$$

This is a sum of two independent normals, and therefore is equal to a normal distribution with variance equal to the sum of the variances which is

$$(\sigma_1 + \sigma_2\rho_{12})^2 + \sigma_2^2(1-\rho_{12}^2) = \sigma_1^2 + 2\rho_{12}\sigma_1\sigma_2 + \sigma_2^2.$$

This means that if $k=2$, we can regard Y as being driven by a single new Brownian motion $W^{(4)}$ and satisfying

$$dY_t = \mu dt + \sigma dW_t^4 \quad (11.19)$$

with

$$\sigma = \sqrt{\sigma_1^2 + 2\rho_{12}\sigma_1\sigma_2 + \sigma_2^2}. \quad (11.20)$$

Note that if $\sigma_1 = \sigma_2$ and $\rho_{12} = -1$, then $\sigma = 0$, and Y_t has become deterministic – the two Brownian motions are perfectly inversely correlated and their movements cancel each other. Note also that if $\rho_{12} = 1$ then

$$\sigma = \sigma_1 + \sigma_2. \quad (11.21)$$

We can interpret (11.20) geometrically. Suppose we regard a Brownian motion as being a vector times a one-dimensional Brownian motion. Perfect correlation means the vectors point the same way, perfect negative correlation means they point the opposite way, and zero correlation means they are orthogonal. In (11.20), the first vector has length σ_1, and the second length σ_2. When we add two vectors, v_1, v_2, the square of the length of the resultant vector is

$$||v_1||^2 + 2\cos(\theta)||v_1||.||v_2|| + ||v_2||^2,$$

where θ is the angle between the vectors. If we interpret the correlation coefficient as being the cosine of the angle between the two Brownian motions, then this means that the new volatility is just the length of the vector obtained by summing the vectors for each Brownian motion.

248   *Multiple sources of risk*

More generally, we could construct a Brownian motion from any vector

$$\alpha = (\alpha_1, \ldots, \alpha_k)$$

with $\sum \alpha_i^2 = 1$, by taking $\sum_{j=1}^{k} \alpha_j X^{(j)}$.

When we have a process driven by $k > 2$ Brownian motions, we obtain a similar expression to (11.20). The volatility becomes

$$\sqrt{\sum_{i,j=1}^{n} \sigma_i \sigma_j \rho_{ij}}$$

and we can write

$$dY_t = \mu dt + \sigma dW_t, \tag{11.22}$$

for a Brownian motion W constructed from the old one. We can similarly regard our processes as vectors in $\mathbb{R}^n$ which add according to their directions.

11.4 The higher-dimensional Girsanov theorem and risk-neutral pricing

A key component of martingale pricing in one dimension was the ability to change the drifts of stock prices via a measure change on the space of Brownian motion paths. An important converse was that this was all that a measure change could do.

The situation in higher-dimensions is very similar. Once again we can change drifts. Once again we cannot do anything else. This second statement has some new aspects however. When we are dealing with correlated Brownian motions, this means that the correlations between them cannot be changed via measure change. In financial terms, this ensures that the correlation between two assets affects the price of a derivative contract written on both of them which is what we might expect. However, as before, the real-world drifts have no effect on prices.

Our sample space is larger than in the one-dimensional case. The space of paths is the space of continuous paths in $\mathbb{R}^n$ rather in $\mathbb{R}$. The set of information, $\mathcal{F}_t$, available at time t is the behaviour of all n Brownian motions up to time t rather than just one. The information for each of the individual Brownian motions is still contained in $\mathcal{F}_t$, it is just that $\mathcal{F}_t$ contains a lot more information. The measure change will be a change of measure on the larger sample space and we can change the drifts of the individual Brownian motions by differing amounts.

Our martingale condition is the same as before:

$$\mathbb{E}(X_t | \mathcal{F}_s) = X_s. \tag{11.23}$$

11.4 The higher-dimensional Girsanov theorem

We want to change the drifts so that this holds for all the assets. This is trickier than in the one-dimensional case because correlation may cause the effective dimensionality of the Brownian motion to be lower than immediately apparent. For example, suppose we have a three-dimensional Brownian motion $W^{(j)}$, and three stocks, $S^{(l)}$. Suppose also that we have

$$dS^{(l)} = \mu_l S^{(l)} dt + S^{(l)} \sigma_l dZ^{(l)}, \tag{11.24}$$

where $Z^{(1)} = W^{(1)}$, $Z^{(2)} = W^{(2)}$ and $Z^{(3)} = \frac{W^{(1)} + W^{(2)}}{\sqrt{2}}$.

The Girsanov theorem allows us to change the drifts of $W^{(j)}$. If we change the drift of $W^{(j)}$ by ν_j then for $j = 1, 2$, we have

$$dS^{(j)} = (\mu_j + \nu_j \sigma_j) S^{(j)} dt + S^{(j)} \sigma_j dZ^{(j)}, \tag{11.25}$$

and

$$dS^{(3)} = \left(\mu_3 + \frac{1}{\sqrt{2}} \nu_1 \sigma_3 + \frac{1}{\sqrt{2}} \nu_2 \sigma_3\right) S^{(3)} dt + S^{(3)} \sigma_3 dZ^{(3)}. \tag{11.26}$$

We want to choose ν_j so that each $S^{(j)}$ has drift r. Equation (11.25) determines ν_1 and ν_2 immediately. The measure-changed drift of $S^{(3)}$ is then already determined and is equal to

$$\mu_3 + \frac{1}{\sqrt{2}} \nu_1 \sigma_3 + \frac{1}{\sqrt{2}} \nu_2 \sigma_3.$$

For a risk-neutral measure this has to be r. So a risk-neutral measure exists if and only if

$$\mu_3 - \frac{1}{\sqrt{2}} \frac{\mu_1 - r}{\sigma_1} \sigma_3 - \frac{1}{\sqrt{2}} \frac{\mu_2 - r}{\sigma_2} \sigma_3 = r. \tag{11.27}$$

How can we interpret this? The fact that there are only two sources of uncertainty driving the stock prices means that we can only assign two drifts arbitrarily in the real-world measure without causing arbitrage.

If (11.27) does not hold then we can hedge the movements in $S^{(3)}$ by suitable holdings in $S^{(1)}$ and $S^{(2)}$, and achieve a riskless portfolio which does not grow at the riskless rate. That is we can achieve an arbitrage.

Note that our correlation matrix in this case is

$$\begin{pmatrix} 1 & 0 & 1/\sqrt{2} \\ 0 & 1 & 1/\sqrt{2} \\ 1/\sqrt{2} & 1/\sqrt{2} & 1 \end{pmatrix}$$

which has a pseudo-square root

$$\begin{pmatrix} 1 & 0 & 0 \\ 0 & 1 & 0 \\ 1/\sqrt{2} & 1/\sqrt{2} & 0 \end{pmatrix}$$

The important thing to note about these matrices is that they are not of full rank. Clearly, $(0, 0, 1)$ is in the null-space of the second matrix. The null-space of the first matrix is the set of scalar multiples of $(1, 1, -\sqrt{2})$.

In general, we can prove

Proposition 11.1 *If C is a $n \times n$ symmetric matrix, and A is a pseudo-square root of C then A and C have null spaces of the same dimension.*

Proof We show that the set of vectors orthogonal to $\mathrm{Im} A$ is the null-space of AA^T. Let $\langle a, b \rangle$ denote the inner product of a and b. If

$$AA^T u = 0$$

then we have that

$$\langle A^T u, A^T u \rangle = \langle AA^T u, u \rangle = 0.$$

So

$$A^T u = 0,$$

and hence

$$\langle Av, u \rangle = \langle v, A^T u \rangle = 0,$$

which means that u is orthogonal to $\mathrm{Im} A$.

If u is orthogonal to $\mathrm{Im} A$, then u is orthogonal to $\mathrm{Im} AA^t$ as it's a subset of $\mathrm{Im} A$. In particular this means that u is orthogonal to $AA^T u$ which implies that

$$\langle A^T u, A^T u \rangle = \langle u, AA^T u \rangle = 0.$$

Clearly $A^T u = 0$ implies that $AA^T u = 0$. We thus have proved that the null-space of AA^T is equal to the null-space of A^T and is equal to the set of vectors orthogonal to the image of A.

If A is an $n \times n$ matrix, then we have that

$$\dim \mathrm{Ker} A + \dim \mathrm{Im} A = n,$$

and

$$\dim \mathrm{Im} A^\perp + \dim \mathrm{Im} A = n.$$

The result is now immediate. □

11.4 The higher-dimensional Girsanov theorem

The upshot is that the correlation matrix is of full rank (i.e. rank n) if and only if n independent Brownian motions are needed to drive the process. If it is of lower rank then we can discard Brownian motions in the null-space of A^T.

We return to the issue of the existence of risk-neutral measures. Let our correlation matrix be C and A be a pseudo-square root. Suppose we have independent Brownian motions Z_j, driving our n stocks via

$$dS_j = \mu_j S_j dt + \sigma_j S_j \sum_{k=1}^{n} A_{jk} dZ_k. \qquad (11.28)$$

If we perform the measure change $Z_k = Z'_k + \nu_k$, then we get

$$dS_j = \left(\mu_j + \sigma_j \sum_{k=1}^{n} a_{jk} \nu_k \right) S_j dt + \sigma_j S_j \sum_{k=1}^{n} A_{jk} dZ'_k. \qquad (11.29)$$

To achieve a risk-neutral measure, we need to have

$$\sum_{k=1}^{n} a_{jk} \nu_k = -\frac{\mu_j - r}{\sigma_j}. \qquad (11.30)$$

If we set $\alpha_j = -\frac{\mu_j - r}{\sigma_j}$, $\boldsymbol{\alpha} = (\alpha_1, \ldots, \alpha_n)$, and $\boldsymbol{\nu} = (\nu_1, \ldots, \nu_n)$, then we can rewrite our equation as

$$A\boldsymbol{\nu} = \boldsymbol{\alpha}. \qquad (11.31)$$

Note that if some σ_j is zero, then either $\mu_j = r$, and no measure change is needed for that asset, or there is a riskless asset which is not growing at the riskless rate. The latter would of course imply an arbitrage. We can therefore assume that all the σ_j are non-zero.

If A is invertible this is easy to solve. Otherwise, $\boldsymbol{\alpha}$ has to be in the image of A. However, as in the special case we examined above, the non-solvability will correspond to the existence of arbitrage. We prove this in the special case where A has been found by Cholesky decomposition, and therefore has the property that if a_{jk} (with $k > 1$) is non-zero then $a_{i,k-1}$ is non-zero for some $i < j$. We then solve iteratively for ν_j, with j increasing. When solving for the drift of each asset, we find either that there is a ν_j entering the drift which has not yet been specified, or that the assets' risk can be hedged precisely by a linear combination of the assets whose drifts have already been fixed. In the latter case, this means that the drift of the asset must already be r, or we can create a riskless asset that grows at a rate other than r, and hence there is an arbitrage opportunity. In the former case, we simply solve for the not yet determined ν_j.

In conclusion, when pricing derivatives we can assume the existence of a risk-neutral measure in which all the stocks grow at the riskless rate, or there will be an arbitrage opportunity just from trading the stocks.

We can proceed to risk-neutral pricing for multi-asset derivatives in much the same way as for single asset derivatives. We change measure so that all asset prices discounted by the numeraire are martingales. The arbitrage-free price for the derivative is then the initial value of the numeraire times the expectation of the ratio of pay-off to numeraire. The same argument that this price is arbitrage-free holds; if all assets are martingales then any portfolio involving trading them must have a chance of being worth less if it can be worth more, so no arbitrages can occur.

As in the single asset case, the derivatives' price must satisfy a PDE, the difference being, of course, that the PDE will be higher-dimensional which reflects the price dependence on each of the stocks. To get the PDE, we can either use a hedging argument as in the Black–Scholes derivation in the one-dimensional case, or we can just use Ito's lemma to compute the drift in the risk-neutral measure. We do the latter.

Thus suppose we have j assets S_j such that

$$dS_j = \mu_j S_j dt + \sigma_j S_j dW^{(j)}, \tag{11.32}$$

and $W^{(j)}$ is correlated with $W^{(k)}$ with correlation coefficient ρ_{jk}. We shift to a risk-neutral measure in which all the assets have drift r. Let the derivative be D. We set

$$D(S_1, \ldots, S_n, t) = e^{rt} \mathbb{E}_{\text{RN}}\left(\frac{D(T)}{e^{rT}}\bigg|\mathcal{F}_t\right), \tag{11.33}$$

where, as usual, T is after all payoffs and the value of $D(T)$ incorporates any previously generated cashflows rolled up to time T. As De^{-rt} is a martingale by construction, it must have zero drift.

We compute; by Ito's lemma we get

$$d(De^{-rt}) = e^{-rt}(dD - rD). \tag{11.34}$$

We have

$$dD = \frac{\partial D}{\partial t}dt + \sum_{j=1}^{n}\frac{\partial D}{\partial S_j}dS_j + \frac{1}{2}\sum_{i,j=1}^{n}\frac{\partial^2 D}{\partial S_i \partial S_j}dS_i dS_j. \tag{11.35}$$

Using the fact that

$$dS_i dS_j = \rho_{ij} S_i S_j \sigma_i \sigma_j dt, \tag{11.36}$$

we conclude that the drift of D is

$$\frac{\partial D}{\partial t} + \sum_{i=1}^{n}\frac{\partial D}{\partial S_j}rS_j + \frac{1}{2}\sum_{i,j=1}^{n}\rho_{ij}S_i S_j \sigma_i \sigma_j \frac{\partial^2 D}{\partial S_i \partial S_j}.$$

The drift condition therefore gives us

$$\frac{\partial D}{\partial t} + \sum_{i=1}^{n} \frac{\partial D}{\partial S_j} r S_j + \frac{1}{2} \sum_{i,j=1}^{n} \rho_{ij} S_i S_j \sigma_i \sigma_j \frac{\partial^2 D}{\partial S_i \partial S_j} = rD. \tag{11.37}$$

This is the higher-dimensional Black–Scholes equation.

11.5 Practical pricing

We can now apply the same techniques to pricing that we developed in the one-dimensional case. We can attempt to solve the PDE analytically, or use a numeric grid method to solve it. If we work with the expectation, we can attempt to evaluate analytically, or approximate it via numerical integration or Monte Carlo simulation.

In fact, it is in higher dimensions that Monte Carlo becomes an important method. The reason is that with a grid-based numerical integration or PDE method, the number of grid points grows exponentially with dimensions. For example, suppose we have n stocks, S_j, and we decide to integrate over the region

$$50 < S_j < 150, \quad j = 1, \ldots, n.$$

If we take 1 as our grid size, then the number of hypercubes in the grid is 100^n, as each cube is of the form

$$[S_1, S_1 + 1] \times [S_2, S_2 + 1] \times \cdots \times [S_n, S_n + 1],$$

with the value of S_j any integer from 50 to 149 and the value of each S_j totally independent of the others. If we were doing a 10-dimensional problem, we would then have 100^{10} cubes which would clearly be prohibitive. We would have the same problems with a grid-based PDE method.

Monte Carlo on the other hand is a lot less affected by high dimensions. We proceed in much the same way as before. We generate paths, evaluate the value of the derivative on each path and then average. How would we actually carry out the simulation? Suppose the payoff of our derivative, D, depends on the values of $S_j(t)$ for $t = t_1, \ldots, t_k$. Suppose that the correlation of S_i with S_j is ρ_{ij}. Let A be a pseudo-square root of the matrix (ρ_{ij}). Let $t_0 = 0$.

We need nk normal random draws. We denote them as Z_{jl}. We first create correlated normal variates by

$$W_{il} = \sum_{j=1}^{n} a_{ij} Z_{jl}. \tag{11.38}$$

We then put

$$\log\left(S_j(t_l)\right) = \log\left(S_j(t_{l-1})\right) + \left(r - \frac{1}{2}\sigma_j^2\right)(t_l - t_{l-1}) + \sigma_j \sqrt{t_l - t_{l-1}} W_{jl}. \tag{11.39}$$

For each j, we iterate through the values of l. In practice, we might absorb the volatilities and times into the correlation matrix to form a covariance matrix, and take the pseudo-square root of that instead, which would have the advantage of reducing slightly the number of computations necessary during each run of the simulation.

Having generated the path, we can then compute D's pay-off for that path. If D generates cashflows at varying times then we discount each one according to when it occurs. We now just repeat this algorithm for each path and average as usual.

11.6 The Margrabe option

The Margrabe option pays off at time t_1 the maximum of $S_2(t_1) - S_1(t_1)$ and zero. To price it by Monte Carlo is straightforward: we apply the method of the previous section using two normal draws for each path. However, this option is sufficiently simple that there are other tractable approaches.

The PDE to be satisfied is

$$\frac{\partial D}{\partial t} + r \sum_{i=1}^{2} S_i \frac{\partial D}{\partial S_i} + \frac{1}{2} \sum_{i=1}^{2} S_i^2 \sigma_i^2 \frac{\partial^2 D}{\partial S_i^2} + \rho S_1 S_2 \sigma_1 \sigma_2 \frac{\partial^2 D}{\partial S_1 \partial S_2} = rD. \quad (11.40)$$

It is possible to solve this by making some educated guesses. If the reader is practiced at such guesses, we urge him to try.

There is however a more financially appealing method of getting the solution. The payoff of the option is

$$\max(S_2(t_1) - S_1(t_1), 0) = S_1(t_1) \max(S_2(t_1)/S_1(t_1) - 1, 0).$$

This means that if we take S_1 as numeraire then the price of the option at time zero satisfies

$$\frac{D(0)}{S_1(0)} = \mathbb{E}\left(\left(\frac{S_2(t_1)}{S_1(t_1)} - 1\right)_+\right), \quad (11.41)$$

where the expectation is taken in the risk-neutral measure associated to S_1. We need to find the drifts of S_1, and S_2 in this measure. As in Example 6.1, the drift of S_1 will be $r + \sigma_1^2$, in order to ensure that the ratio of the money-market account to the numeraire has zero drift. The process for $1/S_1$ is therefore

$$d\left(\frac{1}{S_1}\right) = -\frac{r}{S_1}dt - \frac{\sigma_1}{S_1}dW_1. \quad (11.42)$$

If the risk-neutral drift of S_2 is μ_2 then applying Ito's lemma, we have

$$d\left(\frac{S_2}{S_1}\right) = \frac{S_2}{S_1}(\mu_2 - r - \rho\sigma_1\sigma_2)dt + \frac{S_2}{S_1}(-\sigma_1 dW_1 + \sigma_2 dW_2). \quad (11.43)$$

11.6 The Margrabe option

If we therefore put

$$\mu_2 = r + \rho\sigma_1\sigma_2, \tag{11.44}$$

then S_2/S_1 is driftless and hence a martingale.

To price the option we need to compute

$$\mathbb{E}\left(\left(\frac{S_2(t_1)}{S_1(t_1)} - 1\right)_+\right).$$

The ratio S_2/S_1 has zero drift and effective volatility

$$\bar{\sigma} = \sqrt{\sigma_1^2 - 2\rho\sigma_1\sigma_2 + \sigma_2^2}. \tag{11.45}$$

We can now directly evaluate this expectation by the same methods we used to derive the Black–Scholes formula. Indeed, one can make it into the same expectation by substituting variables. One obtains

$$D(S_1, S_2, t) = S_2 N(d_1) - S_1 N(d_2) \tag{11.46}$$

where

$$d_j = \frac{\log(S_2/S_1) + (-1)^{j-1}\frac{1}{2}\bar{\sigma}^2(t_1 - t)}{\bar{\sigma}\sqrt{t_1 - t}}. \tag{11.47}$$

We can see some interesting facts about the formula for D. It does not involve r. The price of a Margrabe option is therefore independent of the prevailing interest rates. Why does this happen? The pay-off after division by the numeraire S_1, is a function of S_2/S_1. This means, using the martingale representation theorem, that D's pay-off can be replicated purely by trading in S_1 and S_2. We never need to have any cash holdings which means that interest rates are irrelevant.

Suppose now that the market consists only of the stocks S_1 and S_2. Since we have shown that we can replicate D's pay-off by trading purely in S_1 and S_2, the non-existence of the money-market account should not matter. However, our derivation of the martingale measure drifts of S_1 and S_2 depended upon the existence of the riskless bond. If we do not have the bond then S_1 has arbitrary drift – there are an infinite number of equivalent martingale measures. The drift of S_2 is still determined by that of S_1 by a simple relationship, but in any case, we do not actually need to know the drift of S_2 since S_2/S_1 is always driftless. Thus whichever equivalent martingale measure we pick, the expectation of the pay-off of D is always the same.

What feature of the Margrabe option makes this work? The crucial point is that the payoff is homogeneous (of degree one) in the assets; that is

$$f(\lambda S_1, \lambda S_2) = \lambda f(S_1, S_2), \tag{11.48}$$

which implies that

$$f(S_1, S_2)/S_1 = f(1, S_2/S_1); \tag{11.49}$$

so the martingale representation theorem lets us represent $f(S_1, S_2)/S_1$ as an integral against S_2/S_1, which means that the pay-off can be synthesized using S_2 and S_1.

Clearly, there is nothing special about 2 in this argument, and the method can be adapted to work for any number of assets provided the pay-off is homogeneous in all of them, that is

$$f(\lambda S_1, \lambda S_2, \ldots, \lambda S_n) = \lambda f(S_1, S_2, \ldots, S_n). \tag{11.50}$$

One interesting aspect of the Margrabe option is that it is really an option on correlation. In (11.45), we can observe σ_1 and σ_2 from the prices of vanilla options. We can even use vanilla options to Vega-hedge our exposures to them. Thus the only quantity we do not have a concrete handle on is ρ. This means that when pricing a Margrabe option we are really making a market on ρ, in the same way that for a vanilla option we are making a market on the volatility. Note that if we know $\overline{\sigma}$ then we know ρ. This means that if we are pricing more complicated options which depend upon ρ, we can use the market prices of Margrabe options to infer a value of ρ to use. We can also use them to hedge our exposure to ρ.

11.7 Quanto options

A quanto option is, roughly, an option that pays off in the wrong currency. In practical terms, this means that some variable quantity is translated into another currency at a pre-determined fixed exchange rate. Usually this fixed exchange rate is just one for one.

The key to understanding quanto option pricing is to keep a firm grasp on what the tradable quantities are. Suppose we are a sterling investor; our unit of account is the sterling money-market account, and we have a quanto option on a US stock. For example, we have a call option on Microsoft, M_t, a US dollar stock struck at K that pays at time T, the sum of

$$\max(M_T - K, 0)$$

pounds not dollars.

The quantity M_t is a dollar tradable but not a sterling tradable. However, we can convert it into a sterling tradable by multiplying by the exchange rate to give it a price in sterling instead of dollars.

To price this option, we first identify the real-world processes involved. Let F_t denote the value of one dollar in pounds at time t. Let B_t denote the sterling money-market account which grows at continuous rate r. The dollar money-market

11.7 Quanto options

account, D_t, grows at a continuous rate d. We assume that F_t and M_t are log-normally distributed in the real-world measure. The processes are therefore

$$dB_t = rB_t dt, \quad (11.51)$$
$$dD_t = dD_t dt, \quad (11.52)$$
$$dF_t = F_t \mu_F dt + F_t \sigma_F dW_t, \quad (11.53)$$
$$dM_t = M_t \mu_M dt + M_t \sigma_M dZ_t \quad (11.54)$$

where W_t and Z_t are Brownian motions correlated with coefficient ρ.

We want to identify the risk-neutral processes associated with taking the sterling money-market account as numeraire. The exchange rate, F_t, is just the value of one unit of the dollar money-market account in sterling, and so is the same as a dividend-paying stock with dividend rate d. Its risk-neutral dynamics are therefore

$$dF_t = (r - d)F_t dt + F_t \sigma_F dW_t. \quad (11.55)$$

The real-world dynamics of $F_t M_t$, the sterling price of Microsoft, will be

$$d(F_t M_t) = dF_t M_t + F_t dM_t + dF_t dM_t,$$
$$= M_t F_t (\mu_F + \mu_M + \rho \sigma_F \sigma_M) dt + M_t F_t (\sigma_F dW_t + \sigma_M dZ_t).$$
$$(11.56)$$

The risk-neutral dynamics of $F_t M_t$ as a non-dividend paying sterling-denominated stock are

$$d(F_t M_t) = r F_t M_t dt + M_t F_t (\sigma_F dW_t + \sigma_M dZ_t). \quad (11.57)$$

We want the risk-neutral dynamics of $M_t = F_t^{-1}(F_t M_t)$. We have from Ito's lemma that

$$d(F_t^{-1}) = F_t^{-1}(d - r + \sigma_F^2) dt - F_t^{-1} \sigma_F dW_t. \quad (11.58)$$

Applying Ito's lemma again, we have

$$dM_t = d(F_t^{-1}(F_t M_t))$$
$$= (dF_t^{-1})(F_t M_t) + F_t^{-1} d(F_t M_t) + dF_t^{-1} d(F_t M_t)$$
$$= (d - r + \sigma_F^2 + r - \sigma_F^2 - \rho \sigma_F \sigma_M) M_t dt + \sigma_M M_t dZ_t$$
$$= (d - \rho \sigma_F \sigma_M) M_t dt + \sigma_M M_t dZ_t. \quad (11.59)$$

The risk-neutral dynamics of M_t therefore involve an adjustment factor to the drift which depends on the correlation between the price of Microsoft shares in dollars and the dollar/sterling exchange rate. The fact that the correlation has some impact is not surprising. Perhaps more surprising is that it only shows up in the risk-neutral drift. If Microsoft shares are totally uncorrelated with the exchange rate then their risk-neutral dynamics are totally independent of the volatility of the exchange rate.

Now that we have the risk-neutral dynamics, we can price simple quanto options. The simplest option is the quanto forward. This pays £$(M_T - K)$ at time T. Its value will be
$$e^{-rT}\mathbb{E}(M_T - K).$$
As M_T is log-normal, we know immediately that its expectation is
$$M_0 e^{(d-\rho\sigma_F\sigma_M)T}.$$
The value of the quanto forward is therefore
$$e^{-rT}\left(M_0 e^{(d-\rho\sigma_F\sigma_M)T} - K\right).$$
The quanto forward strike, the strike that makes the contract have zero value, is therefore
$$\tilde{M}_0 = M_0 e^{(d-\rho\sigma_F\sigma_M)T}. \tag{11.60}$$
The quanto call, $C(t)$, at time 0 will have value equal to
$$C(0) = \mathbb{E}(e^{-rT}C(T)). \tag{11.61}$$
This is equal to
$$e^{-rT}\mathbb{E}\left(\left(M_0 e^{(d-\rho\sigma_F\sigma_M)T - \frac{1}{2}\sigma_M^2 T + \sigma_M\sqrt{T}Z} - K\right)_+\right),$$
where the expectation is taken over the standard normal variable Z. Here we have used the usual solution to the log-normal stochastic differential equation. The quanto call price can be written in simpler form as
$$e^{-rT}\mathbb{E}\left(\left(\tilde{M}_0 e^{-\frac{1}{2}\sigma_M^2 T + \sigma_M\sqrt{T}Z} - K\right)_+\right).$$
This expectation can now be evaluated by the same procedure as for the Black–Scholes call price. It turns out be to equal to
$$e^{-rT}\left(\tilde{M}_0 N\left(\frac{\log\left(\frac{\tilde{M}_0}{K}\right) + \frac{1}{2}\sigma_M^2 T}{\sigma_M\sqrt{T}}\right) - KN\left(\frac{\log\left(\frac{\tilde{M}_0}{K}\right) - \frac{1}{2}\sigma_M^2 T}{\sigma_M\sqrt{T}}\right)\right).$$

11.8 Higher-dimensional trees

A powerful technique in one dimension was the recombining tree. We can adapt trees to higher dimensions with a little work. The implementation becomes more fiddly but conceptually there is little change. There are however some added complications which limits their use.

11.8 Higher-dimensional trees

We could use trees, as we did in the one-dimensional case, to justify risk-neutral valuation. However, we have already justified risk-neutral valuation using stochastic calculus arguments so we focus instead on their utility for implementing option pricing models.

We concentrate on the two-dimensional case as it illustrates the concepts well without being too technical. Our objective is to create a discrete process which converges to the stock price processes as the step-size tends to zero. We have some choices about how to do this.

We have two stocks following correlated log-normal processes. The stocks will be driven by a two-dimensional Brownian motion and the state will be determined by the value of Brownian motion at a given time. We can therefore regard either the Brownian motion or the stocks as the fundamental process. As usual, we also have the choice of whether to work with the log-processes for the stocks. Denote the stocks by S_j and let the correlation between them be ρ. We assume they follow risk-neutral processes

$$dS_j = rS_j dt + \sigma_j S_j dZ_t^j, \tag{11.62}$$

and the correlation between Z^1 and Z^2 is ρ.

Thus suppose our Brownian motion is

$$W_t = (W_t^1, W_t^2). \tag{11.63}$$

We can synthesize Z from W in the usual manner, for example

$$Z_1^t = W_t^1, \tag{11.64}$$

$$Z_2^t = \rho W_t^1 + \sqrt{1-\rho^2} W_t^2. \tag{11.65}$$

The stock price at time t is then, of course, equal to

$$S_j(t) = S_j(0) e^{(r-\frac{1}{2}\sigma^2)t + \sigma\sqrt{t} Z_t^j}, \tag{11.66}$$

which means that the stock price is a function of W_t.

This all means that we only need to construct the tree for W. A simple first approach is to consider the two components of W separately. Let the time step-size be Δt. In each step the increment of a Brownian motion has mean zero and variance Δt, so, copying the one-dimensional case, each component can move up or down by $\sqrt{\Delta t}$.

Just as in the one-dimensional case, the Central Limit theorem guarantees that the process for each coordinate converges to Brownian motion. As the draws for each coordinate are independent of each other, the Brownian motions for each coordinate will also be independent of each other and hence we obtain a two-dimensional Brownian motion. We can now proceed to pricing just as in the one-dimensional case. First, we fill in the pay-offs in all the terminal nodes, and then at each preceding time slice we assign a value depending on the discounted expectation over the daughter nodes and the current stock prices. The discounted expectation is of course just $e^{-r\Delta t}$ times the average of the values at the discounted nodes. As in the one dimensional case early exercise features are easy to account for.

Note that for the first time step, we have four possible states

$$(\sqrt{\Delta t}, \sqrt{\Delta t}), \quad (\sqrt{\Delta t}, -\sqrt{\Delta t}), \quad (-\sqrt{\Delta t}, \sqrt{\Delta t}), \quad (-\sqrt{\Delta t}, -\sqrt{\Delta t}),$$

and at the jth step we have j^2 states, since we have j states for each of the two components. The number of points in our tree is therefore growing quite rapidly even in two dimensions. In n dimensions, we would have j^n possible states.

One curious aspect of the tree we have constructed in two dimensions is that it implies an incomplete market as at each step the state can move to four possible new states whilst we have only three hedging instruments. This does not matter in that we are only using the tree to approximate the already arbitrage-free risk-neutral measure rather than to deduce the uniqueness of risk-neutral prices. However, it does suggest that we ought to be able to approximate by a complete discrete process, which would only involve three daughter nodes rather than four which would result in a smaller total number of states.

The key to reducing the number of nodes is to consider both components simultaneously. We need a random variable $X_{\Delta t}$ such that in a small step, the mean in each component is zero, the variance in each component is Δt, and the covariance between the components is zero. Let's look for a solution in which the three daughters nodes are assumed with equal probabilities. As we have three daughter nodes, we are really talking about a triangle. The condition on the mean says that the centroid (i.e. the centre of gravity) of the triangle is the origin. As everything is reasonably symmetric the obvious thing to try is an equilateral triangle centred at the origin.

Thus we try, for a time step Δt, the points

$$\alpha(0, 1), \quad \alpha\left(\frac{\sqrt{3}}{2}, -\frac{1}{2}\right), \quad \alpha\left(-\frac{\sqrt{3}}{2}, -\frac{1}{2}\right),$$

where α is to be solved for, and each point is to be taken with equal probability. Let X_j be the jth coordinate. The expectation of each X_j is clearly zero. We have

$$\mathbb{E}(X_1^2) = \alpha^2 \frac{1}{3}\left(\frac{3}{4} + \frac{3}{4}\right) = \frac{\alpha^2}{2}, \tag{11.67}$$

$$\mathbb{E}(X_2^2) = \alpha^2 \frac{1}{3}\left(1 + \frac{1}{4} + \frac{1}{4}\right) = \frac{\alpha^2}{2}, \tag{11.68}$$

$$\mathbb{E}(X_1 X_2) = 0. \tag{11.69}$$

Thus taking $\alpha = \sqrt{2\Delta t}$, we have mean zero, and variance Δt and covariance zero. This is all we need.

Across each time step, now let the asset evolve according to the three possible moves. The tree will still recombine as the ordering of the moves does not affect the final location. We can therefore proceed as before without problems. We let Y^j be a sequence of independent two-dimensional random variables with the distribution of (X_1, X_2) for all j. The discretization of the process for the stocks, $(S^{(1)}, S^{(2)})$ is therefore given by the process

$$\log S^{(i)}_{(j+1)\Delta t} = \log S^{(i)}_{j\Delta t} + \left(r - 0.5\sigma_j^2\right)\Delta t + \sigma_i\left(a_{i1}Y_1^j + a_{i2}Y_2^j\right), \quad (11.70)$$

where (a_{ij}) is any Cholesky decomposition of the correlation matrix. To see this process is a discretization of two-dimensional Brownian motion, we simply invoke the two-dimensional Central Limit Theorem, just as we did in the one-dimensional case.

This technique can be extended easily to higher dimensions by using a tetrahedron, or higher-dimensional analogue, with the appropriate scaling to get the variances correct. As the number of dimensions becomes higher it will soon become impractical, however, because of the huge number of nodes needed to fill out space.

11.9 Key points

In this chapter, we have extended the Black–Scholes theory from a single uncertain to several correlated assets. We have seen that whilst the details are more complex, the fundamental theory is essentially the same.

- Many derivatives have a pay-off which is dependent on the evolution of several stocks.
- A Margrabe option is an option to exchange one stock for another.
- A quanto option is an option that whose pay-off is transformed into another currency at pre-determined rate.
- A multi-dimensional Brownian motion is a vector of processes which have jointly normal increments and is a Brownian motion in each dimension.
- Correlated Brownian motions can be constructed by adding together multiples of one-dimensional Brownian motions.
- The Ito calculus goes over to higher dimensions with the additional rule $dW_j dW_k = \rho_{jk} dt$ where ρ_{jk} is the correlation between W_j and W_k.
- When adding correlation Brownian motions we can find the volatility of the new process by treating the original processes as vectors.
- We can change the drift of a multi-dimensional Brownian motion by using Girsanov's theorem.

- No arbitrage will occur if and only if the discounted price processes can be made driftless by a change of measure.
- We can price by risk-neutral expectation.
- Monte Carlo has the advantage in high dimensions that the rate of convergence is independent of dimension.
- An analytic formula can be developed for the Margrabe option which does not involve interest rates.
- Quanto options can be priced analytically using a modified Black–Scholes formula.
- Trees can be adapted to higher dimensions by placing the nodes on a triangle or tetrahedron.

11.10 Further reading

A recent paper discussing mathematical finance from the point of view of homogeneity is [63].

The original paper on Margrabe options is [91].

A recent paper on higher-dimensional trees is [98]: the authors discuss paradigms for assessing the appropriateness of a given discretization and suggest an icosohedral method in three dimensions.

11.11 Exercises

Exercise 11.1 Suppose we are a dollar investor. The stock we wish to buy is priced in pounds. How would we price a call option on the stock which has a strike in pounds?

Exercise 11.2 Suppose we are a dollar investor. The stock we wish to buy is priced in pounds. How would we price a call option on the stock which has a strike in dollars?

Exercise 11.3 Suppose

$$dS_j = \mu_j S_j dt + S_j \sigma_j dW_t,$$

with the same Brownian motion for $j = 1, 2$. The riskless bond grows at a constant rate r. What relation must hold between μ_j and σ_j to prevent arbitrage?

12

Options with early exercise features

12.1 Introduction

When discussing options, we have so far concentrated on the case of a European option which allows the purchase of the underlying asset on a specific date at an agreed price, or more generally we have considered path-dependent exotic options which do not involve any choice on the part of the holder. The problem of valuing an option when the exercise date is not fixed is considerably trickier. Recall that an option is said to be *American* if it can be exercised at any time before expiry. An option is said to be *Bermudan* if it can be exercised on any one of a fixed set of dates. Whilst Bermudan options are not common in the equity and FX markets, they are very common in the interest rates markets.

We have seen that it is never optimal to exercise an American call option on a non-dividend paying stock before expiry as the value of a European option must always exceed its intrinsic value – that value obtained by exercising immediately were it possible. However, this is not true for a put option nor for a call option on a dividend-paying stock. Pricing must therefore take into account the extra exercise rights and will involve the question of how to make an exercise decision.

Making an exercise decision is simple once the value of an American option is known. A rational investor will exercise if and only if he makes more money by exercising the option. One has to be a little careful about what one means by making more money. No arbitrage implies that the value of an American option will be at least as much as the intrinsic value. Otherwise, one would just buy the American option and exercise it immediately, making an instantaneous profit on the difference in values. If the American option's value is greater than the intrinsic value, one would clearly not exercise as one could sell the option in the market for more money than is to be had by exercising it. This leaves us with the case where the two values are equal. In these cases, we exercise and it would be an error not to do so. The reason is that once the option has been exercised, we hold some cash which will grow at the risk-free rate whereas the rights granted by the option will

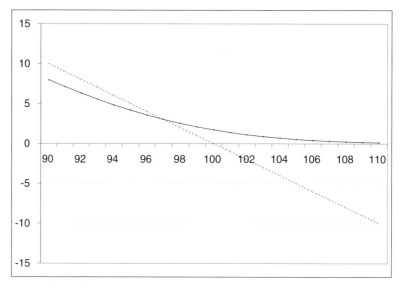

Fig. 12.1. The value of a six-month European put option struck at 100 and the value that would be obtained by exercised today were that possible.

decrease with time. So although the values are equal the time derivatives are not. An alternative way of looking at this is that once one has made the decision not to exercise for a certain very short period of time, then the value of the option need no longer be more than the intrinsic value as one has given up the rights that enforce this no-arbitrage inequality.

Unfortunately, the fact that knowing the value of an American option implies knowledge of when to exercise is not very helpful since computing the value typically depends on knowing when to exercise. To illustrate this, let's consider a simple Bermudan option. Suppose we have a one-year put option but with the opportunity to early exercise after six months. We work in a simple Black–Scholes world with volatility 10%, spot 100, strike 100, and continuously compounding interest rate of 5%. After six months, we clearly do not early exercise if spot is greater than 100. If it is less than 100 then we have a choice: do we prefer the cash we receive by exercising today or do we keep the option hoping to make more money that way. In this case, we can simply compute the two values. The value by exercising today is simply 100 minus spot, whereas the value by not exercising is the price of a six-month put option struck at 100 with today's spot. We exercise according to whichever is bigger. The cross-over point in this case is just below 97: we exercise if spot is below the cross-over point and hold otherwise. See Figure 12.1.

Note the general principle here that the value of an option at an exercise opportunity is the maximum of the exercised value and the unexercised value.

How do we value the original option? We know the value after six months as a function of spot: it is just the maximum of the exercised and unexercised prices. So we can treat the option as a European contingent claim with six-month expiry and payoff equal to the value at six months of the Bermudan. The price can therefore be found by any of the methods for valuing a European contingent claim.

A slightly more complicated option having two early exercise dates after four and eight months can be valued similarly. After eight months the value is the maximum of the intrinsic and a four-month option. We can then, treating this as a European contingent claim, back out the value after four months. The value at four months is then the maximum of the intrinsic and the unexercised option. Note that we will have to compute the price of the unexercised option for every possible value of spot at four months. We then solve back to time zero to get the original value.

This procedure can easily be extended to any number of exercise dates. The crucial point to note is that it is a backwards method. First one computes the values at the final exercise date, then the second final date, then the third final and so on. The reason for this backwardsness is that one cannot make an exercise decision without knowledge of the unexercised value which requires knowing the values for the later-dated exercise opportunities.

The consequence of this in practical terms is that it is more natural to price American and Bermudan options using trees and PDE methods than to use Monte Carlo simulations. To emphasize this point, consider using a risk-neutral evolution of the spot to price an American option. We divide time into lots of little steps. After each step, we either exercise or we do not. If we exercise then we store the exercised value discounted back to today, otherwise we proceed to the next step. We repeat this until we reach the final maturity of the option. The problem is that we have to give our computer an exercise strategy. For example, our strategy could be "exercise if and only if the option is in the money" or "exercise if and only if the exercised value is greater than the price of a European option with the same maturity." Thus to each strategy, we associate a price. But our price should not be strategy-dependent – after all if we have sold an option how can we be sure the purchaser is using the same strategy.

The price of the American option should therefore be the maximum of the prices obtained by any *admissible* exercise strategy. What do we mean by admissible here? Consider the exercise strategy, "exercise when the spot reaches its maximum value on the path." Clearly this is unreasonable as the holder of the option would not be able to know at a given time whether the maximum had been reached. So the decision to exercise should be based solely on the information available at the time of exercise. In mathematical terms, the time of exercise is a *stopping time*. The price of an American option, O, should therefore be the maximum over the set of stopping times of the risk-neutral expectation of the pay-off at the stopping time. If we let B_t denote the continuously-compounding money market account and K be the strike; then we can write this as

$$O = \max_{\tau} \mathbb{E}\left(B_\tau^{-1}(S_\tau - K)\right), \tag{12.1}$$

where the maximum is taken over the stopping times τ. (The reader who knows the difference between a supremum and a maximum should use supremum here.) We are of course evolving S in a risk-neutral log-normal fashion.

12.2 The tree approach

Unfortunately, this has not bought us a huge amount in terms of actually pricing the option! We cannot run a Monte Carlo simulation for every possible exercise strategy. We will return to the issue of Monte Carlo pricing below but now turn to the nicely adapted method of binomial trees. As trees are a backward method, the problems of exercise immediately disappear. We exercise if and only if the exercised value is greater than or equal to the unexercised value which has already been computed. What does this mean in practical terms? We develop a binomial tree in the same way as for a European option except that at each node, rather than setting the value to the weighted average of its daughters nodes suitably discounted, we set the value to the maximum of the intrinsic value and the same weighted average as before. The additional computational difficulty is tiny.

For example, consider a two-period model with initial spot 100, continuously compounding interest rates of 10%, and volatility 20%. We take each step to be one year. Our spot values after one year are then 132.3 and 88.7. After two years they are 175.1, 117.4 and 78.7. We take our option to be a put with strike 106. See Figure 12.2.

The values in the terminal nodes for both European and American options are 0, 0, 27.3. In the middle layer, the European values are 0, 15.0 whereas for the intrinsic values are 0, 17.3. So the American values are 0, 17.3.

The value at time 0 is then the maximum of 6, the initial intrinsic value, and 8.27, the discounted unexercised value. We conclude that immediate exercise is not optimal and the initial value of the option is 8.27.

Our procedure for a general American option is now clear. We build an n-step tree. Working backwards, we assign to each node in the final layer the intrinsic value. In each previous layer, we assign to each node the maximum of the intrinsic value and the discounted expectation of the values of the daughter nodes in the succeeding layer. The value at the base node is then our American option price. In practice, as the tree is an approximation to the underlying Brownian motion which becomes more accurate with increasing n, we would want to compute the price

12.3 The PDE approach to American options

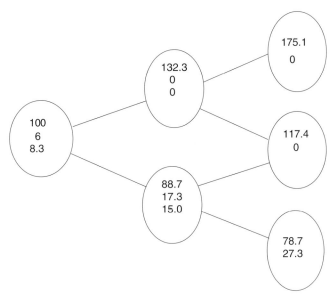

Fig. 12.2. A two-step tree for an American option.

for several values of n to make sure that the price has converged as a function of n. One advantage we have for American options is that we can use the European price as a control. The idea is similar to the control variate approach to Monte Carlo. We know the correct price of the European option, so we know how large the error is for the European price on the tree. We assume that the American price is wrong by the same amount and adjust accordingly. Whilst this will not make the American price totally correct, it will make it more accurate. We illustrate this for an American put option in Figure 12.3.

If we take a finely branched tree and mark the nodes according to whether exercise has occurred, we find that the spot-time domain is neatly divided into two regions. On one side of a smooth curve, we exercise on the other we do not. The area where exercise would occur is known as the *exercise region*. The dividing line is called the *exercise boundary*.

Note that the dividing curve will always pass through the strike at the expiry time, as the decision to exercise there is trivial. Its distance from the strike will increase as a function of time to maturity as the value of optionality increases with time to maturity – the earlier we exercise the more rights we give up.

12.3 The PDE approach to American options

Given that PDE approaches are backwards, they are a natural way to price an American option. In the domain of non-exercise, there is no difference from the vanilla case. If we Delta-hedge the option to cancel the Brownian part, then the

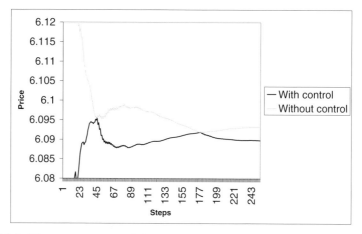

Fig. 12.3. The convergence of the price of an American put option price on a binomial tree with smoothing by averaging odd-step prices with even ones. The graphs are with and without the European put as a control variate. Spot and strike are 100. Interest rates are 5%. Expiry is one year and volatility is 20%. The correct price is 6.090.

same arguments work and the value of our option must satisfy the Black–Scholes equation. The difference is in the boundary conditions. The boundary condition at expiry is as before. However, we now have a second boundary, the edge of the exercise domain. And we do not even know where this boundary is.

Our problem is therefore to identify the location of the boundary and to identify the appropriate boundary conditions on it. The first obvious boundary condition is that the unexercised value should equal the exercised value there. Once the boundary is known this is enough to fix a solution to the Black–Scholes equation and hence a price. Of course, we do not know the location of the boundary so this is of little help.

The boundary is determined by a second boundary condition, which is that the Delta of the option must be continuous across the boundary. Inside the exercise domain, the Delta is trivial to calculate: it is just the derivative of the pay-off. Thus for a put it is -1 and for a call $+1$. Why is this the correct condition? We focus on the put case for simplicity. Let X be the point of exercise at time t and f the solution of the Black–Scholes equation. If the derivative of the unexercised value is lower than -1 on approach to X from above, we can write

$$f(X + \epsilon) = f(X) + f'(X)\epsilon + \mathcal{O}(\epsilon^2). \tag{12.2}$$

Of course, $f(X) = K - X$, so we have

$$f(X + \epsilon) = K - X + f'(X)\epsilon + \mathcal{O}(\epsilon^2) < K - (X + \epsilon) - \delta\epsilon - \mathcal{O}(\epsilon^2), \tag{12.3}$$

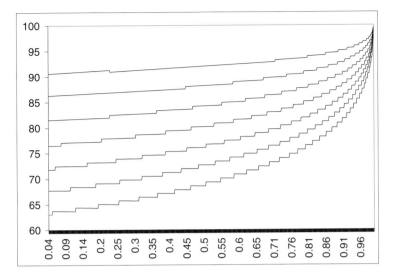

Fig. 12.4. The exercise boundary for an American put deduced from a binomial tree for various volatilities. Spot is 100, strike is 100, r is 5%, maturity is 1 year, and volatilities ranges from 12% to 36% going downwards. The graininess arises from the method of estimation.

where we have put $f'(X) = -1 - \delta$, with $\delta > 0$. For ϵ sufficiently small, this implies that $f(X + \epsilon) < K - (X + \epsilon)$, which means that the exercised value is greater than the unexercised value. This is impossible as we are inside the domain of non-exercise. We therefore conclude that $f'(X) \geq -1$.

What if $f'(X)$ is greater than -1? We show that a smaller value of X would lead to greater value at X. The solution of the equation will vary continuously with X. Let $X_\epsilon = X - \epsilon$, and let f_ϵ be the solution associated with moving the boundary smoothly, so that at time t it is $X - \epsilon$. Then for ϵ sufficiently small we will have $f'_\epsilon(s) > -1$ on the interval $[X - \epsilon, X]$. Hence there exists a positive δ such that $f'_\epsilon(s) > -1 + \delta$ on that interval.

It follows from the mean-value theorem that

$$f_\epsilon(X) > K - X + \epsilon + \epsilon(-1 + \delta) = K - X + \epsilon\delta > K - X. \quad (12.4)$$

Moving the boundary back by ϵ has therefore increased the value of the option at X. This shows that the boundary was not in an optimal position. We therefore conclude that for the boundary to be optimally placed we must have $f'(X) = -1$.

To summarize, we have shown that the price of an American option satisfies the Black–Scholes equation in the domain of non-exercise. It also satisfies three boundary conditions. At expiry it must agree in value with the pay-off for all values of spot. On the boundary of the domain of non-exercise, it must agree with the pay-off and it must have derivative equal to the derivative of the pay-off also. If one knew the location of the boundary this would be too much data, as specifying the value at the boundary is enough to fix the solution uniquely. Instead we have a free boundary-value problem – the boundary is determined by being that boundary for which the solution has derivative equal to that of the pay-off. Whilst there is a theory of such free boundary-value problems, it is not sufficiently well-developed (or more likely it is not possible to develop the theory sufficiently) to give us an analytic solution. Indeed, developing an analytic solution for the price of an American put option is one of the great unsolved problems of mathematical finance. (Warning to PhD students – do not take this as inspiration!)

We therefore must proceed numerically. In fact, the solution can then be found quite simply. We use essentially the same procedure as in the tree case. Using an explicit finite difference method, the value of the option is simulated on a grid and backwards inducted. This is the same procedure as in the European case. The only difference now is that, as in the tree method, at each point of the grid we take the maximum of the exercised and unexercised values, before proceeding to the next layer. Here, as in the tree case, the crucial point is that all the future values are known before the exercise decision is decided, which is possible provided we are using a method which is backwards in time. If one uses implicit finite difference, the algorithm becomes trickier.

We refer the reader who is further interested in PDE approaches to solving American options to [118] or [119].

12.4 American options by replication

An alternative method for carrying out the numerical integration is to approximate the American pay-off by using solutions of the Black–Scholes equation. From a theoretical point of view, this constitutes approximating the American option by a portfolio of European options.

Suppose our American put option has expiry at time T. We divide time into segment $t_0 = 0 < t_1 < t_2 < \cdots < t_n = T$. At time t_n, the value is just the pay-off. Our first approximation is therefore the European option with the same pay-off and expiry. Call this option O_n. At time t_{n-1}, we can compute the value of O_n precisely: it is just the solution of the Black–Scholes equation with expiry time $t_n - t_{n-1}$.

Let $\mathrm{BS}(S, t, \sigma, K)$ denote the solution of the Black–Scholes equation for a put option with strike K, spot S, volatility σ and expiry t.

We numerically find the point where the value of O_n becomes less than the exercised value. That is the value of spot where

$$\mathrm{BS}(S, t_n - t_{n-1}, \sigma, K) = K - S.$$

12.4 American options by replication

We take this point, X_{n-1}, to be the exercise boundary at time t_{n-1}. In the unexercised domain, the value of our American put will be well approximated by the value of O_n but in the exercised domain we want it to be $K - S$.

If we imagine actually hedging an American option then the important thing is not that the value of the American option is $K - X_{n-1}$ in the exercised area since it will be exercised as soon as the domain is entered. Instead the important issue is that the derivative of the American put at the exercise boundary is -1 as we showed in the previous section. We therefore adjust our portfolio in such a way as to make this be true. We therefore add a European put option struck at X_{n-1} with notional equal to 1 plus the Delta of the European option. That is chosen so that the sum of the negative of the notional and the Delta of the European option expiring at time t_n, is -1. Denote this new European option by O_{n-1}. The Delta of the portfolio at $t = t_{n-1}$, and $S = X$ is of course not defined since the pay-off of a vanilla option is not differentiable at the strike. However the Delta of the portfolio as S converges to X from below is -1, as desired.

The portfolio P_{n-1}, consisting of O_n and O_{n-1}, is easy to price at any previous time as the sum of two European options. We can therefore repeat this operation at time t_{n-2}. We find the point X_{n-2} where the value of P_{n-1} becomes less than $K - S$. We compute the Delta of P_{n-1} there and add in an option O_{n-2} with the appropriate notional to make the Delta of the extended portfolio, P_{n-2}, equal to -1 as S tends to X from below.

We now just iterate this procedure back to $n = 0$. The value of our American put is then taken to be the value of O_0 at today's spot at time 0. All the Greeks are also just the Greeks of the portfolio.

The remarkable aspect of this approach is the speed of convergence. A very accurate price can be obtained with only 8 steps as opposed to about the 50 or more that are required for a tree or a finite difference method. The reason for this effectiveness is that we are making good use of the structure of the Black–Scholes equation. Our approximating functions are already solutions so we not need to do small stepping to compute their values. We quote some results from [70]. For a two-year American call option, spot 100, strike 105, volatility 11.35%, domestic rate 4.25%, and dividend rate 6.5%:

> 2-step replication   2.776;
> 4-step replication   2.853;
> 8-step replication   2.879;
> PDE price (501 × 507 grid)   2.878.

The approach given here is of course related to the replication method for pricing continuous barrier options in Chapter 8. Here, as there, an important aspect of the method is the fact that we only replicate up to the exercise boundary and not beyond it. We are able to do so by a non-arbitrage trading strategy argument. We start off holding the approximating portfolio; we continue to hold it until the spot crosses the exercise boundary or the expiry time is reached. If the spot touches the exercise boundary, the American option is exercised and the portfolio is liquidated. Assuming our method works reasonably, these two values are almost equal and in the theoretical limit as the step-size goes to zero they are equal. In any case, the portfolio no longer exists behind the boundary so replication there is both unnecessary and irrelevant.

It is worth noting that we are really only using two properties of the Black–Scholes model for this argument. The first is the continuity of the paths that are followed by the spot price which allows us to liquidate on the boundary. The second is the deterministic nature of the pricing function. This means that as well as knowing the value of a vanilla option for any spot, strike and expiry today, we also know today what the value of a vanilla option will be at any time in the future for any spot, strike and expiry.

Both these properties are retained by models allowing the volatility to be a deterministic function of time and spot. However, the second property fails for fully stochastic volatility models, as future prices will depend on the stochastically-evolved value of volatility. On the other hand, for jump-diffusion models, the first property fails as the spot can jump across the boundary. A jump-diffusion model in which the jumps were only one-way and could not cross the boundary would, however, satisfy both properties.

12.5 American options by Monte Carlo

After having demonstrated above the great superiority of backwards methods for pricing American-type options, let us now look at the use of the classic forward method Monte Carlo simulation. Why would we want to? When working with sufficiently complicated multi-asset models, it is not clear how to apply tree and PDE methods. For example, suppose we have two assets with different time-dependent instantaneous volatility curves and an American 'selector' put option which allows us to sell either one of the two with strike K at a time of our choice. It thus has pay-off,

$$\max(K - S_1, K - S_2, 0). \tag{12.5}$$

We therefore need to simulate both assets' price movements simultaneously.

The problem with a tree method is that the time-dependence of the instantaneous volatility curves destroys the recombining nature of the binomial tree. To see this consider a single asset. Let the asset follow a risk-neutral log-normal process,

$$\frac{dS}{S} = r\,dt + \sigma(t)\,dW \tag{12.6}$$

which we approximate by discrete binomial steps. Suppose that in the first time step the volatility is σ_1 and in the second it is σ_2. Let a time step be Δt long.

After an up-move and then a down-move we have

$$\log\left(S^{ud}_{2\Delta t}\right) = \log(S_0) + \left(2r - \frac{\sigma_1^2 + \sigma_2^2}{2}\right)\Delta t + (\sigma_1 - \sigma_2)\sqrt{\Delta t}, \tag{12.7}$$

whereas after a down-move and then an up-move we have

$$\log\left(S^{du}_{2\Delta t}\right) = \log(S_0) + \left(2r - \frac{\sigma_1^2 + \sigma_2^2}{2}\right)\Delta t + (\sigma_2 - \sigma_1)\sqrt{\Delta t}. \tag{12.8}$$

So $(S^{du}_{2\Delta t})$ equals $S^{ud}_{2\Delta t}$ if and only if σ_2 equals σ_1. Thus the time-dependence of volatility stops the tree from recombining. The number of branches of a non-recombining tree grows exponentially with the number of time steps, which renders them impractical in general.

For a single asset option, there is a trick which removes this problem; we simply vary the size of our time steps so that $\sigma\sqrt{\Delta t}$ is constant. The tree then recombines and the problem disappears. However when valuing an option dependent on two assets with different time-dependence in their volatilities, we no longer have this way out; rescaling time can only flatten one of the volatility curves, not both. Thus to value the 'selector option' we need an alternative approach. In case the reader feels this option is a little contrived, I want to stress that this sort of problem arises very naturally when evaluating interest-rate options with early exercise features.

Now that we are convinced that there is some point to pricing an option with early exercise opportunities via Monte Carlo, let us return to the simple American put option. Once an exercise strategy is chosen, valuation is simple. We simply divide time into a large number of steps. Evolve the asset along the time steps. At each time step, we consult the exercise strategy, if it says exercise we return the exercised value at that time step suitably discounted back to time 0, otherwise we proceed to the next step. If we reach expiry then we return the final pay-off suitably discounted. We do this for a large number of paths and take the average.

The American price is of course the maximum over all possible exercise strategies and it is not possible (let alone practical) to try them all. However, we do not need to try them all. We only need to find the best one. We do not even really need to find the best one, just one that's good enough to give the correct price to a high degree of accuracy. This is not so unreasonable. In addition, we are helped by the fact, from elementary calculus, that the behaviour of a function around a critical point is quadratic so a small error ϵ in strategy translates into a negligible error of order ϵ^2 in price.

For the American put, we know that the exercise strategy is of the form 'exercise at time $t < T$ if and only $S(t) \leq X(t)$' where $X(t)$ is a smooth curve. We also expect that $X(t)$ should increase with t as fewer rights are given up by early exercise the less time there is left to expiry. It is also clear that the exercised value must be greater than the price of a European put with the same expiry and strike.

This all means that we can guess a parametrization form for the curve $X(t)$ and see how well it does. If expiry is at time T, we might therefore try

$$X(t) = \alpha - \beta(T - t)^\gamma, \qquad (12.9)$$

with α, β, γ all positive. We can then use an optimization method to find the best exercise strategy of this form.

If we were to practically implement such a method, we would probably want to generate a reasonable number of paths, say around ten thousand, and store them. We would then using that set of paths, find the average realized price as a function of α, β and γ. Moving α, β and γ around we would search for the optimal parameters for the exercise for that fixed set of paths. Using the fixed set of paths reduces the random noise one would otherwise obtain when trying to assess sensitivities to changes in α, β and γ. Having found an optimum, we would then run a second Monte Carlo simulation with different paths using the optimal strategy and see whether the predicted optimal value was actually realizable. If it is, then this will be a reasonably good predictor for the price. If it is not then the original set of paths was probably either too small or somehow biased and we need to run with a larger base set of paths.

The implementation of the optimization is slightly tricky in that if one moves α, β and γ by very small amounts then it is possible that none of the ten thousand paths change time of exercise, and we get zero derivative. If we use an optimization method relying on differentiation then we have to be sure that our finite differencing width is sufficiently large to stop this occurring. Alternatively, one could such a simplex-type method which does not require the derivatives but might converge more slowly. See [104] for implementations of optimization methods including the simplex method.

Of course, we cannot be sure by exercise strategy estimation that we have found the right price, we can only be sure of having found a lower bound for the price. Nevertheless, in settings where other methods are impractical, Monte Carlo simulation can provide a good way of estimating the price of an option with early exercise features. We discuss these techniques further in Chapter 14.

12.6 Upper bounds by Monte Carlo

The method of the previous section can be very effective but can only give lower bounds for the price of an American option. In this section, we develop a method for finding upper bounds for American and Bermudan option prices by Monte Carlo due to Rogers [110].

The arbitrage-free price for an American option is equal

$$\sup_{\tau} \mathbb{E}(e^{-r\tau} f(S_\tau)),$$

where the supremum (or maximum) is taken over all stopping times τ. If we increase the set of random times then the price can only go up. If we allow all random times, we get a higher price. One random time clearly gives the highest price: the time defined by being the point of optimal exercise along the path when foresight is allowed. Thus we have that

$$\mathbb{E}(\max(e^{-rt} f(S_t)))$$

is an upper bound.

Unfortunately, it is not a very good upper bound. Allowing the holder to see the future means that the price is much higher and the estimate is not particularly useful. How can we tighten the upper bound? If we take a martingale, M_t, of initial value zero, that is the discounted price process of a portfolio of initial value zero, we can subtract before taking the maximum. Because M_t is a martingale the initial value of the American option is equal to

$$\sup_{\tau} \mathbb{E}(e^{-t\tau} f(S_\tau) - M_\tau),$$

and when we pass to exercising with maximal foresight we still get a higher number, so we have that

$$\mathbb{E}(\max(e^{-rt} f(S_t) - M_t))$$

is still an upper bound for the price.

A slight subtlety here arises from the fact that we must take the expectation of M_t at some point. However, it is possible not to exercise an American option at all. One solution is to take f to be the positive part of the pay-off so for a call we take the pay-off at maturity time to be

$$(K - S_t)_+,$$

rather than $K - S_t$. This means that we can assume that the American option is always exercised at some point; for if it has not been exercised before time T its value will be zero which is the exercised value there.

So whatever martingale, M_t, with M_0 equal to zero, we pick we get an upper bound. We therefore search for a best choice. It is a theorem of Rogers that there exists a choice which makes the upper bound equal to the price of the option. Unfortunately, his proof is non-constructive and depends on knowing the price process for the American option which means we cannot directly apply it. We can interpret Rogers' result to say that we can hedge an American option perfectly by investing its value at time zero and trading appropriately. This hedging will be effective even if the option holder is exercising with maximal foresight.

This is a little surprising: even if the holder can see the future we can hedge our exposure to his exercise strategy. However, the seller's price ought to be enough to cover against any exercise strategy or we are not truly hedged. As in option pricing we only live once, we have to be hedged against the the possibility that the buyer's ineptitude by luck imitates seeing the future. Thus a seller's price that did not allow hedging against maximal foresight would not be sufficient.

Whilst Rogers' result is non-constructive, we can optimize by picking a family of portfolios $M_t(\alpha)$ depending upon a parameter or parameters, α. As in the lower bound case, we generate a set of paths and then carry out an optimization of the price by varying α. What sort of portfolio would we choose for α? For an American put option, the obvious M_t is a European put option minus cash bonds equal to its initial value. We can make α the notional of the contract and then optimize. This approach can generate tight upper bounds. As with the lower bounds, the main advantage of this technique lies in its applicability to interest-rate options, particularly Bermudan swaptions, and to multi-asset options rather than for the pricing of single-asset options. We discuss the implementation of this method for a Bermudan swaption in Section 14.10.

12.7 Key points

- An American option can be exercised at any time before expiry.
- A Bermudan option can be exercised on any one of a finite number of dates.
- An American option is always worth at least as much the underlying European option.
- Exercise strategies are interpreted mathematically as stopping times since they must depend on the information available at the time.
- Trees and PDEs are natural methods for pricing American options as they are backwards methods.
- Backwards methods are better for pricing American options because they naturally incorporate the unexercised value of the option.
- The PDE problem for the American option is a free boundary problem where the boundary is not determined but instead the value of the function and its derivative are determined at the boundary.
- We can get lower estimates for American option prices by picking an exercise strategy and then pricing by Monte Carlo.

- We can get upper bounds for American options by allowing exercise with maximal foresight on portfolios consisting of the American option minus a European option.

12.8 Further reading

The tree approach to American options was developed by Cox, Ross & Rubinstein, [36].

For further discussion of the theory in the PDE approach we refer the reader to [119].

The replication approach discussed here was introduced in [70].

A good overview of various methods for pricing American options on multiple assets is Chapter 11 of [27]. Tilley develops a method based on the bundling together of paths in [115]. Broadie & Glasserman develop two different methods in [25] and [26].

There are quite a few approaches to early exercise under Monte Carlo. The two we have given have the virtue that we can be sure that they bracket the price, whereas many other methods have biases but it is not always clear in which direction. A good overview is Chapter 16 of [45]. See also [48] for numerical comparisons of various methods. There is also discussion of various aspects in [68].

The upper bounds method we have developed here is due to Rogers, [110]. A similar approach is developed in [56].

12.9 Exercises

Exercise 12.1 Suppose two options, A and B, have the same pay-offs but A is exercisable on all the dates B is and more. Prove that A is worth at least as much as B. Give an example where they have the same value.

Exercise 12.2 Consider a forward which gives the right and obligation to buy a stock at a fixed price K during a period $[t_1, t_2]$. Thus is if the option has not been exercised before t_2, it must be exercised at t_2. How will the price of this derivative compare to that of an ordinary forward? How will it compare to the price of a American call option exercisable across the period $[t_1, t_2]$? How would you carry out the practical pricing of this option?

Exercise 12.3 Does put-call parity hold for American options in general?

Exercise 12.4 Show that if an American and European option with the same payoff are priced on the same tree then the estimated price for the American option is always at least as high as the estimated price for the European option.

Exercise 12.5 Show that if $\sigma(t)$ is the instantaneous volatility function for a stock S then it is possible to choose uneven time steps in such a way as to make the tree for $\log S$ recombine.

Exercise 12.6 An American–Asian call option pays the positive part of the running average value of spot across a discrete number of dates at the time of exercise. How would you price this option?

Exercise 12.7 The *perpetual American call option* is a call option that can be exercised at any time in the future and never expires. What will its value be in a positive-interest-rate world? What will its value be in a zero-interest-rate Black–Scholes world?

Exercise 12.8 Suppose we have American options, A and B, and B has half the notional of A but is otherwise identical. Consider a portfolio, C, consisting of two contracts of type B. Show that it carries more rights than A. How will its price compare to the price of A?

Exercise 12.9 Suppose we have a forward contract with one year expiry with the additional property that either party can cancel the contract after six months. How much will this contract be worth?

13

Interest rate derivatives

13.1 Introduction

One of the many forms of risk a company has to manage is interest rate risk. A modern company is typically financed by a mixture of debt and equity. In other words, it first raises money from investors by issuing shares and then typically borrows heavily to provide the rest of its funding. This is often called *leveraging* or *gearing*. The idea is that the shareholders get much more 'bang for a buck' because the company has several dollars to play with for every dollar invested.

On the other hand the company has to pay interest on the debt and in general eventually repay it too. If the interest rate varies according to the prevailing rates in the market, it is said to be *floating*. The interest payments can be a severe burden on a fledgling company and if they go up can cripple it. The company may well therefore want a fixed-rate loan, that is a loan for which all the interest payments are fixed in advance. This however introduces extra complications. Suppose rates fall during the period of the loan; the company will then want to refinance. However, the lending bank will not be so keen. It has a fixed stream of interest payments coming in at above the prevailing rate and will not want to break the contract early as that will simply mean losing money. Indeed, the bank may well have matched the fixed stream of payments with another fixed stream of payments going out, and thus will be neutral to interest rate changes. The bank will therefore charge the company a breakage fee equal to the loss. This fee will precisely match any gains from refinancing the loan, and the company is therefore stuck with the high interest rates.

The company, being aware of all this, might ask the bank to include a clause in the loan contract allowing it to break the contract. From the bank's point of view, this is equivalent to granting the company an option to swap a fixed rate of interest for a floating rate of interest at a time of the company's choice. Would the bank be willing to include such a clause? Yes – for the right price. All the bank does is charge a fee, or increase the interest rate to cover the cost of the option. The bank may well cover the option by buying an identical one in the market place.

This option is called an *American swaption* as it gives the right to swap interest payments at an arbitrary time.

Rather than going straight to a bank for a loan, a company may instead issue bonds in the market. Investors buy the bonds from the company and typically receive a fixed interest payment once a year called the *coupon*, and at the expiry of the bond, their original investment, the *principal*, is returned. The pricing of such a bond will depend on the prevailing interest rates and the credit-worthiness of the company. Here, we will stick to studying the former rather than the latter.

As with a straight loan, the company may worry about falling interest rates. It might therefore issue a *callable bond*, which the company can *call*, that is repay early. Typically the right to call is restricted to a set of pre-specified dates. Thus the buyer of the bond is really granting the company a *Bermudan swaption* – the right to swap a fixed stream of interest payments for a floating one on a discrete set of dates. The price of the bond in the market will therefore reflect this fact and go down accordingly. The purchaser of the bond will need to price the option to decide whether the bond is a good buy.

At a more mundane level, consider a homeowner with a mortgage. In the UK, interest payments typically float with prevailing interest rates. In the US, interest payments are typically fixed. However, fixing the rate is becoming more popular in the UK. The homeowner is then protected against rises in interest rates, and is faced with a wholly predictable stream of payments. Suppose the homeowner wishes to sell his house or refinance the mortgage, or, for whatever reason, wishes to early terminate a fixed-rate mortgage. If interest rates have fallen in the meantime, the bank is taking a hit. As in the case of a company loan the bank is effectively granting an option on a swap if it does not make a charge. Yet in the US, that is precisely what happens, mortgages can be early terminated at no cost to the borrower. In the UK, there has typically been a fee for early termination equal to the loss the bank suffers from early termination. However, there is a widespread public perception that such fees are unfair, and borrowers have on occasion managed to avoid them by publicly bleating about the unfairness. The solution is probably for banks to restrict sales of fixed-rate mortgages to ones with early termination allowed, but to price in the additional cost of the option which is what implicitly happens in the US.

At this stage, it is worth examining what a bank actually does! The principal business of a bank is borrowing and lending money. Money is made on the difference between the rate it pays to borrow and the rate it receives for lending. The spread is of course dented by those creditors who go bankrupt in the meantime, and an inherent part of the spread is that the bank is taking on risk.

The bank can borrow in a number of different ways. The simplest is to take money on deposit from savers, paying them interest in return for the use of their money. The bank can also issue bonds just like any other company. Another easy but generally more expensive way is to borrow on the interbank lending market. For each currency, there are a number of interest rates, called LIBOR rates, at which the bank can borrow. The different rates correspond to different lengths of borrowing. Thus there is typically a one-month rate, a three-month rate and a six-month rate. LIBOR stands for London Interbank Offer Rate.

What do we mean by a rate here? The reader should now forget everything to do with continuous compounding we have said so far in this book! Interest rates are never quoted continuously in the market and are generally simple rates. Whilst continuous compounding is a convenient approximation for pricing options in the equity and FX markets, it is positively misleading in the interest rate markets.

Thus if the three-month LIBOR rate is 5% on sterling, a bank can borrow a million pounds today in return for the obligation to repay a million pounds plus 5% of a million pounds times the accrual period. The accrual period in this case is a quarter, as three months is a quarter of a year. The quoted rate of 5% is a rate of 5% a year, but is only available for the period of three months. Note that if we compounded this borrowing, and interest rates did not change, we would end up paying more than 5% for the year because of the interest on the interest.

In equation terms, if we borrow £N for a period τ and the τ-period LIBOR rate is f then the interest payable is $Nf\tau$, and it is payable at time τ. The cashflows are receiving £N today and paying £$N(1 + f\tau)$ at time τ.

Interest rates are generally quoted in this way, and such rates are called *annualized rates*.

What a bank does is continuously borrow and lend. An important consideration for the bank is that it wants to match its receivables and obligations. For example, suppose most of the bank's money comes from short-term deposits which can be withdrawn at any time and the interest payable floats with short-term interest rates. Suppose the bank makes a long-term fixed-rate loan to a company. The maturity mismatch exposes the bank to risks. It may have a short-term liquidity risk if too many of its depositors want their money back at once. It is also exposed to interest rate risk – if the floating rate it pays the depositors rises above the fixed rate received from the company then the bank is making a loss. The bank will therefore try to match maturities in money it receives and pays, in order to avoid these problems, and will use interest rate derivatives when appropriate to reduce risks.

13.2 The simplest instruments

13.2.1 Zero-coupon bonds and present valuing

We have talked about swapping a stream of fixed interest payments for a stream of floating ones. This sort of contract is one of the most widely traded and simplest products to price mathematically. Indeed, it can be perfectly hedged in a static model-independent fashion. In this section, we define and price swaps. In general, the best way to analyze an interest rate derivative is in terms of the cashflows involved, and we illustrate this here.

All pricing of interest rate derivatives assumes the existence of a continuum of zero-coupon bonds which can be freely bought and sold, including short-selling as necessary. We will return to why it makes sense to make this assumption later but for now note that the bonds in question do not exist. The zero-coupon bonds will always have notional one. They will almost always be worth less than their face value as a bigger value is equivalent to negative interest rates, and so their values will be less than 1. In fact, as a function of maturity we will obtain a function which is monotone decreasing and ranges from 1 at $T = 0$ to 0 as T becomes infinite.

First, let's establish some notation. We denote by $P(T)$ the value today of a zero-coupon bond expiring at time T. When we want to think of its value at time t, we write $P(t, T)$.

The importance of the zero-coupon bonds, $P(T)$, is that we can write any deterministic set of cashflows as a linear multiple of such bonds. For example, suppose it is agreed to lend a company a principal N at time 1 year and the company is to pay a fixed six-monthly annualized-rate of 10% for five years, and then at time 6 years to return the principal. We can write this transaction in terms of zero-coupon bonds:

$$-NP(1) + \sum_{j=1}^{10} 0.05 N P(1 + 0.5j) + NP(6). \tag{13.1}$$

The important thing is that, upon substituting the values of the bonds, we have a value for the entire transaction. If this number is zero, the loan is at fair value. If it is positive, the lender makes a profit and if negative a loss. Note that this is the value today of the entire transaction – the value a year from now might be totally different but this would not matter as all the cashflows could have been hedged with zero-coupon bonds, making all interest-rate changes irrelevant. This technique for valuing trades is very standard and is called *present-valuing* or *PVing*.

13.2.2 Forward rates

Of course, the value of $P(T)$ is closely rated to prevailing interest rates. In particular, there is a one-to-one correspondence between LIBOR rates for the periods over which they exist and the value of zero-coupon bonds. Suppose the T-period rate is f. This means I can loan £1 today and receive £$(1 + fT)$ at time T. Equivalently, I can loan £$(1 + fT)^{-1}$ today and receive £1 at time T.

However, the right to receive £1 at time T is equivalent to owning a zero-coupon bond with expiry T. The value of $P(T)$ must therefore be $(1 + fT)^{-1}$. Inverting this relationship we have

$$f = \frac{P(T)^{-1} - 1}{T}. \tag{13.2}$$

13.2 The simplest instruments

The simplest form of interest rate derivative is a *forward-rate agreement*, or FRA. This is an agreement to pay a fixed simple rate of interest on a fixed sum of money between two fixed dates in the future. We take the sum of money, usually called the *notional*, to be 1 since it will simply multiply through everything. Let the two times be T_1 and T_2. If the fixed rate of interest is f then our cash flows are 1 at time T_1 and $-(1+(T_2-T_1)f)$ at time T_2. If we take the PV of these cash flows we have

$$P(T_1) - (1+(T_2-T_1)f)P(T_2). \tag{13.3}$$

The fair rate of interest is then the value that makes the transaction have zero value. Rearranging, we then have that

$$f = \frac{\frac{P(T_1)}{P(T_2)} - 1}{T_2 - T_1}. \tag{13.4}$$

We say that f is the *forward rate from time T_1 to time T_2*. Note that the values of any two of f, $P(T_1)$, $P(T_2)$ fix the value of the third. Note also that the forward rate says absolutely nothing about what the interest rate will be at time T_1; it simply says what the no-arbitrage rate is. In general, we would not expect the forward rate to be in any way constant; rather it will vary with time. Note that the fact that cashflows were synthesizable using zero-coupon bonds means that it is possible to perfectly statically hedge the forward-rate agreement, so the implied forward rate is enforced by no-arbitrage considerations.

Example 13.1 We wish to know the forward rate for putting money on deposit from 1 year to 1.5 years. We therefore look up the discount factors for 1 year and 1.5 years and find

$$P(1) = 0.956, \quad P(1.5) = 0.92.$$

The forward rate is therefore

$$\frac{1}{0.5}\left(\frac{0.956}{0.92} - 1\right) = 0.0783.$$

Once we have entered into a forward-rate agreement, it will soon cease to have zero value. We will want to know what value it has. Suppose the strike of a FRA is K and the current forward rate is f. What is the value of the contract? The value will depend on which side of the contract you hold. Suppose we are the borrower. The PV of the cash flows is

$$P(T_1) - (1+(T_2-T_1)K)P(T_2). \tag{13.5}$$

We can rewrite this as

$$(1+(T_2-T_1)f)P(T_2) - (1+(T_2-T_1)K)P(T_2) = (f-K)(T_2-T_1)P(T_2). \tag{13.6}$$

Thus the value of a forward-rate agreement is the difference between the prevailing rate and the strike multiplied by the accrual period, payable at the end of the agreement.

Now suppose we have three times, T_1, T_2, T_3; we could either enter into a forward-rate agreement from time T_1 to time T_3 or we could enter into two successive rate agreements: one from T_1 to T_2 and then the second from T_2 to T_3. These two transactions ought to be equivalent. Let f_{ij} denote the forward rate from time T_i to time T_j. We have, of course, that

$$f_{ij} = \frac{\frac{P(T_i)}{P(T_j)} - 1}{T_j - T_i}. \tag{13.7}$$

In the first case, we invest a £1 at time T_1 and receive £$(1 + f_{13}(T_3 - T_1))$ at time T_3. In the second case, we receive £$(1 + f_{12}(T_2 - T_1))$ at time T_2 and then immediately give it up again, in order to receive £$((1 + f_{23}(T_3 - T_2))(1 + f_{12}(T_2 - T_1)))$ at time T_3.

If there is to be no arbitrage then we must have

$$1 + f_{13}(T_3 - T_1) = (1 + f_{23}(T_3 - T_2))(1 + f_{12}(T_2 - T_1)). \tag{13.8}$$

Fortunately, this follows immediately from (13.7) since it is equivalent to

$$\frac{P(T_1)}{P(T_3)} = \frac{P(T_1)}{P(T_2)} \frac{P(T_2)}{P(T_3)}, \tag{13.9}$$

which is certainly true.

The important thing about (13.8) is that it expresses compounding effects. Putting money on deposit for six months at an annualized rate of 5% and then rolling it into another six-month deposit at the same rate is not the same as putting money on deposit for one year at an annual rate of 5%.

13.2.3 Swaps

Whilst the forward-rate agreement involves a principal, its close relative, the interest rate swap, does not. Rather than agreeing with a counterparty to put some money on deposit for a fixed period of time in the future at a fixed rate, we instead enter into a contract to pay him the floating rate of interest on a notional, whilst he pays us a rate fixed in advance. As the interest payments go both ways, there is no need to exchange principals. Thus whilst the contract has a notional which multiplies the cashflows, the total sum of money exchanged will actually be quite small in comparison. We take the notional of the swap to be 1 in what follows, as the notional will just multiply through everything.

Unlike a FRA, a swap is generally a multi-period arrangement. It may last several years and will involve payments at regular intervals. These intervals are typically three or six months long. However, from a mathematical point of view, there

13.2 The simplest instruments

is little to be lost be considering unequal interval lengths – indeed in the real world the intervals will vary by one or two days in length in any case. Suppose we have fixed dates

$$T_0 < T_1 < \cdots < T_n.$$

Let $\tau_j = T_{j+1} - T_j$. The fixed part of the swap then involves making payments at the times T_{j+1} for $j < n$ of value $X\tau_j$, where X is the pre-agreed swap rate. The present value of these cashflows is then

$$\sum_{j=0}^{n-1} X\tau_j P(T_{j+1}).$$

If we are the person making the fixed payments, the swap is said for us to be a *payer's swap*. If we are the person receiving the fixed payments, we are said to be in a *receiver's swap*. The set of floating payments is called the *floating leg* and the fixed ones are called the *fixed leg*.

Let f_j be the forward rate from T_j to T_{j+1}. The value of the cashflows on the floating leg is

$$\sum_{j=0}^{n-1} \tau_j f_j P(T_{j+1})$$

as we can turn each of the individual floating rates into a fixed rate by using a FRA. We therefore have that the unique arbitrage-free swap rate X satisfies

$$X \sum_{j=0}^{n-1} \tau_j P(T_{j+1}) = \sum_{j=0}^{n-1} \tau_j f_j P(T_{j+1}). \qquad (13.10)$$

Dividing through, we get

$$X = \sum_{j=0}^{n-1} w_j f_j, \qquad (13.11)$$

where

$$w_j = \frac{\tau_j P(T_{j+1})}{\sum_{i=0}^{n-1} \tau_i P(T_{i+1})}. \qquad (13.12)$$

The weights w_j have the interesting property that $\sum_{j=0}^{n-1} w_j = 1$. This means that the swap rate is a weighted average of the forward rates. It must therefore always lie between the lowest forward rate and the highest one. We shall also see when modelling rate movements that most of the change in X comes from the f_j in (13.11) whilst the w_j are comparatively constant.

Of course, f_j is expressible in terms of zero-coupon bonds and in particular we have that

$$\tau_j f_j = (T_{j+1} - T_j) \frac{\frac{P(T_j)}{P(T_{j+1})} - 1}{T_{j+1} - T_j} = \frac{P(T_j)}{P(T_{j+1})} - 1. \qquad (13.13)$$

Thus we can rewrite the swap rate as

$$X = \frac{P_0 - P_N}{\sum_{j=0}^{n-1} \tau_j P(T_{j+1})}. \qquad (13.14)$$

This reflects the fact that investing a sum at a floating rate over multiple periods is just the same in PV terms as buying a zero-coupon bond with the same termination date.

Example 13.2 Our discount curve is

0.0	1
0.5	0.975609756
1	0.949991019
1.5	0.923971427
2	0.898032347
2.5	0.872449235
3	0.847375747
3.5	0.822893781
4	0.799043132
4.5	0.775839012
5	0.753282381
5.5	0.73136604
6	0.710078203
6.5	0.689404609
7	0.669329748
7.5	0.649837584
8	0.630911969
8.5	0.61253689
9	0.594696597
9.5	0.577375683
10	0.560559119
10.5	0.544232274
11	0.52838092
11.5	0.512991226

13.2 The simplest instruments

We wish to find the swap rate for a swap starting in year with six-monthly payments over 3 years. We thus have that

$$t_j = 1 + 0.5j,$$

for $j = 0$ through 6. We need the discount factors for $P(t_j)$ for each j. The rate is equal to

$$\frac{P(1) - P(4)}{\sum_{j=1}^{6} 0.5 P(1 + 0.5j)}.$$

We substitute the numbers from the curve to get

$$\text{swap rate} = 0.0585.$$

The underlying forwards are

$$0.0563, \ 0.0578, \ 0.0586, \ 0.0592, \ 0.0595, \ 0.0597,$$

which straddle the swap rate as we would expect from the fact that it is a weighted average of the forward rates. ◇

As with a FRA, the fact that a swap is initially worthless does not guarantee that it will be worthless in the future. In fact, it will almost certainly not be. Let B equal $\sum_{j=0}^{n-1} \tau_j P(T_{j+1})$, which is sometimes called the *annuity of the swap*. If the swap is struck at K and the current implied swap-rate is X, then the value of the swap will be

$$(X - K)B.$$

The simplest way to see this is that X is the swap rate such that the floating and fixed legs are of equal value. Thus if the fixed rate is K, we can write $K = (K - X) + X$, which means that if the fixed leg has rate K, the swap is equivalent to a swap with a fixed leg at rate X together with a contract to pay the fixed amount $(K - X)$ times B. The value of paying this amount is therefore simply $(X - K)B$ since we are paying, and the swap struck with strike X is valueless by definition. Note that the definition of B has already taken all the discounting into account.

Sometimes in complicated transactions, there is need for a swap with variable notional. That is the paying and receiving amounts on the jth period are multiplied by a pre-specified sum N_j, for each j. A similar argument, which we leave to the reader, then shows that the arbitrage-free swap rate is

$$X = \sum_{j=0}^{n-1} f_j w_j, \tag{13.15}$$

where

$$w_j = \frac{N_j \tau_j P(T_{j+1})}{\sum_{j=0}^{n-1} N_j \tau_j P(T_{j+1})}. \tag{13.16}$$

Note that the weights still add up to 1.

13.3 Caplets and swaptions

13.3.1 Definitions

The first obvious consequence of trading FRAs and swaps is that there will be a market for options on them. An option on a FRA is a called a *caplet* or a *floorlet* depending on whether it's a call on the rate or a put on the rate. The option's expiry is typically the start of the forward-rate agreement. Thus if the FRA runs from time T_1 to T_2, at time T_1 we have the option to receive a sum equal to the present value of the FRA which is, of course,

$$(f - K)(T_2 - T_1)P(T_2),$$

where f is the prevailing forward rate from T_1 to T_2 at time T_1, and K is the strike of the FRA which is also called the *strike* of the caplet. Thus the pay-off of a caplet is

$$(f - K)_+(T_2 - T_1)P(T_2).$$

Similarly, the payoff of a floorlet is

$$(K - f)_+(T_2 - T_1)P(T_2).$$

Typically caplets and floorlets are sold in large bundles which are called *caps* and *floors*. A ten-year cap would consist of forty three-month caplets; the cap would allow the purchaser to be sure of never having to pay more than the strike rate of the cap, as three-month interest rate on its borrowings, for the entire ten years. A cap is so-called because it means that the borrowing rate for the holder is capped at the cap's strike for the period of the contract. Similarly, a floor would guarantee a minimum interest rate for money on deposit.

Options on swaps similarly exist. They are called *swaptions*. An option on the right to pay the fixed rate in a swap is called a *payer's swaption*. An option on the right to pay the floating rate, i.e. receive the fixed rate, is called a *receiver's swaption*. The expiry of the option is normally the start of the swap. If K is the strike of the swap and B is the annuity then the payoff of a payer's swaption is

$$(X - K)_+ B,$$

where X is the swap rate at expiry, i.e. the prevailing swap rate at the start of the swap. This is immediate from the fact that the swap's value is $(X - K)B$ at expiry.

13.3 Caplets and swaptions

Similarly, the pay-off of a receiver's swaption is

$$(K - X)_+ B.$$

Such swaptions, having one exercise time, are called *European swaptions*.

Often swaptions arise in the sense of the right to break a swap. However, the right to break a swap is mathematically equivalent to the right to enter a swap in the opposite direction at the swap strike. For example, suppose we are in a payer's swap with strike K which lasts 30 years and we have the right to break once a year. The right to break is then equivalent to having the choice after a year to enter a receiver's swap with strike K of length 29 years, or to retain the right to break in the future. After two years, if we have not already exercised, we have the right to enter a swap of length 28 years. Thus the right to break is really a swaption with multiple exercise dates. The big difference from stock options is that the length of the swap decreases with time, so the asset on which we have an option is not always the same. Swaptions of this type are known as *Bermudan swaptions*. The fact that they represent the right to break a long stream of fixed payments means that they are very heavily traded. The early exercise rights make Bermudans tricky to price, and their pricing is still the object of much active research.

13.3.2 Black formulas

First, we need to learn how to price caplets, floorlets and European swaptions. If we take the forward rate to be log-normally distributed, this is easy once we take the right framework. The trickiness in applying risk-neutral valuation is that the forward rate is not a traded asset. However, a forward-rate agreement is a tradable. If the current forward rate is f, and f covers the period from T_1 to T_2, the value of the forward-rate agreement is

$$(f - K)P(T_2),$$

where K is the strike. Since $P(T_2)$ is tradable and K is constant, we have that $fP(T_2)$ is tradable. (Alternatively, we could consider another FRA with zero strike.) If we take $P(T_2)$ to be the numeraire and pass to the risk-neutral measure then we have that the ratio of any tradable to the numeraire is a martingale. This says that

$$f = \frac{fP(T_2)}{P(T_2)},$$

is a martingale.

We know from our study of martingales and measure changes in Chapter 6 that the only effect a measure change can have on a log-normal variable is a change of drift. Thus, if in the real-world measure,

$$df = \mu(f, t)dt + \sigma f dW, \tag{13.17}$$

then in the risk-neutral measure, since all martingales are driftless, we have

$$df = \sigma f dW. \tag{13.18}$$

Let C denote the value of the caplet; then

$$\frac{C_0}{P(0, T_2)} = \mathbb{E}\left(\frac{C_{T_1}}{P(T_1, T_2)}\right), \tag{13.19}$$

which is equivalent to

$$C_0 = P(0, T_2)\mathbb{E}((f - K)_+)(T_2 - T_1). \tag{13.20}$$

We have thus reduced the problem to computing $\mathbb{E}((f-K)_+)$. However, this is just the same as valuing a call option on a non-dividend paying stock in a zero-interest-rate world. (Note that the analogy with a zero-interest-rate world is because the underlying is driftless.) We thus have that the value of the caplet is

$$C_0 = P(0, T_2)(T_2 - T_1)Q(f_0, K, \sigma, T_1), \tag{13.21}$$

where

$$Q(f_0, K, \sigma, T) = f_0 N(d_1) - K N(d_2), \tag{13.22}$$

and

$$d_j = \frac{\log\left(\frac{f_0}{K}\right) + (-1)^{j-1}\frac{1}{2}\sigma^2 T_1}{\sigma\sqrt{T_1}}, \tag{13.23}$$

and f_0 is, of course, the initial value of the forward rate. This formula, (13.21), is generally called the *Black formula* as a variant was first developed by Black when pricing options on futures [18].

We remark that, as with a stock option, the formula remains valid for variable volatilities simply by replacing σ with the root-mean-square volatility. In common with the FX and equity markets, caplets and the caps which they make up are typically quoted in terms of volatility rather than price. The use of the Black formula is assumed by both sides even if neither believes it. Another curious aspect of this is that if one calls a market-maker and asks for a price on a cap, he will quote a single volatility to be used for all the caplets. However, he will have arrived at this volatility, by assigning a different volatility to each caplet according to how much he thinks it is worth, converting these individual volatilities into prices, adding them up, and then converting back into the single constant volatility which makes the cap have the summed price.

13.3.3 LIBOR-in-arrears

The pricing of the caplet was facilitated by the choice of numeraire; in particular the fact that the forward rate times the numeraire is a tradable meant that the forward rate was a martingale in the risk-neutral measure and this made life easy. In general, we will not be so lucky – our good fortune here is really caused by the fact that we are considering an option on a tradable: the forward-rate agreement. Once the pay-off is no longer defined in terms of an option on a tradable, pricing becomes much trickier.

To illustrate this point, consider a contract which is beguilingly similar to a caplet: the LIBOR-in-arrears caplet. The only difference is that at time T_1, instead of receiving the right to receive

$$(f - K)_+(T_2 - T_1)$$

at time T_2, we receive the money immediately at time T_1. In other words, the pay-off has changed from

$$(f - K)_+(T_2 - T_1)P(T_2)$$

to

$$(f - K)_+(T_2 - T_1)P(T_1).$$

Whilst this change seems small, it destroys our previous argument. If we take $P(T_2)$ as numeraire then we still have that f is a martingale but we no longer need the same expectation. Let D_t be the value of the contract at time t. We now have

$$\frac{D_0}{P(0, T_2)} = \mathbb{E}((f_{T_1} - K)_+(T_2 - T_1)P(T_1, T_1)P(T_1, T_2)^{-1}). \quad (13.24)$$

We have that $P(T_1, T_1) = 1$ and $P(T_1, T_2) = (1 + f_{T_1}(T_2 - T_1))^{-1}$. Letting $\tau = (T_2 - T_1)$, we get

$$\frac{D_0}{P(0, T_2)} = \mathbb{E}((f_{T_1} - K)_+(1 + f_{T_1}\tau)\tau). \quad (13.25)$$

It is not clear how to evaluate this expectation for log-normally distributed f. We could, of course, just perform a numerical integration but we have lost our nice closed-form answer.

Another attack is to say that the problems are coming from the fact that the pay-off is at time T_1, whilst we are using $P(T_2)$ as numeraire; this mismatch in times is what messes us up, so clearly we should use $P(T_1)$ as numeraire instead. We then obtain, as in the vanilla caplet case,

$$D_0 = P(0, T_1)\mathbb{E}((f_{T_1} - K)_+)\tau, \quad (13.26)$$

and life looks good. However, there is a hidden thorn: the expectation is to be taken in the risk-neutral measure associated with the numeraire $P(T_1)$. We do not know that f is a martingale in this measure so we need to compute its drift. All we have done is shifted the problem. We shall return to this issue when we examine the pricing of exotic options; for now we stress the fact that this numeraire-mismatch problem is central to the pricing of exotic interest rate derivatives.

13.3.4 Pricing a swaption

We can attack the pricing for a swaption in a similar way to the pricing of a caplet. The key point is to pick a numeraire which makes the swap rate driftless. Thus suppose the swap rate X is log-normal so we have

$$dX = \mu(X,t) X dt + X \sigma dW. \tag{13.27}$$

From our arguments with caplets, we know that if we can find a numeraire, Q, such that XQ is a tradable asset, then X is a martingale with respect to Q's risk-neutral measure and thus X will have zero drift.

In fact, for any quantity that is a rate we can expect to be able to find such an asset. A rate is after all a ratio. Thus if we take the denominator of the ratio as our numeraire then the rate should become a martingale. For a swap rate we have

$$X = \frac{P_0 - P_N}{\sum_{j=0}^{n-1} \tau_j P(T_{j+1})}. \tag{13.28}$$

The numerator and the denominator of this fraction are both tradables – they are just linear multiples of zero-coupon bonds. The swap rate is just the ratio of the values of the floating leg and the annuity. Thus taking the denominator, which is the annuity of the swap rate, as numeraire, the swap rate becomes driftless. Let B_t denote the value of the annuity at time t. If O_t denotes the value of a payer's swaption with strike K at time t then we have that

$$\frac{O_0}{B_0} = \mathbb{E}\left(\frac{O_T}{B_T}\right) = \mathbb{E}\left((X - K)_+\right). \tag{13.29}$$

The value of the expectation is just given by the Black formula, precisely as for a caplet, and B_0 is immediately observable as the appropriate multiple of the zero-coupon bond prices. We thus have the price of a swaption. Similarly, we can value a receiver's swaption, simply by replacing the Black formula for a call with the Black formula for a put. Just as with caplets and equity options, it is typical to quote swaption prices in terms of volatilities, the Black model being assumed by both sides in the transaction. As for equity and FX options, swaptions with different strikes are quoted with different volatilities reflecting traders' lack of belief in the log-normal model.

As with caplets, the argument we have presented here is heavily dependent on the choice of numeraire; the numeraire is the unique tradable, B, such that XB is a tradable. Thus it is the unique numeraire such that X is a martingale in the implied risk-neutral measure. There will be circumstances in which we will want to evolve multiple swap rates simultaneously, or forward rates and swap rates simultaneously; we will then need to understand swap rate drifts under different numeraires. We will return to this point when developing pricing models for exotic options.

The sharp reader will have noticed that there is a slight inconsistency. We price caplets by assuming forward rates are log-normally distributed and we price swaptions by assuming swaps are log-normally distributed. However, forward rates determine swap rates in a complicated way. Thus if a forward rate is log-normal, the swap rate cannot be log-normal. However, the inconsistencies involved in assuming joint log-normality are small, and it is market practice to ignore them. (See for example [108].)

13.4 Curves and more curves

13.4.1 Introduction

Our discussion of swaps and forward-rate agreements has been dependent on the existence of zero-coupon bonds which can be freely bought, sold and even shorted. This is a distortion for a number of reasons but is nevertheless a reasonable way to proceed: in this section we explain why. There are, in fact, many yield curves for each currency whose levels depend on the riskiness of the instruments involved. We discuss the curves for sterling but the issues are essentially the same for the euro and US dollar curves. We will generally talk about constructing discount curves rather than yield curves, as the discount curve is just the price of a zero-coupon bond as a function of maturity which is what we generally want. On the other hand, the yield curve is a notional measure of the effective annual interest rate which we would receive for investing in such a bond. The yield curve is useful from a qualitative point of view as it strips out redundant information by converting everything to interest rates, but to work mathematically with the yield curve is simply annoying. With all the discount curves, one thing to bear in mind is that the theoretical curve will not actually represent a price one can obtain in the market. If we call a market-maker and ask to buy or sell, he will always quote a pair of prices straddling the theoretical price; thus there is always a spread around the theoretical curve.

13.4.2 Gilts

The lowest yielding instruments are, of course, the riskless ones – for example, UK government bonds which are generally known as *gilts*. The UK government does not generally issue zero-coupon bonds so all we can observe in the market is the price of coupon-bearing bonds. However, a coupon-bearing bond is decomposable into a sum of zero-coupon bonds. This is clear if we remember that the bond is really just a sequence of cashflows. The cashflows are the coupon at each coupon-payment date and the repayment of the principal at maturity. Any cashflow is just a zero-coupon bond with expiry equal to the timing of the flow and notional equal to the size of the cashflow.

This means that we can attempt to fit a theoretical discount curve for zero-coupon bonds to the observed prices of UK gilts. The curve should be reasonably smooth and reproduce the prices of all traded gilts exactly. Curve fitting is more a tricky programming issue than a conceptual one so we will not explore precisely how to do this. The main idea is that one should build up the short maturities first using the prices of the short-dated gilts and then gradually bootstrap up to the longer ones, using the fact that the initial coupons now have well-defined present values. This will only give us certain points on the gilt curve associated to gilt payments. Typically, one would then interpolate to find the intervening points. Generally, the interpolation is carried out in a log-linear fashion; that is the logs of two neighbouring points are taken and then interpolated linearly to provide the logs of the intervening points.

13.4.3 Repos

The problem with the gilt curve for pricing options is that a gilt cannot be truly shorted. If we sell a gilt we do not own, then the purchasers are exposing themselves to a credit risk – we might go bankrupt and never actually provide the gilt. However, it is possible to borrow some money, use it to buy the gilt and then sell it, which is effectively the same as short-selling. The lender will require some form of collateral to cover any obligations. Typically, the borrower will have to provide the market value of any instruments borrowed plus a little bit more. The little bit more is sometimes called the *haircut*.

These agreements are generally called *repo* deals, which is short for repurchase rather than repossession. The reason is that they are generally phrased in terms of an agreement to sell the collateral to a broker and then repurchase it at a higher price in the future. The difference between sale and purchase prices is the interest paid to the broker. The implied discount curve from these interest rates is called the *repo curve*. Since the borrowing is collateralized, the riskiness is small but it is not non-existent and so the interest rates are higher. Thus the implied yields from the repo curve are higher and the discount factors correspondingly lower than for gilts.

13.4.4 The LIBOR curve

The third main curve is the LIBOR curve and this is the most important one. (Remember that with this is the rate at which banks can freely borrow and lend to each other and that LIBOR is an abbreviation for London Interbank Offering Rate.) The short end of the LIBOR curve is constructed from the market rates for borrowing. However, LIBOR rates are not available for long-term borrowing – one cannot go to the LIBOR market and ask for a five-year loan. The longer-dated discounts are therefore inferred from the prices of other instruments, in particular from the prices of the most liquid of instruments, swaps. As interbank borrowing is riskier than collateralized borrowing, the LIBOR yield will be higher than the repo and gilt curves whilst the discount factors are lower. However, as it is quite rare for a bank to go under, the spread is not very large.

Although we have priced swaps using zero-coupon bonds, in fact it is the opposite process that is carried out by the market. Swaps are liquidly traded, their rates observed and the resultant discount factors inferred. This is more a tedious task than a theoretical problem and we leave it to the exercises. The consistency of the prices of different swaps is then checkable by repricing the swaps from the curve. For example, if the one-to-two-year-rate is X, and the two-to-three-year rate is Y, there will be a unique compatible rate for the one-to-three-year period.

For us, the LIBOR curve is the most important curve for the pricing of exotic interest-rate options, as it's the curve at which the bank can effortlessly trade swaps that allows it to take positions on interest rates and to freely hedge.

As well as the three curves discussed, there is also a host of other curves associated to the cost of corporate bonds. The price of a corporate bond is dependent on the credit status of the issuer. The rating agencies assign a credit rating to each bond issue which is the main determinant of the spread required over gilts. Thus to each credit rating a discount curve is associated. These range from the AAA curve which yields between gilts and LIBOR to the C curve which can easily have a 5% spread above gilts. These curves are important to the pricing of credit options but play little role in the pricing of exotic interest rate derivatives so we do not examine them in detail.

13.5 Key points

- Interest rate derivatives are used to manage risks arising from exposure to interest rates.
- A forward-rate agreement, or FRA, is the right and obligation to borrow or deposit a sum of money for a fixed rate for a fixed period time in the future.
- Forward rates are quoted in annualized terms over discrete periods.
- Forward contracts can be perfectly replicated by zero-coupon bonds and forward rates are therefore uniquely determined.
- Swaps are contracts to swap a fixed stream of interest rate payments for a floating stream.
- Swaps do not involve exchange of principals.

- The swap rate is determined by no-arbitrage considerations.
- A caplet is call on a forward rate.
- A floorlet is a put on a forward rate.
- If forward rates are log-normal then caplets and floorlets can be valued by the Black formula.
- Swaptions are options on swaps and can be valued by using the Black formula provided swap rates are taken to be log-normal.
- There are many different discount curves depending upon the riskiness of instruments involved.

13.6 Further reading

A comprehensive text on interest-rate derivatives is *Interest Rate Option Models* by Riccardo Rebonato [105].

A good overview of available products in the market is [117].

13.7 Exercises

Exercise 13.1 If the six-month LIBOR rate is 5% and the one-year rate is 5%, what is the forward rate from six months to one year?

Exercise 13.2 If the six-month LIBOR rate is 4% and the one-year rate is 6%, what is the forward rate from six months to one year?

Exercise 13.3 If the six-month LIBOR rate is 6% and the one-year rate is 5%, what is the forward rate from six months to one year?

Exercise 13.4 A LIBOR-in-arrears FRA pays $(f - K)\tau$ at the reset time of the forward-rate f. If f is log-normal, derive a formula for its price.

Exercise 13.5 Suppose
$$0 = t_0 < t_1 < t_2 < \cdots < t_n,$$
and the swap rate, X_j, runs from t_0 to t_j, for each j. Show that the discount factors $P(t_j)$ can be deduced from the rates X_j. Such rates are said to be *co-initial*.

Exercise 13.6 Suppose
$$0 = t_0 < t_1 < t_2 < \cdots < t_n,$$
and the swap-rate, X_j, runs from t_j to t_n, for each j. Show that the discount factors $P(t_j)$ can be deduced from the rates X_j. Such rates are said to be *co-terminal*.

Exercise 13.7 Show that the process for a swap-rate is not log-normal if the underlying forward rates are log-normal.

Exercise 13.8 A *trigger FRA* is a FRA that comes into existence if and only if the forward rate is above H at the start of the FRA. Develop an analytic formula for its price if the forward rate follows geometric Brownian motion.

14

The pricing of exotic interest rate derivatives

14.1 Introduction

The critical difference between modelling interest rate derivatives and equity/FX options is that an interest rate derivative is really a derivative of the yield curve and the yield curve is a one-dimensional object whereas the price of a stock or an FX rate is zero-dimensional. One might be tempted to think that as most movements of the yield curve are up and down it is unnecessary to model the one-dimensional behaviour. However, the yield curve can and does change shape over time, and we shall see that for certain options these changes are the source of most of the option's value. From time to time, yield curves also undergo qualitative changes in shape. For example, the UK yield curve changed from being upward-sloping to being humped in the early 1990s.

The fact that we are modelling the changes of a curve makes life considerably more complicated but also much more interesting. One important thing to realize is that just because most of the movements in the yield curve are up and down does not mean that most of the value of a given derivative comes from these up and down movements. To try and illustrate this point and some others in pricing interest rate derivatives, we introduce an option which to my knowledge is not traded but is very good at demonstrating some of the trickier issues involved. Suppose we have a contract consisting of two forward-rate agreements which span contiguous segments of times, and that we are long the first contract and short the second. That is, we pay fixed on the first one and receive fixed on the second. Thus the first forward-rate agreement runs from time t_1 to time t_2 and the second runs from time t_2 to time t_3. We shall call such a contract a *reversing pair*. Pricing a reversing pair is trivial: we just decompose it into a sum of two forward-rate agreements, which we already know how to price, and we are done.

The interesting thing about the reversing pair is that its value is very insensitive to changes in the overall level of the yield curve. If interest rates go up by 1%,

14.1 Introduction

then we gain on the first forward-rate agreement but lose a similar amount on the second. If, however, the shape of the yield curve changes so that the first rate goes down and the second rate goes up, then we lose on the first and lose on the second. Thus the value of the reverse contract reflects changes in the shape but not the level of the yield curve. In particular, the reverse contract is sensitive to the slope of the curve: a change in slope means money won or lost. We can extend the reverse contract to a double reverse contract by taking two reverse contracts over adjacent periods of time which go in opposite directions. We are then in the situation of being neutral to both the level and the slope of the yield curve. However, we are still sensitive to changes in the shape, in particular the curvature will be the main impact on our profits and losses.

Whilst there are no problems with pricing the reverse and double-reverse contracts because of their decomposability, if we now consider options upon them then the situation becomes considerably more complex. Call an option on a reverse contract a *reverse option*. Note that the reverse option is not the sum of two options of forward-rate agreements because it is only the right to enter into the two forward-rate agreements simultaneously. Thus the pricing is not as trivial as that of the underlying contract. All the value of the reverse option comes from divergence in the neighbouring forward rates, so the crucial thing we need to understand is the correlation between them.

When pricing, we must try to use all the information we have available. We can observe the initial values of the forward rates in the market. We can also observe the cost of caplets and floorlets on the forward rates. If we believe a log-normal model for forward-rate movements then the market price tells us what volatilities to ascribe to each of the forward rates. Or does it? Forward rates generally move according to a time-dependent volatility curve with a peak around two years. Very close to expiry not much happens, far from expiry not much happens but with a couple of years to go the rate moves around a lot.

If the volatility is not constant then the market price reflects the root-mean-square volatility over the period from now until the start of the forward rate. Thus if we are pricing a reverse option, we have the r.m.s. volatility of the first contract over the period until the start of the option, but for the second forward rate, we have the r.m.s. volatility until its own start, which is not the expiry of the reverse option.

This means that every time we need the volatility of a forward rate over a period other than from now till expiry, we need to assume a shape for the instantaneous volatility curve. That is we need to define a function for the instantaneous volatility with the property that the r.m.s. volatility from now till expiry is equal to the observed market value.

Returning to the reverse option, if we have chosen a shape for the volatility function then we can compute the r.m.s. volatility up to the reverse option's expiry.

Now that we have the two forwards' r.m.s. volatilities what can we do with them? Suppose we decide to employ risk-neutral valuation and run a Monte Carlo simulation. First, we have to pick a numeraire. It will not be possible to do so in such a way that both the forward rates become driftless, since they have different finishing times and only the zero-coupon bond maturing at the finishing time will imply a driftless rate. We will therefore need to compute the drift of at least one of the forward rates.

Supposing we know the drifts, the next issue for our simulation is that we have to evolve both the forward rates simultaneously. To run such a multi-asset simulation, we will need to know the correlation between the two rates. In fact, we have to be careful about what we mean by correlation, since two quantities which are instantaneously perfectly correlated can, in fact, become decorrelated by the shape of their volatility curves. To see this, suppose we have a Brownian motion B_t and two assets X_t and Y_t following the processes,

$$dX_t = \sigma_X(t) dB_t \qquad (14.1)$$
$$dY_t = \sigma_Y(t) dB_t. \qquad (14.2)$$

If we let $\sigma_X(t)$ be equal 1 for $t < 0.5$ and 0 thereafter, whilst we let $\sigma_Y(t) = 1 - \sigma_X(t)$, then we have that

$$X_1 - X_0 = B_{0.5} - B_0, \qquad (14.3)$$
$$Y_1 - Y_0 = B_1 - B_{0.5}. \qquad (14.4)$$

Since B is a Brownian motion, it has independent increments so $X_1 - X_0$ and $Y_1 - Y_0$ are totally uncorrelated despite being driven by the same Brownian motion. Thus two assets with perfect instantaneous correlation can be totally uncorrelated because of differences in the shape of their volatility curves. This means that if we are to run a Monte Carlo simulation then we will need to know both the shapes of the volatility curves and the instantaneous correlations.

Once we know the curve shapes, the correlations, and the drifts in the risk-neutral measure, we can run the simulation and obtain a price. For the reverse option, the price will be largely dependent on the choices we have made for the curve shapes and for the instantaneous correlations, since we have seen that all the value comes from changes in the slope, which will not result from correlated movements.

How can we estimate these factors? One approach is to observe historical movements and use these to infer likely future behaviour. A second approach is to try and use the market prices of more instruments as a guide. The advantage of the second approach is one could then use these instruments as a hedge against having the wrong values. So, for example, consider a forward caplet. This is a caplet for which the exercise time is not the start of the forward rate but instead some earlier time. The pricing will proceed in precisely the same way as for an ordinary caplet

14.1 Introduction

except that the expiry time in the Black formula will be the time of exercise, and the volatility will be the root-mean-square volatility from now to the expiry time. Thus if one knew the prices of forward caplets for all possible expiries then one would know the r.m.s. volatilities for all expiries, and one would be able to deduce the shape of the instantaneous volatility curve in a hedgeable way. Unfortunately, forward caplets are not sufficiently liquid to be able to infer the volatilities in a useful manner so we are stuck with historical measures.

Our study of how to price the reverse option has yielded a programme of attack for pricing any interest rate derivative which we now outline.

(i) Write the instrument's payoff in terms of the behaviour of forward rates.
(ii) Observe the current forward rates and their implied volatility in the market.
(iii) Choose a numeraire which makes the rates as driftless as possible.
(iv) Compute the drifts of the forward rates in the risk-neutral measure.
(v) Choose or infer instantaneous volatility curves for the forward rates.
(vi) Choose or infer instantaneous correlations between different forward rates.
(vii) Run a Monte Carlo simulation.

This approach is commonly called the BGM model after Brace, Gatarek & Musiela who published an early paper on the method. It is also known as BGM/J as Jamshidian published another early paper. Since the model relies on evolving LIBOR market forward rates, it is often called the LIBOR market model. Whilst we have have tried to present the approach in such a way that it appears natural and inevitable, historically it was a latecomer. The earlier interest rate models depended on the notion of an evolving short rate – the continuously compounding rate which we used in equity and FX models became a stochastic quantity following its own process. The price of a bond or forward rate was then inferred by the expected quantities obtained by rolling up this short rate to maturity. Heath, Jarrow & Morton introduced a new approach, now known as HJM, in the late 1980s which essentially said that one should model the yield curve directly rather than thinking about a compounding short rate. There was initially a great deal of resistance to such a change in viewpoint but in the longer term the approach has become very popular and use of BGM, which can be viewed as a discretitation of the HJM approach, in particular has become widespread.

Ultimately, the big attraction of the BGM model is that it incorporates in a very natural fashion all the market-observable information. We do not have to fiddle or develop complicated calibration procedures to price the caplets correctly; instead we just plug their volatilities in. The main downside of the approach is that it depends inherently on Monte Carlo simulation, which, as we recall from Chapter 12, is tricky to apply to options with early exercise opportunities.

14.2 Decomposing an instrument into forward rates

The first stage of our BGM programme was to write the instrument to be priced in terms of the values of forward rates; in this section we carry this out for various instruments.

Decomposing a swap or a swaption

Whilst there is little point to pricing a swap using BGM since there is a simple static model-independent price, it is instructive nevertheless to examine how one might do so. It is also worthwhile for testing an implementation of the model; the fact that the price is model-independent means that the model ought to recover it and if it does not then the model is seriously wrong. There are in fact two different ways to approach modelling a swap. The first is to evolve the yield curve up to the time when the swap starts, and then value the swap as its value at that point using the usual formula – this has the advantage of being easily adapted to the case of a swaption. The second method is to evolve all the way through the swap, generating all the cashflows.

The second method is more illustrative so we examine it first. The swap is associated to a set of dates $t_0 < t_1 < t_2 < \cdots < t_n$. Suppose the swap is to pay fixed at rate X. Let $f_j(t)$ denote the forward rate from t_j to t_{j+1} as realized at time t. At time t_{j+1} for $j < n$, we then receive a payment of $((f_j(t_j) - X)(t_{j+1} - t_j))$ pounds.

We are trying to compute an expectation of the ratio of the swap value to the numeraire. Suppose we have chosen as numeraire, B_n, which is the zero-coupon bond expiring at time t_n. The value of B_n at time t_{j+1} is then

$$\prod_{k=j+1}^{n-1} \frac{1}{1 + f_k(t_{j+1})(t_{k+1} - t_k)},$$

as investing this sum at the prevailing interest rates, which could be hedged not to float, we would receive £1 at time t_n. Thus the ratio of the payment at time t_{j+1} to the numeraire is

$$Q_j = (f_j(t_j) - X)(t_{j+1} - t_j) \prod_{k=j+1}^{n-1} 1 + f_k(t_{j+1})(t_{k+1} - t_k).$$

We therefore have for a given realized path that the value of the ratio is $\sum_{j=0}^{n-1} Q_j$.

We have used as numeraire the zero-coupon bond B_n as it's the natural bond which does not expire before the swap does. Now we could use a shorter bond for

14.2 Decomposing an instrument into forward rates

example B_1, the zero coupon bond which expires at time t_1, for the evolution up to time t_1 but we would then have to be careful about what to do after that time. One way out of this is to regard the numeraire as an asset with a trading strategy. So at time t_1, we use the £1 we receive from our maturing bond to buy more bonds! In particular, at time t_1 we could buy bonds, B_2, maturing at time t_2, at time t_2 buy those maturing at time t_3 and so on. When we get to t_n, the numeraire is then a number of B_n bonds. This numeraire is called the discretely-compounding money-market account. How would it work in practical terms?

At time t_1, the numeraire is one B_1 bond and the swap payoff is $(f_0(t_0) - K)(t_1 - t_0)$. The first component of our sum is therefore

$$R_1 = (f_0 - K)(t_1 - t_0).$$

The numeraire is reinvested in B_2 bonds which will cost $(1 + f_1(t_1))^{-1}$ each – we therefore buy $(1 + f_1(t_1))$ of them.

At time t_2, we receive the payment, $(f_1(t_1) - K)(t_2 - t_1)$ and the numeraire is worth $1 + f_1(t_1)$. The second term of our sum of ratios is therefore

$$R_2 = \frac{(f_1(t_1) - K)(t_2 - t_1)}{1 + f_1(t_1)}.$$

The numeraire is now reinvested in B_3 bond which will cost $1 + f_2(t_2)$ each.

So at time t_3, we receive $(f_2(t_2) - K)(t_3 - t_2)$ and the numeraire is now worth $(1 + f_1(t_1))(1 + f_2(t_2))$. The third term is then

$$R_3 = \frac{(f_2(t_2) - K)(t_3 - t_2)}{(1 + f_1(t_1))(1 + f_2(t_2))}.$$

The numeraire is then reinvested and so on.

The reasons for choosing one of these numeraires over the other are mainly technical, and we shall discuss them in the context of computing forward rate drifts.

We now look at the alternate method of valuing a swap in BGM. In this approach, we take as numeraire any of the bonds B_j. Let B_l be the numeraire. We evolve the forward rates up to time t_0. At time t_0, we can value our swap in the conventional way. The discount factors associated to each time t_j as seen from t_0, are easy to compute: in fact they are just the values of the bonds B_j at time t_0. We get

$$P(t_0, t_j) = B_j(t_0) = \frac{1}{\prod_{k=0}^{j-1}(1 + f_k(t_0)(t_{k+1} - t_k))}. \qquad (14.5)$$

We therefore just plug these values into the formula for valuing a swap to get the value of the swap at time t_0. We also have the value of the numeraire, B_l. Thus the realized value of the price ratio is computed and we are done.

Suppose we want to value a swaption. At time t_0, we will exercise if and only if the swap value is positive. We therefore simply proceed as in the second method but rather than taking the ratio of the swap's value to the numeraire, we take the ratio with the positive part of the swap's value.

The trigger swap

A *trigger swap* is a swap that does not take effect until some reference rate passes a trigger level on one of a number of fixed dates. It is similar to a barrier option except that the barrier occurs on the basis of a different underlying than the option. As with barrier options, there can be many sorts: *up-and-out*; *down-and-out*; *up-and-in*; and *down-and-in*. Also as with barrier option, the *in* option plus the *out* option is equal to the vanilla.

In what follows, we concentrate on the 'up-and-in' payer's swap for concreteness leaving the reader to fill in the obvious details for the other cases. In particular, suppose the trigger swap is based on a five-year swap which starts in two years with three-monthly payment dates. Let the swap rate be X. In our notation for swaps above, we have

$$t_j = 2 + j/4, \tag{14.6}$$

where j runs from 0 to 20. We take the reference rate to be three-month LIBOR and the reference dates to be the setting dates of the swap, i.e. t_j. Let $f_j(t)$ be the forward rate from t_j to t_{j+1} as evaluated at time t. Take the trigger level to be K.

To run a BGM simulation, we would simulate all the rates f_t. At time t_0, we check f_0. If it is above K, we value the swap as seen at time t_0 and take the ratio with the numeraire in precisely the same way as for the vanilla swap. If it is below K we continue on to time t_1. At time t_1, if f_1 is above K, we value the remaining part of the swap by the same method and take the ratio with the numeraire.

If $f_1(t_1)$ is below K, we continue on to t_2, and repeat until either we reach t_{19} or the three-month LIBOR rate is above K.

Whilst this method of pricing the trigger swap is intuitively obvious, there is a more rapid way of proceeding. The crucial observation to make is that we only need to know the value of each forward rate on its own reset date. If we now let a forward rate have zero volatility after its reset date, then we need only evolve all the forward rates to a single time horizon: namely, the reset date of the last forward rate.

To price, we then run through the forward rates until the swap is triggered, and use the final values of the forward rates to simulate all the cashflows. Note that the evolution of the forward rates takes a twentieth of the time that the naive approach does; however we do have to be careful that the errors arising from evolving forward rates over long periods are not large. In practice, we would probably do five-year steps in order to reduce such errors.

The Bermudan swaption with a given exercise strategy

A Bermudan swaption is an option to enter a swap on any one of the swap's fixing dates. The theoretical value of the Bermudan will be the maximum attainable by any exercise strategy. One interesting aspect of the 'up-and-in' trigger swap is that we can regard it as a Bermudan swaption with the exercise strategy of "exercise if and only if the 3-month LIBOR rate is greater than the trigger." Thus the price of a trigger swap gives us an immediate lower bound of the price of a Bermudan. We could optimize over all trigger levels to improve this bound. The value of the remaining optionality decreases with time because there are fewer exercise dates left and because the underlying becomes a swap of shorter and shorter length. We will therefore want to let the trigger level depend on time also. Close to the start where the option is worth a lot, we could expect the optimal trigger level to be higher than at the end, when it is optimal to exercise if and only if the swap is in the money.

Suppose we have chosen an exercise strategy which could be based on all of the information available. We could equally well have made it depend on the level of the remaining swap rate which would probably be better, or the level of the discount factors remaining. As all the information available is any case determined by the forward rates, we can regard our exercise strategy as a function from the set of forward rates cross (i.e. Cartesian product) the set of exercise times to the set

$$\{\text{exercise}, \quad \text{don't exercise}\}.$$

We then proceed simply as in the trigger swap case. At each time, we check the value of the exercise function, if it says *exercise*, we value the swap and numeraire as for the trigger swap. Otherwise, we go on to the next time and check again. We repeat until we reach the end, or the strategy tells us to exercise.

Of course, valuing the swaption once the exercise strategy has been chosen is the easy part – choosing it is hard. We discuss exercise strategies in Section 14.9.

The LIBOR-in-arrears FRA or caplet

A forward-rate agreement is associated to two times t_0 and t_1. As usual, let $f_0(t)$ denote forward rate from t_0 to t_1, at time t. The pay-off of the *LIBOR-in-arrears FRA* is $(t_1 - t_0) f_0(t_0)$ and is paid at t_0. The pay-off of the LIBOR-in-arrears caplet is simply the positive part of the pay-off of the LIBOR-in-arrears FRA. To simulate these options, we take the bond expiring at t_0 as numeraire and then evolve to t_0 where the pay-off is immediately computable.

A caption or a floortion

A cap is a sequence of caplets, and similarly for floors. Valuing a cap or floor does not require BGM: we just add together the prices of the caplets or floorlets and we are done. Indeed, caps and floors are more liquid than caplets and floorlets so it really the opposite procedure which is carried out: cap prices are observed and then caplet prices are inferred.

An option on a cap, known as a *caption*, is a more interesting object. As some time, typically the expiry time of the first caplet, we have the option to purchase the cap for a pre-agreed price. As the option is on all the caplets at once, we cannot decompose into the individual caplets. Indeed, an option on a caplet at the expiry of a caplet would be a rather boring instrument – it's just a caplet. How do we approach the valuation of the caption? If we assume that volatilities are deterministic, which is an explicit assumption of BGM in any case, then at any time, given the state of the yield curve, we can rapidly price all the caplets by using the Black formula with the residual volatility for each caplet.

Our procedure is now easy. We take as numeraire the zero-coupon bond with the same expiry as the caption. We evolve the rates underlying the caplets to the expiry time of the caption. We then compute the caption's payoff by pricing the caplets using the Black formula, subtracting the strike price and taking the maximum with zero.

Similarly for *floortions*. Note that we should however be slightly careful with compound options because of the possible change in volatilities during the life of the option. We refer the reader to Chapter 18 for further discussion of this point.

The cash-settled swaption

True swaptions are actually less liquid than a closely related variant, the *cash-settled swaption*. A swaption in its original form is the right to enter into a swap. However, the purpose of holding a swaption is often to hedge certain risks, possibly volatility risk. When this is the case, one does not wish to actually enter into a swap when the swaption is in the money, rather one desires just to receive the cash value of the swap. The problem is in agreeing this cash value: the value depends on the discount factors which make up the swap's annuity. These discount factors are not a market observable and the swap counterparties are unlikely to agree them precisely. Of course, the swaption holder could just trade a swap in the market to balance

14.2 Decomposing an instrument into forward rates

the swap coming from the swaption. However, this would expose the holder to bid-offer spread, the necessity to manage the two swaps over what could be a long period of time, and counterparty credit risk.

To cope with these problems, the cash-settled swaption was devised. In computation of the annuity, the swap rate itself is used as interest rate across each period. Thus if the period of the swap is τ and there are N periods, the cash-settled swaption's payoff is

$$(\text{SR} - K)_+ \sum_{j=1}^{N} (1 + \text{SR}\tau)^{-j} \tau.$$

To price this using BGM, we simply need to evolve the forward rates underlying the swap up to the expiry time of the swaption, and then compute the swap rate.

What numeraire would we choose? It's important to realize that we cannot use the approximate annuity,

$$\sum_{j=1}^{N} (1 + \text{SR}\tau)^{-j} \tau,$$

as numeraire. Why? It is not the price process of a traded asset. Whilst we can certainly define an asset which pays this amount at time T, its value at previous times will not be equal to $\sum_{j=1}^{N} (1 + \text{SR}\tau)^{-j} \tau$. We therefore use a zero-coupon bond such as the bond maturing at the expiry time of the swaption as numeraire.

What happens in the markets? Although there is no theoretical justification for using the Black formula with the approximate annuity to price a cash-settled swaption, in fact this is often done. Of course, the trader will adjust his volatility in order to take account of the errors.

The constant maturity swap

The *constant maturity swap* (CMS) is a hybrid between a forward-rate agreement and a swap. It is a series of payments each one based on a fixed length of swap rate rather than the prevailing LIBOR rate in an ordinary swap.

Thus at each reset time, t, we compute

$$(\text{SR}(t) - K)\tau$$

where τ is the accrual period, and we receive this sum at $t + \tau$. The rate $\text{SR}(t)$ is a fixed length rate observable on that day; for example the ten-year rate, starting at time t.

Thus every six months we would observe the ten-year swap rate, X, and arrange to pay $0.5(X - K)$ six months later.

To value using BGM, we can do each payment individually. The swap rate is then computable from the forward rates over the time periods underlying the swap. We also need to simulate the discounting from t to $t + \tau$. This means that we also need the forward rate from t to $t + \tau$.

It is important to realize that there's an alternative approach to pricing CMS swaps (and swaptions) which is to observe that both the CMS and cash-settled swaption have pay-offs which are functions of the prevailing swap rate (except for the fact that the CMS pays a little later), which means that one can replicate the CMS pay-off using cash-settled swaptions. We can therefore price by strong static replication provided we take the timing mismatch into account.

The general procedure

Given a new instrument which has no early exercise opportunities, how do we proceed? First, we must identify the times at which we need to know the state of the yield curve, whilst remembering that a forward rate at its own reset time can be observed at any time after that reset. Second, we must recognize which forward rates are necessary to computing the payoff. There are two ways in which a forward rate can be necessary. One is simply in terms of defining the size (or existence) of a cashflow. The second is in terms of its use in discounting cashflows, or equivalently in assessing the ratio of the value of the cashflow to the numeraire. Having decided which forward rates are relevant, we must then write the ratio of the payoff to the numeraire as a function of the forward rates. We give a rather contrived example to illustrate the process.

Example 14.1 Let B_1, B_2 be two zero-coupon bonds expiring at times t_1, t_2 respectively. We take $t_1 < t_2$. Let $B_j(t)$ denote the value of B_j at time t. Let t_0 and t_3 be such that t_j is strictly increasing.

Suppose we have an instrument that pays

$$(B_1(t_0) - B_2(t_0))^2$$

at time t_3. The cashflow only depends upon the state of the yield curve at time t_0. We therefore have one evolution time: t_0. Observing that the value of a bond expiring at t_0 at time t_0 is 1, we can write

$$B_1(t_0) = \frac{1}{1 + f_0(t_0)(t_1 - t_0)}. \tag{14.7}$$

where f_j denotes the forward rate running from t_j to t_{j+1}. Similarly, we have

$$B_2(t_0) = \frac{1}{(1 + f_0(t_0)(t_1 - t_0))(1 + f_1(t_0)(t_2 - t_1))}. \tag{14.8}$$

The size of our cashflow is therefore easily expressed in terms of f_0 and f_1. However, we still need to consider discounting. If we take B_0 (the bond expiring at t_0) as numeraire then we have to multiply the cashflow by

$$\prod_{j=0}^{2} \frac{1}{1 + f_j(t_0)(t_{j+1} - t_j)}.$$

If, however, we take the bond B_3 (the bond expiring at t_3) as numeraire then we do not have to multiply the cashflow. However, we would still need to include f_3 in our simulation since its value will affect the drifts of the other forward rates, as we shall see in Section 14.3. ◇

Which instruments are possible?

Whilst BGM is a general approach, there are instruments that do not fit within the framework. What do we need to make BGM work? The instrument's pay-off should depend upon the value of a finite number of forward rates, and bond prices on a finite number of dates. The restriction to a finite number of rates and bond prices is not really a restriction at all – there are only a finite number of instruments and rates to observe the market prices of, so the pay-off function is inevitably a function of a finite number of them. The second restriction is more of a problem. For example, suppose we have a trigger swap that triggers whenever a certain LIBOR rate passes a reference level. If the trigger occurs at any time rather than on a certain discrete subset of dates, then our finiteness condition is violated and we will not be able to fit the product into the BGM framework. Similarly, an option to exchange two bonds at a time of the holder's choice will not fit into the framework.

14.3 Computing the drift of a forward rate

We have observed that we can only expect forward rates to be martingales in the risk-neutral measure when the numeraire is the zero-coupon bond with maturity equal to the pay-off time of the forward rate. This means that in general we will need to compute the drifts of our forward rates.

Note that any rate is really the ratio of two asset prices, so the rate will be a martingale when the second asset has been taken as numeraire, and it will generally not be a martingale otherwise. This second asset is sometimes called the *natural pay-off* for the rate.

The simplest case to compute is the LIBOR-in-arrears case, that is when the numeraire is the bond B_0 expiring at t_0 and the forward rate f runs from t_0 to

t_1. Let $\tau = t_1 - t_0$. We know that fB_1 is tradable so $f\frac{B_1}{B_0}$ is a martingale in the risk-neutral measure. Thus we have that

$$f\frac{B_1}{B_0} = \frac{f}{1+f\tau}, \tag{14.9}$$

is driftless. We also know that $\frac{B_1}{B_0}$ is driftless, that is $\frac{1}{1+f\tau}$ is driftless.

Recall that given two Ito processes X_t and Y_t with respect to the same Brownian motion, we have

$$d(X_t Y_t) = X_t dY_t + Y_t dX_t + dX_t dY_t. \tag{14.10}$$

If we let $X_t = f$, and $Y_t = (1+f\tau)^{-1}$ then we can compute the drift of $X_t Y_t$, which we know must be zero, and use this to deduce the drift of f.

Let

$$df_t = \mu(f,t)f dt + \sigma(t)f dW_t, \tag{14.11}$$

then

$$d\left(\frac{1}{1+f\tau}\right) = -\frac{\tau\sigma f}{(1+f\tau)^2} dW, \tag{14.12}$$

using the fact P_1/P_0 is a martingale.

Computing and using the fact that $dW^2 = dt$, we find that the drift of $f\frac{P_1}{P_0}$ is

$$\frac{\mu(f,t)f}{1+f\tau} - \frac{\sigma^2 \tau f^2}{(1+f\tau)^2}, \tag{14.13}$$

which must be zero by the martingale property. We therefore conclude that

$$\frac{df_t}{f} = \frac{\tau f}{1+\tau f}\sigma^2 dt + \sigma dW_t. \tag{14.14}$$

Note that the crucial part of this argument was to find a tradable asset, B, such that fB was tradable. If N is the numeraire, then the fact that both B/N and fB/N are martingales in the risk-neutral measure is what allows us to determine the drift of f.

An important aspect of (14.14) is that not only is f not driftless but that its drift is state-dependent, that is, it depends on the value of f. This complicates the running of Monte Carlo simulations considerably as we no longer have a nice closed-form solution for the stochastic differential equation.

We can use similar arguments to compute the drifts of a forward rate with other numeraires. As we have seen in the previous section, we will generally want to evolve a string of forward rates, f_j, which span contiguous intervals of time, $[t_j, t_{j+1}]$. Let B_j be the zero-coupon bond expiring at time t_j. If we choose B_l as numeraire then we have that f_{l-1} is driftless and we have just computed the drift of f_l. How do we compute the drift of f_{l+r}?

We know that $f_{l+r}B_{l+r+1}$ is tradable and that B_{l+r+1} is a tradable. Hence, both $f_{l+r}B_{l+r+1}B_l^{-1}$ and $B_{l+r+1}B_l^{-1}$ are martingales. In terms of forward rates, this says that

$$Z_r = \frac{f_{l+r}}{\prod_{j=0}^{r}(1+f_{l+j}\tau_{l+j})}$$

and

$$Y_r = \frac{1}{\prod_{j=0}^{r}(1+f_{l+j}\tau_{l+j})}$$

are driftless. To infer the drift of f_{l+r} we compute that of Z_r, which must be zero, in terms of Y_r and f_{l+r}. Let

$$df_{l+j} = \mu_{l+j}(f_{l+j})f_{l+j}dt + \sigma_{j+l}f_{l+j}dW_j.$$

Using the product rule and remembering that Y_r must have zero drift,

$$dY_r = -Y_r \sum_{j=0}^{r} \frac{\tau_{l+j}}{1+f_{l+j}\tau_{l+j}} \sigma_{l+j} f_{l+j} dW_j. \qquad (14.15)$$

We have that

$$dW_j dW_r = \rho_{l+j,l+r} dt \qquad (14.16)$$

where $\rho_{l+j,l+r}$ is the instantaneous correlation between the forward rates f_{l+j} and f_{l+r}. Using

$$dZ_r = f_{l+r}dY_r + Y_r df_{l+r} + df_{l+r}dY_r, \qquad (14.17)$$

we have on computing the drift of Z_r that

$$Y_r \mu_{l+r} f_{l+r} - Y_r \sum_{j=0}^{r} \frac{f_{l+j}\tau_{l+j}}{1+f_{l+j}\tau_{l+j}} \rho_{l+j,l+r} f_{l+r} \sigma_{l+r} \sigma_{l+j} = 0. \qquad (14.18)$$

Hence, we conclude that

$$\mu_{l+r} = \sum_{j=0}^{r} \frac{f_{l+j}\tau_{l+j}}{1+f_{l+j}\tau_{l+j}} \rho_{l+j,l+r} \sigma_{l+r} \sigma_{l+j}. \qquad (14.19)$$

We have computed a not very nice, but extremely useful, expression for the drift. We have in general that the drift is state-dependent and depends not just on the forward rate f_{l+r} but also on all the others. This will mean that the interaction amongst all the rates will have to be considered when evolving them in a simulation.

The expression (14.19) holds for numeraires which are too short, that is, the numeraire bond's maturity is before the payment time of the forward rate. We will equally need to evolve forward rates when the numeraire bond is too long: that is, it has maturity after the payment date of the bond. Keeping B_l as numeraire, we want to compute the drift of f_{l-r} for $r > 1$. The drift for $r = 1$ is of course zero. A very similar argument yields a very similar expression with the principal difference being one of sign:

$$\mu_{l-r} = -\sum_{j=1}^{r-1} \frac{f_{l-r+j}\tau_{l-r+j}}{1 + f_{l-r+j}\tau_{l-r+j}} \rho_{l-r,l-r+j} \sigma_{l-r} \sigma_{l-r+j}. \tag{14.20}$$

The reader is encouraged to derive this expression!

Note that if we are allowing volatilities to be time-dependent, which we will need to do later, then the drifts are implicitly time-dependent, as well as state-dependent.

14.4 The instantaneous volatility curves

In our discussion of the reverse option, we saw that there were two important reasons to understand the instantaneous volatility curves of the forward rates. The first was that we will often need to evolve a forward to some time other than its expiry, and we therefore need to know how much volatility has been used up by a previous time. The second is that the shapes of the instantaneous volatility curves affect the degree of correlation between rates and will therefore critically affect the price of highly slope-dependent instruments.

As we have mentioned above the price of a caplet gives us the root-mean-square volatility of the underlying forward. Thus if the forward rate runs from time t_0 to time t_1, and the volatility function is $\sigma(t)$ then we can recover

$$\sigma_{\text{total}} = \int_0^{t_0} \sigma(t)^2 dt.$$

There are clearly lots of functions which give the same value to σ_{total}. In the absence of further market information, we must therefore make a choice.

One assumption we can make to narrow the choice, is that the shape of a forward-rate volatility curve ought to be time-homogeneous. This means that, in future, the forward-rate volatility curves, as a function of time-to-go, should look the same as they do today. We can achieve this by letting the instantaneous volatility of a forward rate be a function of the amount of time until its reset. This means that today's two-year rate will have volatility one year from now equal to the volatility of the one-year rate today.

So if we have forward rates f_j running from t_j to t_{j+1}, we can take the volatility, $\sigma_j(t)$, of f_j to have the form $p(t_j - t)$ where the function p is independent of j. This expresses the idea that the behaviour of the forward rate should be largely

14.4 The instantaneous volatility curves

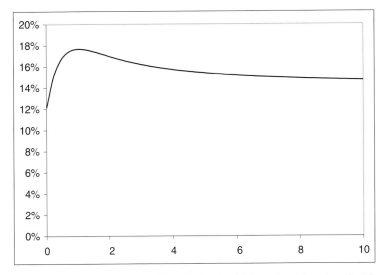

Fig. 14.1. The term structure of implied volatilities of caplets implied by the functional form (14.21), with $a = -0.02$, $b = 0.3$, $c = 2$, and $d = 0.14$.

dependent on the time until its own expiry, rather than on absolute time. This reflects the observation that the volatility of forward rates is low with a long time to expiry, has a high around two years to expiry and then falls again close to expiry.

One still needs to choose a functional form for p. One simple form that works reasonably well is to put

$$p(s) = (a + bs)e^{-cs} + d, \qquad (14.21)$$

where a, b, c, d are the parameters to fit with. (See figures 14.1 and 14.2.) The virtues of this function are that is sufficiently flexible to allow an initial steep rise followed by a slow decay, and that its square has an analytic integral. This virtue is very important – when carrying out any fit, one needs to be able to rapidly evaluate the function a large number of times; doing a numeric integration would simply not be practical.

Thus we take all the observed caplet volatilities in the market and search for the values of a, b, c and d which makes $p(t_j - t)$ fit them all optimally. That is, if the caplet C_j starting at time t_j has observed r.m.s. volatility σ_j, those values such that the quantities

$$\sigma_j^2 t_j - \int_0^{t_j} p(t_j - s)^2 ds$$

are made as close to zero as possible. For this we would fit all the caplets in the market rather than just the ones we need to simulate, since we want to use as much information as possible in inferring the shape of the volatility curves.

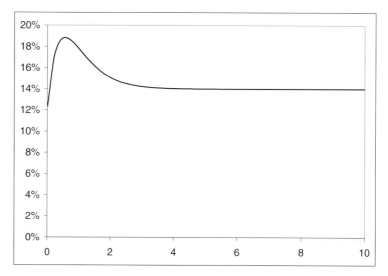

Fig. 14.2. The instantaneous volatility of a caplet as a function of time to expiry implied by the functional form (14.21), with $a = -0.02$, $b = 0.3$, $c = 2$, and $d = 0.14$.

It will, however, be impossible to obtain a perfect fit to all the caplets simultaneously. This means that we then have to rescale p by a different factor, K_j, for each f_j in order to ensure that all the caplets are priced correctly.

Our procedure is therefore to find the a, b, c and d which best match all the caplet volatilities and then letting p be the implied function, we set K_j so that

$$\sigma_j^2 t_j = K_j^2 \int_0^{t_j} p(t_j - s)^2 ds. \tag{14.22}$$

The instantaneous volatility function, $\sigma_j(t)$, for f_j is then $K_j p(t_j - t)$. We have thus achieved a perfect fit to all the caplet volatilities at the cost of losing a little time-homogeneity.

How much time-homogeneity has been lost is reflected by the divergence of the K factors. If they are very close to 1 then our parametric form has matched the market well, if they are far then perhaps our fit is bad, or it may be that our parametric form is simply not suitable for the current market. If a good fit cannot be obtained then it is probably time to consider a different parametric form, or to consider whether there's an economic reason to expect a lack of time-homogeneity.

For example, interest rate volatilities for the millennium period were much higher because of worries about the millennium bug. Similarly, volatilities tend to be much higher around uncertain U.S. presidential elections. If we do believe in a good reason for the lack of time-homogeneity, we might want to try a fitting function of the form $h(t)g(t_j - t)$, with h reflecting the expected information arrival rate, whilst g reflects the sensitivity of the forward rate f_j to the information.

14.5 The instantaneous correlations between forward rates

We have seen that for certain options assessing the amount of decorrelation between neighbouring forward rates is crucial for pricing. We therefore want to estimate the correlation between forward rates. This is a very tricky quantity to get hold of – there's no obvious instrument whose price directly reflects it. We can expect the movement of neighbouring forward rates to be more closely correlated than those that are far apart and since most of the movements of the yield curve are up and down, we can also expect the instantaneous correlations to be high.

An easy simplification we can make is to assume that the correlation between forward rates is purely a function of the number of years that separate them. That is, the correlation, ρ_{ij}, between the rates f_i and f_j should be of the form $\rho(|t_i - t_j|)$. This assumption is arguably dubious in that a nineteen-year rate is probably more correlated with a twenty-year rate than a six-month rate is with an eighteen-month one. However, it's a reasonable first approximation and we shall use it here.

Suppose we have three forward rates, f_1, f_2, f_3, associated to times T_j, such that $T_1 < T_2 < T_3$; now suppose that any movements of f_1 which are uncorrelated with movements of f_2, are also uncorrelated with movements of f_3. This is not an unreasonable assumption in that we would not expect information to arrive which would affect both f_1 and f_3 but not f_2. This assumption may be criticized on statistical grounds but it is not too bad. How do we translate our assumption into mathematical terms? Here we assume that the processes for the rates can be written as follows:

$$df_1 = f_1\mu_1 dt + f_1\sigma_1\left(\sqrt{1 - \rho_{12}^2}dW_1 + \rho_{12}dW_2\right), \tag{14.23}$$

$$df_2 = f_2\mu_2 dt + f_2\sigma_2 dW_2, \tag{14.24}$$

$$df_3 = f_3\mu_3 dt + f_3\sigma_3\left(\rho_{23}dW_2 + \sqrt{1 - \rho_{23}^2}dW_3\right), \tag{14.25}$$

where W_1, W_2 and W_3 are uncorrelated Brownian motions. The fact that W_1 and W_3 are uncorrelated comes from our decorrelation assumption. Note that this is a quite strong interpretation of our condition.

It is then immediate that the instantaneous correlation of f_1 and f_3 is just $\rho_{12}\rho_{23}$. This means that if $\rho_{ij} = \rho(T_i - T_j)$, then

$$\rho(T_3 - T_1) = \rho(T_2 - T_1)\rho(T_3 - T_2). \tag{14.26}$$

This means that ρ must satisfy the functional equation

$$\rho(s_1 + s_2) = \rho(s_1)\rho(s_2). \tag{14.27}$$

This implies that $\log \rho$ is linear and hence that

$$\rho(s) = \exp(\alpha s). \tag{14.28}$$

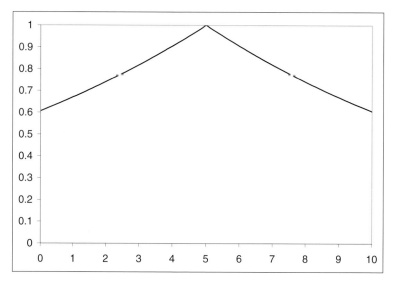

Fig. 14.3. The correlation between a forward rate expiring in five years and the forward rate expiring at time T according to the functional form (14.29) with β equal to 0.1.

Given that ρ must be less than or equal to 1, and that $\rho(0) = 1$, we conclude that

$$\rho_{ij} = \rho(T_i - T_j) = \exp(-\beta |T_i - T_j|), \tag{14.29}$$

for some $\beta \geq 0$.

We will still need to choose β. To do so we can use the additional data available from swaption volatilities. In particular for any given value of β, having chosen our instantaneous forward-rate volatility curves, we can price a swaption by running a BGM Monte Carlo. The price of the swaption will be affected by β since the swap's volatility will increase as the forward rates underlying it become more correlated. Thus we can choose the value of β which best matches all the swaption volatilities. Typically a β of around 0.1 tends to fit the market quite well. (See figure 14.3.)

14.6 Doing the simulation

Having decomposed the product, selected the instantaneous volatility curves, computed the drifts and chosen the instantaneous correlations, we can now run a Monte Carlo simulation and obtain a price.

However, this is not as easy as it sounds – for two reasons. The first is that the shapes of the instantaneous volatility curves will affect the correlations between assets, and the second is that we have state-dependent drifts. If we had infinitely-fast computers there would not be any issues: we would simply divide time into

14.6 Doing the simulation

lots of small steps and evolve over little steps. However, running a multi-asset simulation is time-intensive and if we run lots and lots of little steps, the amount of time required will be prohibitive.

We therefore want a way of accurately evolving over long steps without looking at the intervening values. For the second problem, we can only use approximations, though we can find very good ones. For the first problem, there is a method of solving the stochastic differential equation which involves computing the effective correlation over a long step; we now examine this. We take the drift to be constant for the purpose of this discussion.

Thus suppose we have

$$df_j = f_j \mu_j dt + f_j \sigma_j(t) dW_j, \tag{14.30}$$

where μ_j is constant and the correlation between the Brownian motions W_j and W_k is a known constant quantity, ρ_{jk}. Since f_j is log-normal we have

$$f_j(T) = f_j(0) \exp\left(\mu_j T - \frac{1}{2}\bar{\sigma}^2 T + \bar{\sigma}\sqrt{T} X_j\right), \tag{14.31}$$

where $\bar{\sigma}$ is the root-mean-square value of σ over $[0, T]$, and for each j, X_j is a normally-distributed random variable with mean 0 and variance 1. The remaining question is what the correlation between X_j and X_k is. An obvious first answer would be ρ_{jk} but this is incorrect. The reason is that the instantaneous volatility curves' differing shapes mean that the forward rates will have different sensitivities to a piece of information arriving at a given time, which will cause extra decorrelation. (See (14.4) for a simple example.)

Suppose we have two quantities moving under scaled Brownian motions:

$$g_1 = \sigma_1(t) dW_t^{(1)} \tag{14.32}$$
$$g_2 = \sigma_2(t) dW_t^{(2)}. \tag{14.33}$$

If the instantaneous correlation between $W^{(1)}$ and $W^{(2)}$ is ρ, then, introducing a new Brownian process $W^{(3)}$ which is independent of $W^{(1)}$, we can write

$$W^{(2)} = \rho W^{(1)} + \sqrt{1-\rho^2} W^{(3)}, \tag{14.34}$$

and hence

$$dg_1 = \sigma_1(t) dW_t^{(1)} \tag{14.35}$$
$$dg_2 = \sigma_2(t) \rho dW_t^{(1)} + \sigma_2(t)\sqrt{1-\rho^2} dW_t^{(3)}. \tag{14.36}$$

As W_1 and W_3 are uncorrelated, the correlation between g_1 and g_2 comes purely from the first dW^1 term. We therefore need to understand the correlation between g_1 and g_3 defined by

$$dg_3 = \sigma_2(t) dW_t^{(1)}. \tag{14.37}$$

The processes g_1 and g_3 are perfectly instantaneously correlated; all decorrelation comes from the different shapes of $\sigma_1(t)$ and $\sigma_2(t)$. We suppose that $\sigma_1(t)$ and $\sigma_2(t)$ are piecewise constant. For notational convenience let $\sigma_3(t) = \sigma_2(t)$. We thus have that there exists a strictly increasing sequence of times, t_j, such that on the interval $[t_j, t_{j+1}]$, we have that $\sigma_k(t)$ takes the constant value $\sigma_{k,j}$. For notional convenience write $T = t_n$. Then

$$g_k(T) - g_k(0) = \sum_{j=0}^{n-1} \sigma_{k,j}(W_{t_{j+1}} - W_{t_j}). \tag{14.38}$$

We want the covariance of $g_1(T) - g_1(0)$, and $g_3(T) - g_3(0)$. Since the increments of a Brownian motion are independent, we have

$$\mathbb{E}((W_{t_{j+1}} - W_{t_j})(W_{t_{l+1}} - W_{t_l})) = 0, \tag{14.39}$$

for $j \neq l$. For $j = l$, this expectation is the variance of $W_{t_{j+1}} - W_{t_j}$ which is $(t_{j+1} - t_j)$. This means that

$$\mathbb{E}((g_3(T) - g_3(0))(g_1(T) - g_1(0))) = \sum_{j=0}^{n-1} \sigma_{1,j}\sigma_{3,j}(t_{j+1} - t_j). \tag{14.40}$$

We can identify this sum as $\int_0^T \sigma_1(t)\sigma_3(t) dt$. We conclude that the correlation coefficient of $g_3(T) - g_3(0)$ and $g_1(T) - g_1(0)$ is

$$\frac{\int_0^T \sigma_1(t)\sigma_2(t) dt}{\left(\int_0^T \sigma_1(t)^2 dt\right)^{\frac{1}{2}} \left(\int_0^T \sigma_2(t)^2 dt\right)^{\frac{1}{2}}},$$

using the fact that $\sigma_2(t)$ and $\sigma_3(t)$ are equal.

Of course, we actually wanted the correlation between g_1 and g_2 rather than g_1 and g_3. However, we can write

$$g_2(T) - g_2(0) = \rho(g_3(T) - g_3(0)) + \sqrt{1-\rho^2} Z, \tag{14.41}$$

where Z is an independent normal variable. It is therefore immediate that

$$\rho(g_1(T) - g_1(0), g_2(T) - g_2(0)) = \rho \frac{\int \sigma_1(t)\sigma_2(t) dt}{\left(\int_0^T \sigma_1(t)^2 dt\right)^{\frac{1}{2}} \left(\int_0^T \sigma_2(t)^2 dt\right)^{\frac{1}{2}}}. \tag{14.42}$$

Whilst we have only deduced this expression for piecewise constant σ, a limiting argument shows that it remains true for general volatility functions. Note that the final values of the rates have less correlation that the instantaneous changes. This effect is known as *terminal decorrelation*.

We can now return to (14.31). To do the evolution from zero to time T we populate the covariance matrix for the logs with

$$\rho_{jk} \int_0^T \sigma_j(t)\sigma_k(t)dt,$$

and then proceed as in Chapter 11. More generally we could carry out an evolution between any two times, T_1 and T_2, by taking the covariance matrix of the logs to be given by the integral over the interval $[T_1, T_2]$.

We have been ignoring the state-dependence of the drift, and also the implicit time-dependence of the drift coming from the time-dependence of the instantaneous volatility functions. Taking the latter into account is simple. If we want to evolve from time zero to time T, it is really the integral of the drift that is important, so, ignoring the state-dependence of the drift, we can use as total drift

$$\int_0^T \mu_{l+r}(t)dt = \sum_{j=0}^r \frac{f_{l+j}\tau_{l+j}}{1+f_{l+j}\tau_{l+j}} \rho_{l+j,l+r} \int_0^T \sigma_{l+r}(t)\sigma_{l+j}(t)dt. \qquad (14.43)$$

If the forward rates were constant along the path, this drift would be precisely correct.

However, they are not; the whole point of our model is that they are state-dependent and therefore effectively time-dependent. We can pretend that they are constant and obtain a reasonable approximation for medium-sized time steps, for example one year. But the approximation is too rough for longer steps.

We present an improved approximation method which works well in practice. Since we are evolving the log, the volatility is state-independent and the contribution of the drift is generally small in comparison. We can therefore think of the drift as a perturbation to the driftless equation. Our procedure is first to evolve pretending the drift is constant, recompute the drift at the evolved time, and take the average of the beginning and ending drifts. We then re-evolve using the same random numbers with this evolved drift. This approximation works extremely well, allowing evolution out to ten years for reasonable values of volatility with only small errors. This is a variant of the *predictor-corrector* method of solving ordinary differential equations.

We have completed our programme for developing the BGM model. We can now price any interest rate derivative which does not involve early exercise decisions by running a Monte Carlo simulation. For early exercise, we would have to apply a variant of the Monte Carlo techniques discussed in Chapter 12.

14.7 Rapid pricing of swaptions in a BGM model

We have covered many of the principal ideas involved in implementing a BGM-type approach. In this section, we mention some of the further issues involved in implementing the model. One natural question is why use forward rates? Jamshidian has suggested using swap rates instead. If we use a set of swap rates with the same ending date and the same fixing dates but different starting dates, then they will determine discrete points on the discount curve in the same way that forward rates did. Such a collection of swaps is sometimes said to be *co-terminal*. If they have the same start date but different end dates, they are *co-initial*. We can therefore evolve swap rates instead of forward rates and apply similar techniques. The advantage of this approach is that calibration to the swap-rate volatilities is automatic, and if one is pricing a product that is essentially dependent on swap rates this is an extremely useful feature.

We may, however, wish to calibrate our LIBOR market model to swaption volatilities. To do so, we will have to be able to price swaptions rapidly. One can always run a Monte Carlo simulation for each swaption and compute the price. However, if one is carrying out a calibration, fast repeated calculations will be needed and Monte Carlo is generally not fast enough for that.

Rebonato has suggested an approximation which can be very effective. Recall (13.11),

$$X = \sum_{j=0}^{n-1} w_j f_j, \qquad (14.44)$$

where X is the swap rate. We therefore have

$$dX = \sum_{j=0}^{n-1} w_j df_j + \sum_{j=0}^{n-1} (dw_j df_j + dw_j f_j). \qquad (14.45)$$

This is not very tractable; however if we make the courageous assumption that w_j is constant, we obtain

$$dX = \sum_{j=0}^{n-1} w_j df_j. \qquad (14.46)$$

Computing, this implies that

$$dX = \left(\sum_{j=0}^{n-1} \sum_{k=0}^{n-1} w_j w_k \rho_{jk} \sigma_j \sigma_k f_j f_k \right)^{1/2} dZ, \qquad (14.47)$$

plus possible drift terms, where Z is a Brownian motion. If we want to think of X as log-normal, this becomes

$$\frac{dX}{X} = \left(\sum_{j=0}^{n-1} \sum_{k=0}^{n-1} w_j w_k \rho_{jk} \sigma_j \sigma_k f_j f_k X^{-2} \right)^{1/2} dZ. \qquad (14.48)$$

We can therefore approximate the variance of log X to put in the Black formula for an option of expiry T by

$$X^{-2} \int_0^T \sum_{j=0}^{n-1} \sum_{k=0}^{n-1} w_j w_k \rho_{jk} \sigma_j(t) \sigma_k(t) f_j f_k dt.$$

Here X, w_l and f_l are taken to have their initial values rather than their stochastic values. Whilst, this expression has been derived in a slightly hand-waving fashion it is surprisingly accurate. This means that we can price swaptions quickly and analytically which allows rapid calibration to the swaption market.

This formula involved approximating the derivative of the swap rate with respect to the forward rate by the weight w_j. The formula can be improved slightly by removing this approximation and instead computing the derivative precisely.

14.8 Automatic calibration to co-terminal swaptions

If we are pricing a Bermudan swaption then our primary concern is to correctly price all the European swaptions underlying the product. In particular, if the product has exercise and reset times

$$t_0 < t_1 < \cdots < t_n,$$

then for each j the European swaption associated to the set $t_j, t_{j+1}, \ldots, t_n$ underlies the Bermudan, and its price must be a lower bound for the Bermudan's price – just consider using the exercise strategy of 'exercise at time t_j if in the money and do not exercise at any time otherwise'. These swaptions are also natural hedges for the Bermudan swaption, we can use these swaptions to Vega hedge to reduce our exposure to changes in volatility.

We therefore wish to calibrate to co-terminal swaption prices; one solution is just to use a swap rate market model. However, sometimes we will want to be sure both of caplet and swaption prices, particularly when the product has additional exotic features, so it is useful to be able to calibrate the FRA-based model to the swaption prices. We can do this by extending the ideas of the last section. We think in terms of a change of coordinates – we wish to specify covariances in a swap rate framework but work in a FRA-based framework. This means that we have to understand how the covariance matrix transforms.

Let SR_j denote the swap rate for times $t_j, t_{j+1}, \ldots, t_n$, and f_j the forward rate for t_j and t_{j+1}. Note that we can certainly express each swap rate in terms of forward rates and by a tedious calculation it is possible to do the reverse.

We want to relate the processes for the two sorts of rates. If we ignore drifts, we can write

$$d \log SR_j = \sum_{k=0}^{n-1} \frac{\partial \log SR_j}{\partial \log f_k} d \log f_k \qquad (14.49)$$

which implies

$$d \log \text{SR}_j = \sum_{k=0}^{n-1} \frac{f_k}{\text{SR}_j} \frac{\partial \text{SR}_j}{\partial f_k} d \log f_k. \qquad (14.50)$$

Let

$$Z_{jk} = \frac{f_k}{\text{SR}_j} \frac{\partial \text{SR}_j}{\partial f_k}. \qquad (14.51)$$

Clearly, we have that Z_{jk} is zero for $j < k$, and the diagonal elements, Z_{jj}, will be non-zero so the matrix has non-zero determinant and is therefore invertible. Writing $\log \text{SR}$ for the vector of rates $\log \text{SR}_j$ and similarly for the FRAs we can now write

$$d \log SR = Z d \log f, \qquad (14.52)$$

still ignoring drifts.

This mean that if the instantaneous covariance matrix of the log forward rates (i.e. the product of the instantaneous correlations and the volatilities), is C^f, then the instantaneous covariance matrix of the log swap rates, C^{SR}, is equal to

$$Z C^f Z^t.$$

To see this we consider

$$d \log \text{SR}_i d \log \text{SR}_j = \sum_k Z_{ik} d \log f_k \sum_l Z_{jl} d \log f_l$$

$$= \sum_{k,l} Z_{ik} d \log f_j d \log f_l Z_{lj}^t. \qquad (14.53)$$

Now

$$d \log f_i d \log f_j = \rho \sigma_i \sigma_j dt, \qquad (14.54)$$

where ρ is the instantaneous correlation, and similarly for the swap rates. We therefore conclude that if we cancel dt then we have

$$C^{\text{SR}} = Z C^f Z^t \qquad (14.55)$$

as desired. Since Z is invertible we can equally well go in the opposite direction, and write

$$C^f = Z^{-1} C^{\text{SR}} \left(Z^t\right)^{-1}. \qquad (14.56)$$

However, this is still not particularly useful in that it is an instantaneous expression and we will want to evolve rates across long time intervals, and we will not want to have to recompute Z every time step since it would be rather slow. We therefore make an approximation, by taking Z to be equal to $Z(0)$, and taking the swap-rate covariance matrix across an interval (s, t) to be equal to the

14.8 Automatic calibration to co-terminal swaptions

conjugated FRA-rate covariance matrix across the same interval. In particular, we write

$$C^{SR}(s, t) = Z(0) C^f(s, t) Z(0)^t \qquad (14.57)$$

for any $s < t$.

We have to be a little careful with this expression because we do not want a swap rate to continue to change after its own reset although the FRAs underlying it will continue to change until their own resets. We therefore cut time into intervals $[s_i, s_{i+1}]$ such that for each i such that no swap rates reset in the interior of the intervals. Over each interval we then just work with the restricted set of forward rates and swap rates which have not yet reset.

This means that we can now prescribe a swap-rate covariance structure and use (14.57) to infer a FRA covariance structure and use that to run a BGM simulation. The only difficulty is that it is hard to get a good feeling for the covariances of swap rates. We can therefore adopt a compromise approach where the variances of the swap rates are determined by the market but we infer correlations from the forward rates.

Thus suppose we have calibrated the BGM model to a set of caplets and decided forward-rate correlations and instantaneous volatilities. This means that we have determined $C^f(s, t)$ for all s and t, i.e. we can infer that

$$C^{SR}(0, t_j) = Z(0) C^f(0, t_j) Z(0)^t \qquad (14.58)$$

for each j. If the model is pricing the swaptions correctly then we should have that the implied volatility of SR_j, $\hat{\sigma}_j$, is equal to

$$\sigma_j = \sqrt{\frac{1}{t_j} C^{SR}_{jj}(0, t_j)}.$$

In general, it will not be. We therefore need to adjust appropriately. Let

$$\lambda_j = \frac{\hat{\sigma}_j}{\sigma_j}.$$

Let Λ be the diagonal matrix with entries equal to Λ_j; then if we replace $C^f(s, t)$ by

$$\tilde{C}^f(s, t) = Z(0)^{-1} \Lambda Z(0) C^f Z(0)^t \Lambda (Z(0)^t)^{-1},$$

we certainly have that

$$\tilde{C}^{SR}(0, t_j) = \Lambda Z(0) C^f(0, t) Z(0)^t \Lambda;$$

has the correct variance for SR_j by construction.

This method is highly effective at calibrating to at-the-money swaption prices with an error of only a few basis points, even over twenty years. This method is essentially due to Jäckel & Rebonato, [71], but we have rephrased the argument slightly.

In extreme cases, the method can break down, for example for a 30-year deal with sharply varying yield curve and volatilities. However, in that case we can use it as a first approximation. In particular, suppose we have calibrated as above, and we have obtained a vector of prices p_j, or equivalently a vector of implied volatilities, σ'_j. We can create a new scaling matrix Λ' by setting

$$\lambda'_j = \lambda_j \frac{\hat{\sigma}_j}{\sigma'_j}. \qquad (14.59)$$

We now replace Λ by Λ', and our calibration will be accurate to about a basis point even in extreme cases.

Why does this method work? Although a swap rate is not log-normal in a FRA-based model, it is approximately log-normal and the at-the-money Black formula is approximately linear. By rescaling we are effectively multiplying the volatility by the ratio of the prices, and the price scales accordingly.

14.9 Lower bounds for Bermudan swaptions

In this section, we look at the problem of pricing a Bermudan swaption using a LIBOR market model. The trickiness arises from the fact that Monte Carlo is the only tractable method for pricing using market models, but as a forward method it is ill-adapted to pricing options which feature early exercise opportunities. We can, however, apply the methods of Section 12.5 to obtain lower bounds, and Section 12.6 to obtain upper bounds. We now look at lower bounds and leave upper bounds to the next section.

Whilst we concentrate on the case of a Bermudan swaption for concreteness, we stress that these techniques could be applied to many callable structures. Early exercise opportunities generally arise from the ability to break a contract. Thus if the bank has entered into a swap with a counterparty, the right to break that swap is equivalent to the right to enter into a swap in the opposite direction. In general, any contract with the right to break can be rephrased as an unbreakable contract plus the right to enter into a contract with the opposite cashflows.

We assume that our swaption is associated to times

$$t_0 < t_1 < t_2 < \cdots < t_n.$$

Let $\mathrm{SR}_j(t)$ denote the swap rate for t_j through t_n at time t. At each t_j, for $j < n$, the holder of the swaption has the right to enter into a swap associated to the times $t_j < t_{j+1} < \cdots < t_n$, provided the right has not already been exercised.

14.9 Lower bounds for Bermudan swaptions

Thus for a payer's Bermudan swaption struck at K, the holder has the right to receive

$$A_{t_j} = (\text{SR}_j(t_j) - K)_+ \sum_{i=j}^{n-1}(t_{i+1} - t_i)P_{i+1}(t_j)$$

provided the option has not already been exercised. We write the positive part here so we can assume that the option is always exercised at the final time slice, t_{n-1}, even if out-of-the-money.

If we take P_n as numeraire, then the price of the Bermudan swaption is equal to

$$P_n(0) \sup_\tau \mathbb{E}(P_n^{-1}(\tau)A_\tau),$$

where the supremum (maximum) is taken over all stopping times τ. The stopping time τ encapsulates the exercise strategy for the Bermudan swaption. The decision to exercise must be made on the basis of information available at the time which is why we require τ to be a stopping time.

Once the stopping time τ has been chosen, pricing is straightforward: we just run a BGM simulation as usual. Our problem is therefore to find the optimal choice of stopping time. Note that if we take $\tau = t_j$, then we have the European swaption associated to the times $t_j, \ldots, t_n$. This means that these European swaptions will always be worth less than the Bermudan swaptions.

At the last two time horizons, the decision of whether to exercise is simple. At time t_{n-1}, we exercise if and only if in the money. At time t_{n-2}, we have a choice between a caplet associated to the times t_{n-1} and t_n and exercising into a swap. We can value the caplet by using the Black formula since the final swap rate, i.e. the forward rate, is log-normal and the swap is simply priceable in the usual fashion. We therefore exercise at time t_{n-2} if and only if the swap is worth more than the caplet.

It is at previous times that the decision of whether to exercise becomes trickier. The decision at each time is based on whether the money received by immediate exercise is less than the value of the Bermudan swaption over the remaining times. As the Bermudan swaption over the remaining times effectively has a smaller underlying, it is a balance between whether the extra optionality is worth more than is being given up by the change of underlying. If the swap is out-of-the-money or barely in-the-money then the remaining optionality will be worth more; but if it is deeply in-the-money it will not since the value of the additional optionality falls away as an option gets more and more in-the-money.

This suggests an approach to pricing. For each j, we set a trigger level, $R_j > K$. At time t_j, we exercise if and only if SR_j is greater than R_j. In other words, our exercise strategy has turned the Bermudan swaption into a trigger swap which triggers on the remaining swap-rate. For each choice of triggers, we are choosing a different exercise strategy and therefore obtain a different lower bound for the Bermudan swaption. We therefore optimize over the trigger levels to get the best possible price.

How do we carry out the optimization? There are two different ways. One is to to carry out a global optimization adopting a functional form for the triggers. The second is to iterate backwards storing nominal values for the unexercised portion as we go. We discuss the global optimization first.

Global optimization

We have certain properties for our array of triggers. The final trigger will be at the strike. The level of trigger will decrease with exercise time. A simple strategy will therefore be to trigger on the basis of a swap level which varies linearly. Thus we set the trigger level at time t_j to be

$$R_j = K + \alpha \frac{t_j - t_{n-1}}{t_0 - t_{n-1}}. \tag{14.60}$$

We have one unknown parameter here, α. We can therefore optimize over α to get a best lower bound. As behaviour is quadratic near a maximum, the value of the trigger as a function of α will roughly look parabolic. This means that the optimization is simple to carry out. As usual, we would generate a set of paths to carry out the optimization and determine α. We would then generate a second set of paths to correctly estimate the expectation for this optimal α.

This method is not optimal in the sense of price obtained; however, it does get close, is very simple to implement and tends to be quite robust. One simple improvement that can be made is to allow a more complicated functional form. The exercise boundary for the American put tended to show more a square-root type behaviour as a function of time-to-go. We could therefore allow the trigger level to vary as a function of a power of time-to-go. For example, we could take

$$R_j = K + \alpha \left(\frac{t_j - t_{n-1}}{t_0 - t_{n-1}} \right)^\beta, \tag{14.61}$$

and then optimize over α and β. This will generally give a slightly higher price.

Local optimization

Rather than adopting a functional form and then solving globally, an alternative approach is to solve for the exercise boundary at each time slice and iterate backwards. How would this work? We first generate a set of paths. For each path, at each exercise time we store the relevant information for our exercise strategy, and the exercised value divided by the numeraire. In this case, the relevant information

14.9 Lower bounds for Bermudan swaptions

at each slice would be the swap rate for the remaining period, but it could be other rates or quantities.

At the final time slice, the exercise strategy is clear: exercise if and only if in the money. At the second last time slice, we need to develop the price as a function of trigger level and then optimize. We make the approximation that the unexercised value divided by numeraire on a given path is the value divided by numeraire at the next time slice on the *same* path. Thus for each trigger level, we estimate the value of this two-time-slice product by taking for each path the exercised value (divided by numeraire) at time t_{n-2} if the swap rate is above the trigger, and the exercised value (possibly zero) (divided by numeraire) at time t_{n-1} otherwise. This gives us a function of trigger level, and we can therefore optimize over trigger level to get a best lower bound.

In what follows, all values should be regarded as discounted, that is all values will have been divided by the numeraire. Having determined the second last trigger level, we move on to the third last trigger. First, we store for each path a nominal value at time t_{n-2}, this will be the value on that path of the exercise value at time t_{n-2} if the swap rate is above the trigger, and the value at time t_{n-1}. We now carry out the same procedure. For each trigger level, we take the value on each path to be the exercised value if the rate is above the trigger level and the unexercised value otherwise. We then average over all paths, and we have a value for that trigger level. We optimize over the time t_{n-3} trigger level to get the best value as before.

We now just repeat this procedure iterating back to the beginning; this gives an estimate of the price. There is a slight upwards biasing arising from the interaction between choice of trigger level and the precise sample drawn. Also, the use of the value at the next time slice for the unexercised value is an approximation. We therefore run a second simulation using a different set of random numbers, and using the optimized trigger levels, to estimate the expectation for this exercise strategy. As usual, we can be sure that this second expectation will give a lower bound for the Bermudan price.

Note that one could first carry out the global optimization to estimate the trigger levels, and then carry out the local backwards optimization using the global estimates as a starting point for the optimization.

A more powerful strategy

When we use the swap rate to trigger an exercise strategy, we are effectively collapsing the multi-dimensional yield curve onto a one-dimensional quantity. Whilst the swap rate is the right quantity in that it encapsulates well the level of the yield curve and also how far the swaption is in-the-money, our dimensional collapse necessarily throws away information. An obvious extension is to collapse onto a two-dimensional space. The question is, which two-dimensional space? Exercise will depend upon whether the current intrinsic value is greater than the next unexercised value. Thus the two main obvious rates are the swap rate for the entire period and the rate for the next period. So at time, t_j, we consider SR_j and SR_{j+1}. However, our projection is less clean than it could be in that these two rates overlap. If we change SR_{j+1}, we also change SR_j. An alternative projection encapsulating similar information is therefore to use f_j the forward rate from time t_j to t_{j+1} and SR_{j+1}. The level of f_j will express well the difference in the two rates.

In fact, one can determine the efficacy of a projection by using a non-recombining tree. One develops a price in very simple cases, and stores the location of exercise and non-exercise nodes. One can plot the projection of these nodes onto various two-dimensional subspaces. If a choice of projection is good then the two sets of nodes will not overlap. This was carried out by Jäckel, [69], [68], and he found that the projection onto f_j and SR_{j+1} was very good and better than the obvious alternatives.

One can then develop an exercise strategy and iterate backwards as for the trigger strategy. The main question in the two-dimensional case, is what functional form to adopt for the exercise strategy at each time horizon. Jäckel suggests exercising according to the sign of

$$f_j(t_j) - p_1 \frac{SR_{j+1}(0)}{p_2 + p_3 SR_{j+1}(t_j)} - K,$$

where the parameters p_i vary with j and are optimized over. He also makes the additional constraint of never exercise out of the money which is not necessarily forced by this functional form in the way that it was for the trigger strategy. Jäckel's strategy works very well. See figure 14.4.

14.10 Upper bounds for Bermudan swaptions

In the last section, we developed lower bounds for Bermudan swaptions using the BGM model, and now we study the complementary problem of finding upper bounds. We proceed by using the method of Rogers which we introduced in Section 12.6. For concreteness, we restrict ourselves to considering payers' swaptions but the extension to receivers' swaptions is obvious.

Recall that we showed that an upper bound for an early exercisable product is

$$\mathbb{E} \left(\max_j B(t_j)^{-1} (A(t_j) - M(t_j)) \right) B(0),$$

where A is the exercised value at time t_j, M is any portfolio of assets with initial value zero, and B is the numeraire. Here we have only a finite set of exercise times

14.10 Upper bounds for Bermudan swaptions

$t_0, \ldots, t_{n-1}$. If B_j is the annuity, we take the exercise value of A to be the positive part of the value of the payoff at time t_j, that is,

$$(SR_j - K)B_j$$

if the swaption is in the money. If the swaption is out-of-the-money then we can take the payoff to be $-\infty$ since decreasing the payoff at a point where the swaption would never be exercised, cannot affect its value. At the final time slice, we must take the payoff to be

$$(SR_j - K)_+ B_j,$$

however. We need to do this as the argument of Rogers relies on the fact that for any martingale M' we have

$$\mathbb{E}(M'_\tau) = M'_0 \qquad (14.62)$$

for any finite stopping time τ. The important point here is the finiteness: we must take the value of M_τ at one of the sampling dates, and when we do so it is possible that the Bermudan will be out-of-the-money and so its correct value is zero, not the negative value. This roughly corresponds to the fact that the Bermudan may not ever be exercised, and the hedging portfolio will then dissolved at the final sampling date. We can therefore think of not exercising as being the same thing as exercising into zero value.

We need to choose the portfolio M in such a way as to minimize the expected value of

$$\max(B(t_j)^{-1}(A(t_j) - M(t_j))).$$

We therefore want M_t to look as much as possible like A_t. It should therefore grow linearly above the strike and, below it, look like zero. The obvious product with this sort of behaviour is the European swaption.

We could therefore consider a portfolio of European swaptions and optimize over the weights to get the best possible M and an upper bound. However, the European swaptions violate a second important property we require which is that we need to be able to take the value of M at each exercise time, and thus we need the value of the European swaption before its expiry. If swap rates were log-normal this would not be an issue as we could use the Black formula, but if we are using a BGM model this is not the case. One solution is to use a swap-rate market model instead; this was done in [77]. A second solution is to use the approximate swaption volatilities to rapidly evaluate the swaptions, as in Section 14.7. However, whilst the evaluation is rapid it is not instantaneous and does slow things down considerably. Also one obtains additional approximation errors, whilst the approximation is superb at-the-money, it is not so good away-from-the-money.

We do not pursue the path of European swaptions here as we do not wish to develop the swap-rate market model. We therefore use portfolios of caplets instead. Our initial portfolio is a caplet for each exercise time with some notional to be optimized over, together with zero coupon bonds to make the initial value zero. One subtlety is what to do with a caplet after its expiry. The portfolio M can be dynamic so we can carry out a trading strategy as long as we remain self-financing. Thus when a caplet reaches expiry, we receive cash equal to its intrinsic value discounted from the next time slice. We can (and must) now use this money to buy further assets, for example more caplets or zero-coupon bonds. What appears to work best is either to use the money to buy caplets with expiry equal to the next expiry time or to buy zero-coupon bonds expiring at the final time. Buying caplets has the desirable property of giving the portfolio that is most reactive to the level of the swap rate. Note that the caplets can always be priced using the Black formula as the forward rates are log-normal. We thus take as our portfolio

$$M^\alpha = \sum_{j=0}^{n-1} \alpha_j \mathrm{Caplet}_j - \beta P(t_{n-1}) \qquad (14.63)$$

where β has been chosen to make the portfolio initially valueless. The parameter α is the parameter to be optimized over.

We use the zero-coupon bond expiring at time t_{n-1} as numeraire for simplicity. Our procedure to find an upper bound is therefore as follows:

(i) Generate a set of paths of the underlying forward rates. Say 32768 paths.
(ii) For this set of paths, optimize over α to get the possible upper bound.
(iii) Generate a new set of paths with different random numbers to evaluate

$$\mathbb{E}\left(\max_{t_j} \left(A_{t_j} P(t_j, t_{n-1})^{-1} - M^\alpha(t_j) P(t_j, t_{n-1})^{-1}\right)\right) P(0, t_{n-1}),$$

with the optimal α.

Just as for the lower bound, we use a two-pass approach in order to ensure that the optimization procedure has not resulted in biasing arising from exploitation of the microstructure of the paths involved.

This method is highly effective and results in bounds that are of the order of a few basis points and generally small compared to a Vega. We define Vega here to mean the change in value obtained by increasing all the underlying swap-rate volatilities by 1%. See figure 14.4.

One interesting side effect of this method is that the optimal portfolio M^α, having been constructed to be as similar to the intrinsic value as possible, provides a good control variate. In other words, if we want to estimate the lower bound and

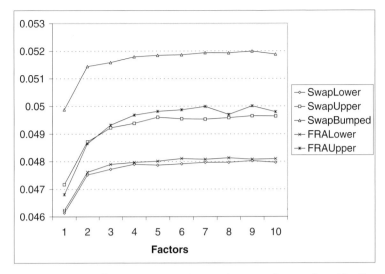

Fig. 14.4. Prices for a five into ten year Bermudan swaption produced by forward-rate based and swap-rated models as a function of the number of factors. We present both lower and upper bounds for each model. To give an idea of scale we also show the lower bound for a swap-rate based model with volatility 1% higher.

run a Monte Carlo simulation to estimate

$$\mathbb{E}\left(\left(A_\tau - M_\tau^\alpha P(\tau, t_{n-1})^{-1}\right)\right),$$

where τ is the optimal exercise strategy, it will converge much faster since we have removed much of the variance. It will give the same price as

$$\mathbb{E}\left(M_\tau^\alpha\right) = M_0^\alpha = 0. \tag{14.64}$$

This fact is of limited utility for computing prices in that lower bounds tend to be faster to compute than upper bounds in any case, since the optimization is easier. Nevertheless if one wishes to compute Greeks one has many simulations to run for the lower bound and it can then be highly useful.

14.11 Factor reduction and Bermudan swaptions

We have so far implicitly assumed that the dimension of the Brownian motion driving our forward-rate process is equal to the number forward rates. In fact, few banks use such a full factor model. Instead they tend to use a small number of factors, typically between 2 and 4. In addition, although we have discussed the predictor-corrector method to allow long-stepping of rates whilst correcting for drift errors, most banks short-step the forward rates and use constant drifts across the small steps. As they take steps of about 3 months, the error in the drift approximation is negligible and there is no need for a more sophisticated approach.

What effect does using fewer factors have? Reducing the number of factors makes the rates more correlated. However, as the rates are being short-stepped it is only the instantaneous correlation matrix which is affected: decorrelation arising from the varying shapes of the instantaneous volatility curves will be totally unaffected. In general, it is the terminal decorrelation that has more effect.

Before presenting an example, we discuss how one would carry out the factor reduction. Given the instantaneous correlation matrix C of the form, say,

$$C_{ij} = e^{-\beta|t_i - t_j|}, \tag{14.65}$$

we compute the eigenvectors, e_j, and (decreasing) eigenvalues, λ_j. The matrix A with column j equal to $\sqrt{\lambda_j} e_j$ is then a pseudo-square root of A. Let $A^{(l)}$ be the matrix with first l columns equal to those of A and the others equal to zero. We can now form a reduced factor matrix equal to

$$C^{(l)} = A^{(l)} (A^{(l)})^t. \tag{14.66}$$

However, $C^{(l)}$ is not a correlation matrix; it will not generally have 1s on the diagonal but will be positive semi-definite and thus be a covariance matrix. We can therefore form $\tilde{C}^{(l)}$ by taking

$$\tilde{C}^{(l)}_{ij} = \frac{C^{(l)}_{ij}}{\sqrt{C^{(l)}_{ii} C^{(l)}_{jj}}}. \tag{14.67}$$

Since $\tilde{C}^{(l)}$ is just the correlation matrix associated to $C^{(l)}$, we can be sure it is a correlation matrix.

Since we are short-stepping we can ignore terminal decorrelation effects, and take the covariance matrix for a short step from t_r to t_{r+1} to be

$$\bar{\sigma}_i \bar{\sigma}_j \tilde{C}^{(l)}_{ij},$$

where $\bar{\sigma}_k$ is the r.m.s. volatility of σ_k across the interval $[t_r, t_{r+1}]$.

Now we know how to carry out a short-stepped, reduced-factor simulation, what effect does the factor reduction have on prices? We have to keep the price of the objects calibrated to constant or the question is meaningless. We present one example. We consider a payer's Bermudan swaption on an annual rate. It starts in five years and lasts for ten years. It is exercisable and pays annually. So we put

$$t_i = i + 5, \tag{14.68}$$

for i running from 0 to 10. The swap pays at the dates t_i for i equal to 1 through 10. It can be exercised at t_i for i equals 0 through 9.

14.11 Factor reduction and Bermudan swaptions

The strike is 5.91%. The natural calibrating instruments are the co-terminal swaptions underlying the Bermudan so we need the volatilities of European swaptions running from t_i to t_{10} for i equals 0 through 9. We take these to be as follows

t_i	Vol
5	11.83%
6	11.50%
7	11.13%
8	10.80%
9	10.55%
10	10.39%
11	10.30%
12	10.28%
13	10.32%
14	10.45%

To apply the calibration method of Section 14.8, we need to know the instantaneous volatilities of the forward rates, their instantaneous correlations and of course the yield curve.

We take the instantaneous correlation matrix before factor reduction to be $e^{-\beta|t_i-t_j|}$, with β equal to 0.1. For simplicity we define the yield curve by a functional form. The rate from time i to $i+1$ is taken to be

$$-1\% e^{-0.3i} + 6\%,$$

for i an integer. We therefore obtain a gently upwards-sloping curve. Other discount factors are to be computed by log-linear interpolation. The instantaneous volatility of a forward rate expiring at time T is taken to be $(a + b(T - t))e^{-c(T-t)} + d$, where

$$a = 5\%,$$
$$b = 10\%,$$
$$c = 50\%,$$
$$d = 10\%.$$

For each number of factors, we factor-reduced and then carried out the calibration procedure of Section 14.8. Upper bounds and lower bounds were then computed using the methods presented above. We present the results in Figure 14.4. For comparison, we also present prices found by a swap-rate based model with the same calibration. To give some idea of scale, we also present the lower bound when volatilities have been increased by 1%.

The graph shows a marked increase from one to two factors. A slight increase from two to three and quite slight increases thereafter. Note that there is a certain amount of noisiness in the prices arising from the fact that the bounds result from an optimization procedure and we cannot always be sure that the optimal bounds have been attained.

14.12 Interest-rate smiles

An important issue in pricing exotic interest-rate derivatives is smiles. Just as in the equity and FX worlds, traders typically use different volatilities to price swaptions and caplets of different strikes which are otherwise identical. If one truly believed forward rates and swap rates were log-normal, then this would make no sense. (They cannot both be log-normal in any case.) However, this strike dependence is, of course, just expressing the traders' belief in a non-log-normal world. One strong reason to be sceptical of log-normality is that there is an absolute component to the movement of interest rates which does not exist in the equity/FX setting. Whilst interest rates do move in smaller amounts when they are low, they do not move proportionately less. Part of the reason for this is that interest rates already express a ratio, and therefore do not have the same invariance of scale that stock prices have. Rebasing an index to be a tenth of its current value has no effect on anything but this has no analogue in the interest-rate world. A 5% interest rate really is different to one of 100%

For example, yen rates are very low at the time of writing, yet they still move around a lot more than would be predicted by a log-normal model. Interest-rate smiles are therefore typically downward-sloping, reflecting the effectively greater volatility for low values of the forward rate.

We can build this effect into our models in a couple of ways. The first is to say that log-normal volatilities go up as the rate goes down so we let the log-normal volatility depend on the rate. A forward rate would therefore follow a process of the form

$$\frac{df}{f} = \mu dt + \sigma f^{\gamma - 1} dW, \quad (14.69)$$

where γ is a constant between 0 and 1. The $\gamma = 0$ case corresponds to normal rates and $\gamma = 1$ case to log-normal. Such a process is called a *constant elasticity of variance process* or more ordinarily a *CEV process*. The main downside of this approach is trickiness of implementation in terms of pricing the vanilla options and running the Monte Carlo.

A model that in easier to work with is the *displaced diffusion* model which models the process as being a mixture of normal and log-normal processes. Where the CEV process is the geometric mean of the processes, the displaced diffusion is the arithmetic mean. In particular, we put

$$df = \sigma (f + \alpha) dW, \quad (14.70)$$

14.12 Interest-rate smiles

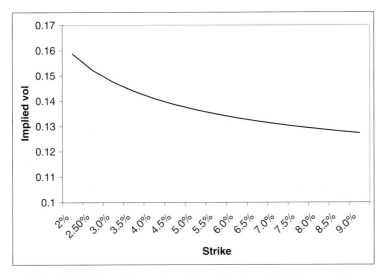

Fig. 14.5. The smile produced by a displaced-diffusion model.

where α is the constant displacement coefficient. As $d\alpha = 0$, we can rewrite the process as

$$\frac{d(f+\alpha)}{f+\alpha} = \sigma dW, \qquad (14.71)$$

which shows that the model is equivalent to saying that $f + \alpha$, rather than f, is log-normally distributed. This has the huge advantage that the Black analysis carries over immediately if we regard a caplet of strike K as being a call option of strike $K + \alpha$ on $f + \alpha$. In particular, one can use the Black formula just by shifting the forward rate and strike by α. Running a Monte Carlo simulation is also no harder than in the vanilla case, with the drift expression only being very slightly more complicated; the displacement must be added to the forward rates in the numerator in (14.19) but otherwise there is no change. One can therefore work in a displaced-diffusion setting for BGM with little difficulty. We display a typical displaced-diffusion smile in Figure 14.5.

One further advantage of the displaced-diffusion setting is the simplicity of calibration for varying values of α. The Black formula is approximately linear at-the-money (see Figure 3.7), and in fact it is very accurately approximated by

$$Cf\sigma\sqrt{T}P\tau, \qquad (14.72)$$

where P is the discount bond for the payment time, τ is the accrual, σ is the volatility, T is expiry and f is the forward rate today. As the displaced Black formula just

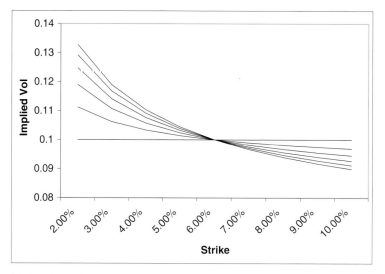

Fig. 14.6. The smiles produced by displaced-diffusion models by varying the displacement but rescaling the volatility to keep the at-the-money (6%) implied volatility constant. The displacements vary from zero (flat line) to 5% (most sloped line).

involves moving rate and strike by α we have a similar approximation

$$C(f + \alpha)\sigma_\alpha \sqrt{T} P \tau, \qquad (14.73)$$

where σ_α is the volatility of $f + \alpha$. We can therefore obtain the same price in the two settings by letting

$$\sigma_\alpha = \frac{f}{f + \alpha} \sigma. \qquad (14.74)$$

Using this formula allows us to pivot the smile about the at-the-money volatility by changing the displacement. See Figure 14.6.

Unfortunately, displaced diffusion cannot be the whole story. If one plots caplet smiles, one notices a slight upkick for high strikes. It is impossible to match this with a displaced diffusion smile, however, as the nature of displaced diffusions means that smiles can only ever be down-sloping.

To go further, one needs a more sophisticated model. There are many such any alternative model for equity/FX evolutions can be adapted to modelling forward-rate changes. One could therefore incorporate the possibility of jumps or of stochastic volatility. Or more generally, since each forward rate has an instantaneous volatility curve, one could make the shape of that curve stochastic. We do not explore these possibilities here but merely suggest the reader bears them in mind when studying the alternative models of stock evolution.

14.13 Key points

- The pricing of exotic interest-rate derivatives depends on the evolution of a one-dimensional object: the yield curve.
- The modern approach to pricing exotic interest rate derivatives is to evolve market observable rates.
- The BGM (or BGM/J) model is based on the evolution of log-normal forward rates.
- Forward rates only have zero drifts in the martingale measure when the numeraire is a bond with the same payoff time as the forward rate.
- In general, the drift of a forward rate is both state- and time-dependent.
- The BGM model can used to price any instrument that is dependent on a finite number of rates at a finite number of times.
- A crucial part of the calibration of the BGM model is choosing the instantaneous volatility functions for the rates.
- Decorrelation between rates can occur both through instantaneous decorrelation and through terminal decorrelation arising from the differing shapes of volatility curves.
- The state-dependence of drifts means that Monte Carlo is the natural method of pricing in the BGM model.
- Forward rates can be evolved over long time intervals by using a predictor-corrector technique.

14.14 Further reading

The approach given here to calibrating the instantaneous volatility curves is due to Rebonato and discussed at length in [106]. A comprehensive discussion of market models can be found in [107]. For a discussion of the historical models predating the BGM approach as well as some discussion of market models see [105]. Musiela & Rutowski, [96], contains a quite extensive discussion of market models including BGM.

Other recent books on the topic of exotic interest rate derivative pricing from a similar point of view are Pelsser, [102] and Brigo and Mercurio, [22].

The original paper suggesting that one should evolve a yield curve rather than a short rate is by Heath, Jarrow & Morton and such models are often called *HJM models*. See [60].

Some early papers on market models were by Brace, Gatarek & Musiela, [20], Jamshidian [67], and Musiela & Rutowski, [97].

The predictor-corrector technique discussed here was introduced by Hunter, Jäckel & Joshi in [65].

The use of trigger swaps to price Bermudan swaptions was introduced by Anderson, [3]. Extensive discussion of the use of non-recombining trees to design exercise strategies and in particular the Jäckel strategy of Section 14.9 is in [68].

The upper bound for Bermudan swaptions method discussed here was introduced by Joshi & Theis in [77]. An alternative approach for upper bounds for Bermudan swaptions based on similar theory but different practicalities was introduced by Andersen & Broadie in [8].

In [86], Longstaff, Santa-Clara & Schwartz argued that banks were throwing away billions of dollars by using low-factor models for pricing Bermudan swaptions. In [7] Andersen & Andreasen argue that if a model is calibrated to both all caplet prices and all swaption prices, then the factor dependence is a myth.

The equivalence between displaced-diffusion and CEV models is due to Marris, [92].

We have not examined short-rate models at all. They are covered in [105], [17] and [12] as well as many other places. Indeed most books on mathematical finance treat the topic.

Developing smile models in a market model context is an active area of research. Some recent papers on the topic are [4], [50], [51], [76] and [121].

14.15 Exercises

Exercise 14.1 Derive the drift of a forward rate following a displaced-diffusion process when the zero-coupon bond expiring at its reset date is the numeraire.

Exercise 14.2 Show that if the forward rates underlying a swap rate are all the same then the derivative of the swap rate with respect to f_j is equal to w_j. (Notation as in Section 14.7.)

Exercise 14.3 Suppose all the forward rates followed normal processes instead of log-normal processes. What changes would be necessary to implement the BGM approach?

Exercise 14.4 Suppose we decide that all the trouble in the BGM model is caused by the non-tradability of the rates and therefore decide to evolve the values of forward-rate agreements with zero strike instead. What process for the FRAs is equivalent to the log-normal process for forward rates? Suppose we make the FRAs log-normal, what process do we get for the rates?

Exercise 14.5 Every three months, an inverse floater pays $\max(2L - K, 0)\tau - L\tau$, where L is the three-month LIBOR rate for the preceding three months. How would you price this derivative using BGM? How many evolution times would be needed? Could you price it without using BGM?

Exercise 14.6 Suppose we take a forward rate f with P the zero-coupon bond with the same payoff time, and use fP as numeraire. What is the drift of f? Why is it valid to use fP as numeraire?

Exercise 14.7 Develop an analytic formula for a trigger FRA under a displaced-diffusion model (see Exercise 13.8).

Exercise 14.8 Suppose the volatility of f_1 is equal to $0.1 + e^{-2t}$ and the volatility of f_2 is equal to $0.2 + e^{-3t}$. The instantaneous correlation is 0.5. What is covariance of f_1 and f_2 over the time interval from 0 to 1?

15

Incomplete markets and jump diffusion processes

15.1 Introduction

We have mentioned the imperfections of the Black–Scholes model of stock-price evolution. In this chapter, we look at one method of improving it. Our improvement is to add the possibility of the stock price jumping discontinuously. The motivation for this model is that stock markets do crash and during a crash there is no opportunity to carry out a continuously-changing Delta hedge. One consequence of this will be the impossibility of perfect hedging; at any given time the stock price can increase slightly or decrease slightly or fall a lot. It is not possible to be hedged against all of these simultaneously. The impossibility of perfect hedging means that the market is incomplete, that is not every option can be replicated by a self-financing portfolio. The price of a non-replicable option can then only be bounded rather than fixed using no-arbitrage methods. This means that the risk preferences of investors re-enter the picture despite their banishment from the Black–Scholes world. Our secondary purpose in this chapter is to discuss many of the issues, both philosophical and practical, involved in pricing in incomplete markets.

In the equity world, the call and put option prices generally display a steeply downward-sloping smile. One explanation of this smile is that it is caused by strong demand for slightly out-of-the-money put options. A fund manager has his performance reviewed every three months. He wants to be protected against the possibility of a market crash in the mean time. He therefore buys put options which guarantee that his portfolio's value can only fall by a small amount even if the market crashes. Thus he is buying the put option as insurance. Since there are many fund managers doing the same thing, hedging is impossible, and no one wants extra exposure to crashes, the market is all one way, and the price of the put option is bid up. The bank can still make money by selling put options but it is doing so by taking on risk, rather than by charging for the cost of hedging, as in the Black–Scholes framework. The purchase of the put option is therefore a transference of risk from the fund manager to the bank. The market price will settle on a point where the

15.2 Modelling jumps with a tree

banks feel that they are being adequately compensated for taking on the extra risk. A major determinant of the price is therefore risk preferences rather than arbitrage.

Once we have moved to an incomplete market, there are two different issues to be addressed. The first is how to use arbitrage to bound the prices of vanilla options. The second is to determine prices for exotic options which are compatible with both the model and the prices of the vanilla options traded in the market.

Our objective in this chapter is place these ideas in a mathematical context and to examine some of the consequences.

15.2 Modelling jumps with a tree

We can see some of the problems with modelling jumps, by returning to a tree model. Suppose that in each small time-step, the log of the stock price can move up or down a small amount, or else it can fall by 10%. If the current price is S_t, length of the step is Δt, the volatility is σ and the drift of the stock is μ, then after the time step the stock has the possible prices,

$$S_t(1 + \mu \Delta t - \sqrt{\sigma \Delta t}), \quad S_t(1 + \mu \Delta t + \sqrt{\sigma \Delta t}), \quad 0.9 S_t.$$

More generally, we might want a distribution of possible jumps which would add more branches. As we let the time-step get smaller and smaller, the first two prices get closer and closer together but the third price does not change. This reflects the instantaneous nature of a crash; you cannot catch the stock price in between in the way that you could for continuous models. Of course, the probability of a crash occurring would decrease as the size of a time-step decreased but its size would not change. In Chapter 3, we showed that for a binary tree, the probabilities of attaining a particular node was not important; there was a single arbitrage-free price independent of the probabilities. We also demonstrated that for a tree with three branches, there was an infinite number of prices for an option that were arbitrage-free. This means that when modelling a stock with jumps there will always be an infinite number of possible prices.

How can we assess which prices are possible? We can use risk-neutral valuation. For example, suppose interest rates are zero, the stock price is 100 and the stock can take the values 110, 100 or 50 tomorrow. We wish to price a call option struck at 100. What are the possible prices? If we assign probabilities p_1, p_2 and p_3 to 110, 100 and 50 respectively, then the expected price tomorrow is

$$110 p_1 + 100 p_2 + 50 p_3$$

for a risk-neutral measure, and this must equal 100, since interest rates are zero. We must also have $p_1 + p_2 + p_3 = 1$. We solve to obtain,

$$p_1 = 5 p_3, \tag{15.1}$$

$$p_2 = 1 - 6 p_3. \tag{15.2}$$

This means that an equivalent martingale measure can be achieved by taking any value of p_3 between zero and $1/6$. The value of the option can therefore be anywhere between zero and $\frac{5}{6}(110-100) = 8\frac{1}{3}$, without introducing arbitrage. Which is correct? Any of them – it is simply a question of risk-aversion.

We can now define a multi-step model by stringing trees together. However, at each stage there will be three branches and we thus obtain a bound on the possible prices at each node which then has to be propagated backwards.

Although we cannot price an option precisely, we can relate the prices of options to each other. Suppose we have two call options, O_1 and O_2, which are struck at 100 and 90. The first option pays off 10, 0 and 0 in the states 110, 100 and 50 respectively. The second option pays off 20, 10 and 0 in the three states. We can regard these prices as vectors, namely

$$(10, 0, 0) \quad \text{and} \quad (20, 10, 0).$$

We can hedge any product which can be written as a linear combination of the stock,

$$S = (110, 100, 50),$$

and the bond,

$$B = (1, 1, 1).$$

The set of all such linear combinations (i.e. the linear span) of the two options' vectors will be two-dimensional as will the linear span of the stock and the bond. As we are working in three-dimensional space, the intersection must be at least one-dimensional. This means that it must be possible to find a linear combination of the two options which is equal to a linear combination of stock and bond. Solving, we find that

$$S - 50B = 5O_2 - 4O_1, \tag{15.3}$$

as vectors. This equation holds in any of the three final states and therefore must hold initially. Thus although the prices of O_1 and O_2 are not determined, the price of either completely determines the other. Note that this argument has, of course, depended on the fact that there is only one possible jump amplitude and if we allows k jump amplitudes, we would have a $(k+2)$-dimensional state space, and would need k options to force the pricing of all other options. This illustrates the fact that in the general setting that although any individual option price is not fixed, there is a lot of rigidity in the possible overall shape of the volatility surface. By volatility surface, we mean the two-dimensional surface of Black–Scholes volatilities implied by the prices of options for all strikes and maturities. We shall return to the issue of hedging with options once we have developed more theory.

15.3 Modelling jumps in a continuous framework

Suppose we have a stock moving under geometric Brownian motion, with the added possibility of crashes. What properties should crashes have? They should occur instantaneously: the probability of one occurring in a given small time interval should be roughly proportional to the length of a time interval. We can achieve these properties by modelling crashes with a Poisson process.

We recall the characteristics of a Poisson process. The main parameter is the intensity λ and the probability of an event in a given small time interval, Δt, is $\lambda \Delta t$ plus a smaller error. We denote the number of events up to a time t by $N(t)$. We therefore have that $N(t)$ is an integer-valued function which is constant for a while and then jumps up by 1. Another important property is that the probability of an event is independent of the number of events that have already occurred: that is for $t > s$, we have that $N(t) - N(s)$ is independent of the value of $N(s)$.

We find
$$\mathbb{P}(N(t) = j) = \frac{(\lambda t)^j}{j!} e^{-\lambda t}, \tag{15.4}$$

that is, the number of jumps up to time t is Poisson distributed with parameter λt. It is often useful to simulate the times of arrivals between jumps. That is the time to the first jump, and then the time from the first jump to the second jump, and so on. These times are all independent and they are continuous random variables. Denote the random time by X. The probability that $X > t$ is the same as the probability that no jumps occur in the period $[0, T]$ which is $e^{-\lambda t}$. We conclude that X has density function

$$\lambda e^{-\lambda t} H(t), \tag{15.5}$$

where $H(t)$, the Heaviside function, is 1 for $t > 0$ and 0 otherwise. For further discussion of the Poisson process, we refer the reader to Grimmett & Stirzaker, [53].

Returning to our stock, it moves according to a geometric Brownian motion with superimposed jumps. We suppose that at a jump the stock price is multiplied by a random variable J. We could take J to be log-normal, or to be a single constant number, or anything we want.

We can write the process as

$$\frac{dS_t}{S_t} = \mu dt + \sigma dW_t + (J - 1)dN(t). \tag{15.6}$$

Note that when a jump occurs, S changes by $S(J - 1)$, which is equivalent to multiplying S by J. We want to be able to price options under this process. We saw in Chapter 6, that if we find an equivalent martingale measure and set the option prices to the discounted expectations of their values in that measure, then there can be no arbitrage. We therefore look for an equivalent martingale measure. Suppose interest rates are a constant r and the riskless bond is therefore worth e^{rt} at time t. The expected change of S in a small time interval will be

$$\mu S \Delta t + \mathbb{E}(J-1)\lambda S \Delta t.$$

For the ratio S/e^{rt} to be a martingale, we need S to grow at the risk-free rate so we need the expected change to be $rS\Delta t$. That is we need

$$\mu = -\mathbb{E}(J-1)\lambda + r. \tag{15.7}$$

We therefore conclude that an arbitrage-free price for a European option, O, of expiry T is obtained by taking

$$e^{-rT}\mathbb{E}(O(S_T)), \tag{15.8}$$

with S_T evolved according to (15.6) with drift given by (15.7). We can do this change of measure by invoking Girsanov's theorem.

How do we actually use this expression? The first point to note is that if we evolve the log, there is no interaction between the Brownian part and the jump part. The log will follow the process

$$d\log S = \left(\mu - \frac{1}{2}\sigma^2\right)dt + \sigma dW_t + (\log J)dN(t); \tag{15.9}$$

to see this note that if no jump occurs the log will just evolve as in the continuous case, and if a jump occurs then multiplying S by J is equivalent to adding $\log J$ to $\log S$.

We therefore have that the final distribution of $\log S_T$ is of the form

$$\log(S_0) + \left(\mu - \frac{1}{2}\sigma^2\right)T + \sigma\sqrt{T}N(0,1) + \sum_{j=1}^{N(T)} \log J_j,$$

where $N(0,1)$ is a standard normal variable, and J_j are independent variables distributed according to the jump distribution. Note that this expression involves $N(T)$, the number of jumps, which is itself a random variable.

We can now price by Monte Carlo. We just draw the number of jumps, then the size of each of the jumps and the Brownian part to get the final spot, and evaluate the option.

To do better, we need to make some assumptions on the nature of the jumps. Suppose there is only one jump size, really a jump ratio, since we are multiplying by J. We then have that the final distribution of $\log S$ is

$$\log(S_0) + \left(\mu - \frac{1}{2}\sigma^2\right)T + \sigma\sqrt{T}N(0,1) + N(T)\log(J).$$

15.3 Modelling jumps in a continuous framework

We can rewrite this as

$$\log(S_0) + (\mu - r)T + N(T)\log(J) + \left(r - \frac{1}{2}\sigma^2\right)T + \sigma\sqrt{T}N(0,1).$$

If we fix $N(T)$, then this has the same terminal distribution as the log of a standard Black–Scholes evolution with initial spot

$$\log(S_0) + (\mu - r)T + N(T)\log(J).$$

This means that, integrating over the possible values of $N(0, 1)$, we just obtain a Black–Scholes price for the option but with spot $S_0 e^{(\mu-r)T} J^{N(T)}$.

Thus, using the distribution for $N(T)$, the formula for the price of a call option is

$$\sum_{j=0}^{\infty} e^{-\lambda T} \frac{(\lambda T)^j}{j!} \mathrm{BS}\left(S_0 e^{(\mu-r)T} J^j, \sigma, r, T, K\right), \tag{15.10}$$

where $\mathrm{BS}(S, \sigma, r, T, K)$ is the Black–Scholes price of a call option struck at K with spot S, interest rate r, volatility σ, and expiry T. We therefore have a semi-closed-form solution but note that it involves an infinite series. In practice, however we would be able to cut off after a finite number of terms, since the sum of the remaining terms will converge to zero.

Of course, a single jump amplitude is not particularly realistic. We can develop similar but more complicated expressions for any finite number of jump amplitudes. If we want to use a continuous jump distribution, we can develop an expression when the amplitude is itself log-normal, the reason being that we can then absorb the jump into the Brownian part of the Black–Scholes expression. Thus if J is distributed as $m \exp(-\frac{1}{2}v^2 + vZ)$ with Z a standard normal variable then the terminal distribution of $\log S$ is

$$\log S_0 + rT + (\mu - r)T - \frac{1}{2}\sigma^2 T + \sigma\sqrt{T}Z_0 + N(T)\log m - \frac{1}{2}v^2 N(T) + \sum_{j=1}^{N(T)} vZ_j, \tag{15.11}$$

where Z_j are independent draws from $N(0,1)$. We can regroup all the normal distributions together to obtain

$$\log S_T = \log S_0 + rT + (\mu - r)T - \frac{1}{2}\sigma^2 T + N(T)\log(m) - \frac{1}{2}v^2 N(T)$$
$$+ \sqrt{\sigma^2 T + N(T)v^2}\,W, \tag{15.12}$$

with W a standard normal variable. Fixing $N(T)$, this is the terminal distribution of a stock in a Black–Scholes world with volatility $\sqrt{\sigma^2 + \frac{N(T)v^2}{T}}$ and initial spot $S_0 m^{N(T)} e^{(\mu-r)T}$. The price will therefore be given as a weighted infinite sum over Black–Scholes prices with these parameters, as in the single jump amplitude case.

With a little work, the price can be rewritten as

$$\sum_{n=0}^{\infty} \frac{e^{-\lambda' T}(\lambda' T)^n}{n!} \mathrm{BS}(S_0, \sigma_n, r_n, T, K) \qquad (15.13)$$

where

$$\begin{aligned} \sigma_n &= \sqrt{\sigma^2 + nv^2/T}, \\ r_n &= r - \lambda(m-1) + n \log m / T, \\ \lambda' &= \lambda m. \end{aligned} \qquad (15.14)$$

These formulas were originally derived by Merton using a PDE approach.

15.4 Market incompleteness

In the last section, we derived formulas for arbitrage-free prices for call options (and hence puts) on a non-dividend-paying stock in a jump-diffusion world. We did not show that these were the *only* arbitrage-free prices, only that they were arbitrage-free. This arbitrage-free price was arrived at by taking an equivalent martingale measure obtained by changing the drift of the Brownian part of the evolution. The question therefore arises of whether there are other equivalent measures and how to find them. Recall that an equivalent martingale measure has to have the same set of events of probability 0 and 1, but that any other event can change in probability. This means that if all but a set of measure zero of Brownian paths has a certain property before measure change, then all but a set of measure zero will have it after the measure change. (A set of measure zero is another name for a set of probability zero.)

In particular, the amount of jaggedness is the same for almost all paths. What do we mean by jaggedness? To make the concept mathematical, we need the idea of variation. Consider a function

$$f : [0, T] \to \mathbb{R},$$

for example a stock-price path. The first variation is obtained by taking a partition,

$$t_0 = 0 < t_1 < t_2 < \cdots < t_n = T,$$

and taking the sum

$$\sum_{i=0}^{n-1} |f(t_{i+1}) - f(t_i)|;$$

this measures the total amount of up and down moves for this partition. If we make the partition finer and finer, this number can only get bigger (by the triangle inequality). For a continuously differentiable function it will converge. If the limit exists,

15.4 Market incompleteness

this number is called the *first variation* of f. However, for a Brownian motion path, this number will always diverge to infinity (with probability 1.) The easiest way to see this is to consider approximating the Brownian motion by up and down moves. If we divide our interval into N equal length pieces allowing an up or down move of size $\sqrt{T/N}$, at each step, then the total size of moves will always be $\sqrt{TN}$, regardless of which the way the moves are, and the variance of the path is T. Letting N tend to infinity, the paths converge to Brownian motion and the first variation becomes infinite. Since (almost) every path has infinite first variation, this property must be preserved by changes of measure.

How can we measure how infinitely jagged the paths are? If we take the second variation this has a better chance of converging, since we square the differences which being small become smaller. We therefore consider

$$\sum_{i=0}^{n-1} |f(t_{i+1}) - f(t_i)|^2,$$

and let the partition size go to zero. If we again approximate by N up or down moves of size $\sqrt{T/N}$, then this sum will always be T. Letting N tend to infinity, we deduce that Brownian motion paths always have quadratic (or second) variation T over an interval $[0, T]$. This actually holds with probability 1. This is why we can use measure changes to change the drift but not the volatility of a Brownian motion. There is an immense rigidity in the paths caused by the fact that the quadratic variation is always the same. The amazing thing about Girsanov's theorem is not that one can change the drift of a Brownian motion, but the fact that a measure change can do nothing else.

If we consider a stock moving under geometric Brownian motion, this means that the quadratic variation of the log of the stock over an interval $[0, T]$ will always be $\sigma^2 T$. (The part coming from the drift will have finite first variation and disappear in the limit.) We can therefore observe from a given path what the volatility σ is. This is essentially why Black–Scholes pricing works. The cost of hedging is determined by the amount of vibration and the amount of vibration is always the same. We can interpret this by saying that all paths are the same, there are no lucky paths nor unlucky paths; we are totally indifferent to which path occurs.

If we now move to a jump-diffusion process, our indifference disappears. Suppose spot moves according to a jump-diffusion process such that jumps occur with an intensity λ, and the diffusive part has volatility σ. If we examine the quadratic variation over a small, jump-free period then we can measure σ. However, we have no way of measuring λ from within a single path. A given path may have no jumps, one jump or a hundred jumps, whatever the value of λ is. If we take a large sample of paths, then we can start approximating which values of λ are likely to be consistent, but in a single path there is no way to tell. Lucky and unlucky paths do exist in the jump-diffusion setting, and we are not indifferent to how many jumps occur.

The upshot of this is that we change the measure on our space of paths by letting λ be whatever we want as long as it is non-zero. We thus have an infinite number of pricing measures: first we choose a value for λ, and then we enact a drift change on the Brownian motion to make the adjusted measure a martingale. As the implied prices have been arrived at by risk-neutral expectation from an equivalent martingale measure, they are arbitrage-free. We thus have an uncountable infinity of arbitrage-free prices.

In fact, we have even more measures than this. There is nothing forcing λ to be a constant. We could take λ to be function of time or more generally a function of spot and time. There is also the issue of changing the jump distribution. We can change the density of jumps to any other jump density which has the same zero set. This means that if we have a single discrete jump amplitude, measure changes cannot change it at all, but if we have log-normal jumps then we can replace them by any distribution which has non-zero density everywhere on the positive real line.

These infinities of pricing measures reflect the impossibilities of hedging across jumps. We can Delta-hedge to remove the infinitesimal vibrations but we will always be exposed to the possibility of a jump. For this reason, jump-diffusion models tend to be unpopular in the markets as traders do not like the lack of a guaranteed price. This attitude is ostrich-like in that jumps do occur, so to say that we will not use models that incorporate them because we do not like the consequences is ultimately perverse.

15.5 Super- and sub-replication

It is worth relating the range of arbitrage-free prices to the concept of replication. We have seen that if we can replicate the payoff of an option by a self-financing portfolio in the riskless asset and the underlying, then the set-up cost of that portfolio is the unique arbitrage-free price of the option. Since we have a multitude of possible prices in the jump-diffusion world, we can be sure that it is impossible to set up a replicating portfolio. However, replicating portfolios are not useless: we can use them to bound option prices.

In particular, suppose we have an option O with payoff $f(S_T)$. If we can create a self-financing portfolio Q such that whatever path S_r follows, we have

$$O(T) = f(S_T) \leq Q(T),$$

then the portfolio Q is said to be *super-replicating*. We must have $O(0) \leq Q(0)$. Otherwise, the portfolio of $Q - O$ will define an arbitrage. So the price of the option O must be less than or equal to that of any super-replicating portfolio.

15.5 Super- and sub-replication

Similarly, if a self-financing portfolio R is such that

$$R(T) \leq f(S_T) = O(T)$$

then we have that $O(0) \geq R(0)$. Such a portfolio is said to be *sub-replicating*.

Thus the option price lies in the interval bounded by the most expensive sub-replicating portfolio and the cheapest super-replicating portfolio. In fact, in Section 2.8 when we proved model-free arbitrage bounds on the prices of options, we were constructing sub- and super-replicating portfolios.

We should really be a little careful about what we mean by a self-financing portfolio. The self-financing condition is not difficult: we just require

$$d(\alpha S + \beta B) = \alpha dS + \beta dB, \qquad (15.15)$$

as before. The hard bit is determining of which functions α and β are admissible. We previously took them to be functions of S and t. This meant that they were *previsible* functions, that is they were determined by information available before time t. The reason for then, despite the dependence on S_t, was that S_t was continuous, and so

$$S_t = \lim_{r \to t-} S_r, \qquad (15.16)$$

and the right-hand side of this equation is determined by the set of all information available before time t.

In the jump setting, (15.16) no longer holds. If a jump occurs at time t, the path is discontinuous: the left and right sides of the equation differ by the size of the jump. However, the right-hand side of (15.16) always exists and is determined at time t. If we define

$$S_{t-} = \lim_{r \to t-} S_r, \qquad (15.17)$$

then we can consider functions $\alpha(S_{t-}, t)$ and $\beta(S_{t-}, t)$, which are previsible.

Why do we want previsible functions? Essentially, one should only be able to hedge against the possibility of a jump, not the definite occurrence. If we allow our hedging strategy to know the value of $S_t - S_{t-}$ then we are letting it know in advance when the jumps occur, and thus allowing the possibility of hedging in a different fashion at the times of jumps.

One application of this idea of sub-replication is to relate the Black–Scholes price of an option to the possible jump-diffusion prices. (By the Black–Scholes price here we mean the price obtained by keeping the diffusive volatility constant and setting the jump intensity to zero.) We know that the measure associated to any value of λ is valid so we can expect the Black–Scholes price to be arbitrarily

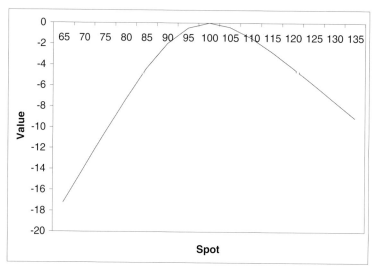

Fig. 15.1. The Black–Scholes value of a portfolio consisting of minus one call option and its Delta hedge.

close to an arbitrage-free jump-diffusion price. However, it is not so obvious where the Black–Scholes price will lie in the range of possible prices. We show that for a vanilla call or put option, the Black–Scholes price is always less than any arbitrage-free jump-diffusion price.

The crucial point we use is that the price of a vanilla call or put option is a convex function of spot in the Black–Scholes world. (See Theorem 4.1.) Suppose our option expires at time T and let $C_{\mathrm{BS}}(t, S_t)$ denote the Black–Scholes price of an option at time t if spot is S_t.

We carry out the Black–Scholes hedging strategy. Our initial portfolio therefore costs $C_{\mathrm{BS}}(0, S_0)$ to set up and at any time we hold $\frac{\partial C_{\mathrm{BS}}}{\partial S}$ units of the stock, with the remaining value in riskless bonds. As long as a jump does not occur then the hedging works perfectly, just as in the Black–Scholes setting. So if no jumps occur throughout the simulation then the option's pay-off is replicated precisely. i.e. the option has been hedged perfectly.

We need to consider what happens across a jump. The fact that $C_{\mathrm{BS}}(t, S)$ is a convex function of S means that any tangent to its graph will always lie below the graph. (See Figure 15.1.) The value of our Delta-hedged portfolio as a function of S has been constructed to be precisely the tangent through the point $(S_t, C_{\mathrm{BS}}(t, S_t))$. Since a jump occurs instantaneously, the effect will be that the value of the portfolio jumps to a point below the Black–Scholes value of the option. Call the difference in values X. We can now continue to Delta-hedge (the amount of the hedge will have changed greatly) but the value of our replicating portfolio will now always be $C_{\mathrm{BS}}(t, S_t) - X$. If there are subsequent jumps, X will of course increase, and X

15.5 Super- and sub-replication

will, of course, grow at the risk-free rate. At time T the value of our portfolio will therefore be

$$C_{\text{BS}}(T, S_T) - Xe^{r(T-t)}.$$

But $C_{\text{BS}}(T, S_T)$ is by definition the payoff of the option and so the replicated value will be less than the payoff of the option.

In conclusion, we can set up a portfolio of initial value $C_{\text{BS}}(0, S_0)$ which is sometimes of the same value as the option at payoff time and sometimes of lower value. The value of the option today must, by no-arbitrage considerations, therefore be greater than $C_{\text{BS}}(0, S_0)$.

Note that this argument depends heavily on the convexity of the Black–Scholes price of a call or put option as a function of spot. A careful examination of the proof shows that this is the only property of the option that we used. It is possible to prove, see Section 15.10, that any option with a convex payoff has a convex Black–Scholes value so our result actually holds for any option with a convex payoff. We therefore have

Proposition 15.1 *Suppose a stock S follows a jump-diffusion process. If there is no arbitrage then the price of any vanilla option on S which has a convex payoff must be greater than the price obtained by taking no jumps.*

The result ceases to be true if one allows general options. For example, consider an in-the-money digital-call option. If interest rates are zero then the price of the option is just the risk-neutral probability of finishing in-the-money. If the diffusive volatility is very low this will be very close to 1. If we now crank up λ, the chance of the option finishing out-of-the-money will increase and the price of the option goes down. An alternate way to see that not all options can be monotone in λ is to observe that a digital call plus a digital put with the same strike is the same as a riskless bond. The price of the digital call going up means that the price of the digital put goes down and vice versa. Another interpretation is that adding in jumps increases volatility so we would expect the prices of options which are monotone in volatility to go up but the prices of others need not increase. See Figures 15.2 and 15.3.

This leads onto the natural question of how to hedge vanilla options in a jump-diffusion world. In order to answer that, one really needs to decide an objective – in a complete market the objective of no-arbitrage forces a unique choice of hedging strategy and achieves a final portfolio of zero variance but in an incomplete market this is impossible so we have to decide what the objective of our hedging is.

One possible objective is to minimize the variance of the final portfolio. There is also the issue of whether one wants to have minimal variance at all times, or just

352   *Incomplete markets and jump-diffusion processes*

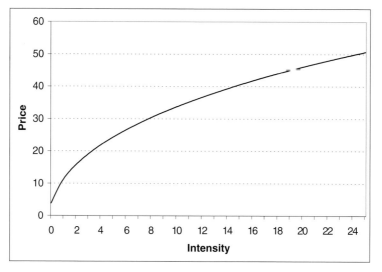

Fig. 15.2. The price of an at-the-money call option as a function of jump intensity for a jump-diffusion model with downward jumps.

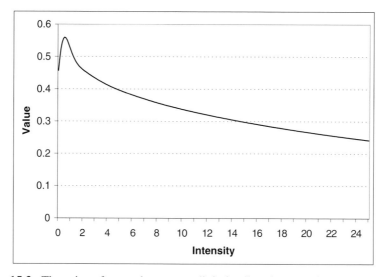

Fig. 15.3. The price of an at-the-money digital call option as a function of jump intensity for a jump-diffusion model with downward jumps. Note the initial increase from the increasingly positive drift.

at the final time. Trading positions are assessed on a day-to-day basis and daily profit and losses are observed. An approach that only minimizes the final variance may look bad en route – or if it does not, its local optimality still has to be demonstrated. Alternatively, one could decide to be perfectly hedged if no jumps occur

15.5 Super- and sub-replication

and bear the risk if one does happen. This may seem slightly feckless but since whatever hedge one chooses, one will do badly in the case of a jump arguably it's pointless to try.

If we decide to hedge only the diffusive part and bear the risk of jumps, then clearly we should Delta-hedge. However, the definition of Delta is not as obvious as it first looks. Delta is only defined relative to a set of parameters. We can therefore choose whether to use the real-world or risk-neutral parameters for hedging. For the purpose of this discussion, we assume that the market is not fickle (see Section 15.6) and the risk-neutral parameters do not change. A third alternative is to Delta-hedge as if there are no jumps, that is, use a jump intensity of zero.

If our objective is solely to minimize variance for paths containing no jumps, then the third of these is correct. If there are no jumps, then we are in a Black–Scholes world, holding the Black–Scholes hedge, which corresponds to the zero jump-intensity case, and we know this strategy yields a final portfolio of zero variance, provided no jumps occur. Thus if no jumps occur, our final portfolio is worth the risk-neutral jump-diffusion price minus the Black–Scholes price. Of course, if a jump does occur we do extremely badly. The no-jump final portfolio value really represents the insurance premium we are taking, in return for taking on jump risk.

Note also that on a day-to-day basis we are not Delta-hedged relative to the market. The market price uses the risk-neutral parameters, so if spot changes, the rate of change of the option's price will not equal the Delta hedge that we hold. This means that the Delta of the option minus our hedge is non-zero, so a change in spot will cause a change in the portfolio value en route, despite the final variance being zero. Thus if we are worried about keeping the change as small as possible in the short term rather than the long term, we should Delta-hedge with the risk-neutral parameters.

If we decide that we wish to hedge the final variance whatever happens, then Delta-hedging is no longer appropriate. We must instead hold a hedge that reduces the damage of a jump. There is some discussion of how to reduce the variance by hedging in [118].

We showed in Proposition 15.1, that the jump-diffusion price is always higher than the Black–Scholes price with the same diffusive volatility for vanilla options with convex options. A related question is whether the jump-diffusion price must be a monotone-increasing function of jump-intensity? We show that it is. We deduce our result from the domination of Black–Scholes prices by jump-diffusion prices.

Suppose $\lambda_1 < \lambda_2$, and the expiry of our option is $T > 0$. Let

$$\tilde{\lambda}_1(t) = 2\lambda_1 \quad \text{for} \quad t \leq T/2,$$

and zero otherwise. Let

$$\tilde{\lambda}_2(t) = \begin{cases} 2\lambda_1 & \text{for } t \leq T/2, \\ 2(\lambda_2 - \lambda_1) & \text{for } t > T/2. \end{cases}$$

Given a volatility σ, a jump distribution J and an interest rate r, the price of an option on a stock with jump-intensity λ_j will be the same as that for one with intensity $\tilde{\lambda}_j$ as the integral of the intensity is the same in both cases.

We therefore take a market in which there are two stocks, S_1, S_2 which move identically up to time $T/2$ with the same diffusive volatility σ driven by the same Brownian motion and with jumps according to the same Poisson process with intensity $\tilde{\lambda}_1$. After the time $T/2$ however, S_1 jumps with intensity zero and S_2 has intensity $\tilde{\lambda}_2$. Each stock has the instantaneous drift at all times required to make the discounted price process a martingale. We take S_1 and S_2 to have the same value initially.

The price of an option with intensity $\tilde{\lambda}_j$ is equal to the discounted expectation of its pay-off as an option on S_j. If we consider a portfolio consisting of being long an option on S_2 and short an option on S_1, then proving its value at time zero is positive is equivalent to saying that the option price is a strictly increasing function of λ.

By the monotonicity theorem, it is enough to prove that the portfolio is of positive value in all world states at time $T/2$. But at time $T/2$, S_1 turns into a nice purely diffusive stock so the price of an option on it is just the Black–Scholes price. However, S_2 is still jumpy so since S_1 and S_2 must be equal at time $T/2$, we invoke our result that the Black–Scholes prices is less than the jump-diffusion price and conclude the portfolio is indeed of positive value in all worlds at time $T/2$ and hence it is of positive value at time zero. Our result follows. We conclude

Theorem 15.1 *The price of a vanilla option which has a convex payoff on a stock following a jump-diffusion process is an increasing function of jump intensity provided we fix the other risk-neutral parameters.*

15.6 Choosing the measure and hedging exotic options

Given that there are an infinite number of pricing measures, how can we choose one and what does that choice mean? Also, given that the price is not unique, what good is it?

The first point to realize is that choosing a given value of the jump intensity is really equivalent to assessing the risk-aversion to jumps. A risk-averse investor will use a greater intensity and a risk-seeking investor a lower one. A trader can therefore assess whether a given piece of jump risk has a price that is greater or

15.6 Choosing the measure and hedging exotic options

lower than what he wishes to pay and therefore trade accordingly. When Merton first introduced jump-diffusion models, he argued that stock-price jumps were a diversifiable risk and that therefore no risk-aversion would be attached to them. This is equivalent to saying the pricing measure should be the one with jump intensity equal to the real world intensity. Unfortunately, whilst this view may have been accurate in 1976 when Merton wrote the first paper, it has not been true since the 1987 crash which demonstrated to market participants that the possibility of a crash was very real, and that the event of a crash was definitely a piece of undiversifiable risk. Indeed since the 1987 crash, equity market option smiles have become much more strongly downward sloping. This reflects an increased risk-aversion to downward jumps amongst market participants.

It is important to remember that there is a large number of vanilla option prices observable at a given time. We can therefore assess whether these prices are compatible with a jump-diffusion model and assess how much of the price is coming from risk-aversion to jumps, and how much from volatility. In practice, this means fitting a jump-diffusion model to the observed market prices and seeing what values of λ and σ are implied. The required values may well be time-dependent. If no good fit is possible, we have a choice between deciding that the market prices of options are arbitrageable and exploiting the opportunity, or abandoning our model.

Assuming we have found a fit, the market has given us its view on λ and σ. The market has therefore chosen a measure for us. To quote Björk, [17],

> Who chooses the martingale measure?

The answer is

> the market.

Given that the market has chosen a measure for us, what can we do with it? We can assess whether we think the market is overpricing certain pieces of risk. We can take the vanilla option prices as a given and use the given measure to price exotic options. We can attempt to hedge by using the vanilla options as a hedging instruments.

To illustrate the issues involved in hedging using vanilla options in a jump-diffusion world, let's consider a simple model with only a single possible jump amplitude. Suppose we use the vanilla call option, O, struck at 100 as a hedging instrument and we want to hedge a call option, C, struck at 110. We can set up a hedging portfolio consisting of -1 units of C, and α units of O and Δ stocks, together with bonds to ensure the self-financing condition holds.

We want the portfolio's value to be invariant under small changes in S, that is, to be Delta-neutral, and to be invariant under jumps. This means we require α and

Δ to be such that

$$-\frac{\partial C}{\partial S} + \alpha \frac{\partial O}{\partial S} + \Delta = 0, \qquad (15.18)$$

and that

$$-C(S,t) + \alpha O(S,t) + \Delta S = -C(JS,t) + \alpha O(JS,t) + \Delta JS, \qquad (15.19)$$

where J is the jump amplitude. We have two equations in two unknowns which we can solve for α and Δ. We are now perfectly hedged and the price of C is guaranteed.

Suppose we have a larger finite number of jump amplitudes, then we can hedge in a similar fashion provided we add in another hedging option for each jump amplitude, we just have to solve a larger linear system. Generally, we will want to have an infinite number of amplitudes. However, by hedging all the jumps in a large finite set, we will expose ourselves to little risk since we only have the small amount of slippage between the actual realized jump and the nearby hedged ones. Thus provided we can use vanilla options to hedge, the hedging problems have disappeared.

Or have they? We have, in fact, made a huge but non-obvious assumption. We have assumed that the hedging call options continue to be priced according to the same measure. This was not an issue in the Black–Scholes world where the pricing measure was unique, but in an incomplete market, the market chooses the measure and the market can change its mind. We can therefore add to the answer to Björk's question above,

and the market is fickle.

This means that in (15.19) even if the equation holds before the jump, the value of O after the jump may not be as expected because the market's risk aversion to jumps will probably be affected by the jump's occurrence. The market may also think a jump back up is more likely after a jump. This means that after a jump the market may price with a different jump-intensity and jump-distribution. Therefore to hedge in a jump-diffusion world really requires a model of how the market chooses the measure, and hedging against the markets' fickleness. Note that the fact that the market's choice of measure has changed will mean that the implied volatilities of options in the market have changed. Thus implied volatilities can and do change from day to day without the real-world volatility and jump-intensity changing, simply because the market has chosen a different measure.

Thus in conclusion if we use a jump-diffusion model to hedge then we are really using a model for the price processes of the hedging options. Note the difference here from the Black–Scholes world: there a process for the underlying was chosen and then all the vanilla option prices were determined. In the jump-diffusion world

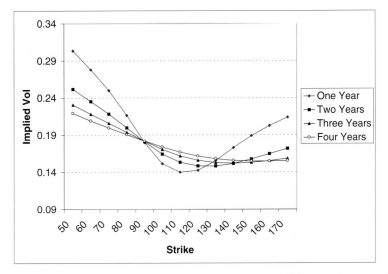

Fig. 15.4. Jump-diffusion smiles with log-normal jumps which have a downward mean.

there are many possible price processes and we therefore have to choose one in order to hedge.

15.7 Matching the market

In trying to decide the appropriateness of jump-diffusion models, one important issue is how well they reproduce the market prices of options. As we have mentioned, the cost of equity put options is bidded up by an excess demand for protection against jumps. If we translate back into Black–Scholes implied volatilities, as the market does, we therefore obtain a steeply downward-sloping curve.

If we price according to a jump-diffusion model and compute the Black–Scholes implied volatilities, then we similarly get a smile. We therefore want to see how accurately these smiles mimic market observed smiles. The jump-diffusion model produces downward-sloping smiles quite easily by taking the expected jump ratio to be less than one and it is fairly easy to match the market smile at a single maturity. (See Figures 15.4 and 15.5.)

However, we do not wish to match the smile at just one maturity. In the market, we can simultaneously observe many maturities. So the issue is, how do the smiles change as a function of the maturity of the option. For both the equity market and the jump-diffusion model, we observe the property that the smile is much steeper for short maturities than for long maturities. In general, the rate of decay of the two smiles need not be the same – the market-observed smiles are driven by traders making prices, not by jump-diffusion smiles. This means that if we want to match the smile at all maturities we may need to use time-dependent parameters.

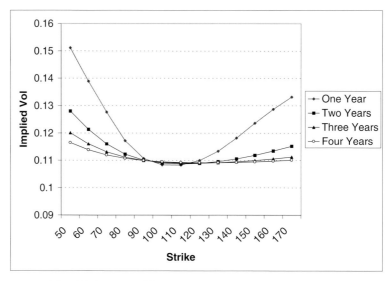

Fig. 15.5. Jump-diffusion smiles with symmetric jumps.

How can we interpret this? One explanation is simply that the jump-diffusion model is wrong. However, it is hard to find a model with time-independent parameters that does reproduce the market smile. Another interpretation is that the jump-intensity really is a function of time to jump. This is not so unreasonable if we remember that the jump-intensity is not a real world quantity but instead reflects the risk-aversion of the market to jumps. So a decreasing jump-intensity says that the market cares more about short-dated jumps than about long-dated ones. That is, there is more demand for short-dated put options than long-dated ones. This is quite believable if one believes that the demand for put options comes from fund managers worried about the short-term performance of their investments.

15.8 Pricing exotic options using jump-diffusion models

One possible use for a jump-diffusion model is to price exotic options. If we have fitted the vanilla market with a jump-diffusion model then we have effectively chosen a measure. We can now use this measure to price an exotic option and be assured that the price obtained is arbitrage-free. How do we actually carry out the pricing?

The simplest approach is to use Monte Carlo simulation. We have to evaluate a risk-neutral expectation and the simplest way to do it is by sampling the paths and averaging over them. Another approach is to use replication techniques; as we have a formula for the vanilla options, we can price accurately any option that is replicable in terms of vanilla options. If we are using only a few jump

15.8 Pricing exotic options using jump-diffusion models

amplitudes we could also use tree techniques though they would tend to be tricky to implement.

In what follows, we assume that λ is a function of time but not spot. Before proceeding to the pricing of exotics, we examine what difference this makes to the pricing of vanilla options. In fact, if λ is non-constant then the number of jumps in a time segment $[0, T]$ has the same distribution if λ is replaced by the average value of λ, which we denote by

$$\bar{\lambda} = \frac{1}{T} \int_0^T \lambda(s) ds. \quad (15.20)$$

If we consider the evolution of the log, we see that the terminal distribution of the Brownian part of S is unaffected by the occurrence of jumps. We essentially have two separate non-interacting processes for $\log S$, one displacing it in a Brownian way and the other in a jump fashion. This means that changing the timing of jumps does not affect the terminal distribution of $\log S$, in neither the risk-neutral density nor the real-world density.

The effect of this is that our pricing formulas hold true in the presence of variable jump-intensity provided we replace λ by its mean over the period of the option.

Pricing a multi-look option

Suppose our option is a multi-look option as we studied for the Black–Scholes model in Chapter 9. Our option therefore pays at time T a function f of the values of spot at times $t_1, t_2, \ldots, t_n$. The two principal examples to keep in mind are the Asian option

$$f_A(S_{t_1}, S_{t_2}, \ldots, S_{t_n}) = \left(\frac{1}{n}\sum_{j=1}^n S_{t_j} - K\right)_+, \quad (15.21)$$

and the discrete-barrier knock-out call option where $f_D(S_{t_1}, S_{t_2}, \ldots, S_{t_n})$ equals $(S_{t_n} - K)_+$ if S_{t_j} is greater than the barrier B for all j, and zero otherwise.

How can we price this option by Monte Carlo? We only need the value of the spot at times S_{t_j}, so we can simply evolve over these times. Indeed, if we work with log-normal jumps, we can use (15.12). Thus for each evolution step from time t_j to time t_{j+1}, we draw the number of jumps using (15.4) with λt taken to be equal to $\int_{t_j}^{t_{j+1}} \lambda(s) ds$. We then take a normal variable W which we plug into (15.12) and store the value of $S_{t_{j+1}}$ and evolve on to the next time up to time t_n. At the end, we just store $e^{-rt_n} f(S_{t_1}, S_{t_2}, \ldots, S_{t_n})$. We then average over all paths as usual.

This procedure is guaranteed to converge to the correct price eventually. As with any Monte Carlo simulation, we may want to consider how to improve the speed of convergence. In the case of a discrete-barrier option, we could immediately abandon any path once the barrier is crossed, reducing slightly the computation time for each path. Interestingly, at time t_{n-1}, both the discrete-barrier option and the Asian option, turn into a vanilla option. For the discrete-barrier if the barrier has not been crossed, it is a vanilla call of duration $t_n - t_{n-1}$ and strike K. For the Asian call, it becomes a vanilla call with a modified strike which is dependent on the realized values. In both these cases, one can therefore just plug the formula value of this option into the expectation, instead of running the final time slice. This will probably have more effect in the case of a discrete-barrier option, where it is the distribution at final time that is all-important, rather than for the Asian option where the final distribution has the same effect as the distribution at any other time slice.

How else can we price these options? The semi-static replication method we presented in Chapter 10 for discrete-barrier options will work with no changes, simply by plugging in the pricing formula for a jump-diffusion option instead of the Black–Scholes formula. To see this one simply has to observe that the only feature used of the Black–Scholes model is the fact that prices are deterministic in the sense that not only do we know the price of all options today, we also know the prices of all options for any give value of spot and time in the future. This is also true in the jump-diffusion model provided we stay within a given risk-neutral measure, which we are implicitly doing when we carry out a risk-neutral expectation in any case.

15.9 Does the model matter?

Suppose we have observed market for vanilla options, and we use this to determine the choice of pricing measure for a jump-diffusion model. We can now use the model to price exotic options. Suppose we calibrated a different model to the vanilla option prices and used that to price the exotic options: would it give the same prices? Let us make the problem more concrete by developing an alternate model which can be used to match any smile perfectly, and then examining the differences between it and a jump-diffusion model.

The model we consider is geometric Brownian motion, but we let volatility be a function of spot and time. We thus take the stochastic differential equation for the stock to be

$$\frac{dS}{S} = \mu(S, t)dt + \sigma(S, t)dW_t, \qquad (15.22)$$

with W as usual a Brownian motion. The Black–Scholes analysis applies equally well in this case; the Black–Scholes equation still holds with the volatility taken to be a function and there is a unique risk-neutral pricing measure. Unfortunately, actually evaluating expectations is tricky because the volatility is state-dependent.

15.9 Does the model matter?

However, one can always run a PDE-solver model to compute the price of an option. It can be shown that any non-arbitrageable smile surface of call options can be exactly matched by this model. We do not develop the method here, but refer the interested reader to [106] for one approach to implementing the method. This model is sometimes called the *Dupire* model, [44], or a *restricted stochastic volatility* model, as the volatility is a deterministic function of the stochastic stock price. An alternative interpretation is that it is a tree with nodes varying according to the local volatility referred to as the *Derman–Kani implied tree*, [41].

Given that we have precisely calibrated the Dupire/Derman–Kani and jump-diffusion models to the same vanilla option prices, to what extent will exotic prices be different? Any option that can be precisely strong statically replicated will have the same price. In practice, this means that the vanilla European options will have the same prices but not a lot more. In particular, recalling the results of Section 6.3, the risk-neutral distributions for the stock price at any given time must be the same for both models since it is directly determined by the vanilla prices. However, the two-dimensional distributions need not, and will not, be the same. So the probability that the stock price lies both in a given interval I_1 at time T_1 and a given interval I_2 at time T_2 will depend on the model.

For example, suppose we have a knock-out put discrete-barrier option struck at 100 which expires at time T_2 and that the option knocks out if the spot is below 110 at time T_1 less than time T_2. Thus for the options to pay off, the price of the stock must fall from above 110 to below 100 in the time segment $[T_1, T_2]$. This fall can occur in the jump-diffusion model via a crash, and in either model via diffusion. If the times T_1 and T_2 are close together, then the probability of a diffusive move over the interval becomes very small. In particular, one can prove that the probability of passing over the interval is an exponentially decaying function of $\frac{1}{T_2-T_1}$. However, the probability of a jump is roughly proportional to $T_2 - T_1$. The effect of this is that when $T_2 - T_1$ is small the probability of the option paying off in a jump-diffusion model is much greater than in the Dupire/Derman–Kani model even if they assign the same prices to vanilla options. This of course means that the Dupire/Derman–Kani price will be much lower than the jump-diffusion price. In effect, the option is pricing the possibility of a jump in the interval $[T_1, T_2]$.

In conclusion, the model certainly does matter when pricing exotic options: the choice of underlying process can have a large impact on the price even when the different models have been calibrated to give the same vanilla option prices. One consequence of this is that if we wish to assess the riskiness of exotic option positions, not only do we need to assess the probability of adverse market moves but also the possibility that our model is wrong. We return to these issues in Chapter 18 where we discuss the philosophical issues and practical problems associated with model choice.

15.10 Log-type models

The Black–Scholes model and the jump-diffusion model have a nice property in common: the distribution of increments of the log of the stock price is independent of the current value of the stock price. This is equivalent to the fact that the process for the log is constant-coefficient. Our purpose in this section is to prove that this property leads to some nice results about option pricing. The results of this section apply equally well to the models of Chapters 16 and 17.

Our fundamental assumption for this section is that the risk-neutral distribution of $\log S_T$ given the value of $\log S_t$ can be written as

$$\Phi_{t,T}(\log S_T - \log S_t) d\log S_T.$$

We include the $d\log S_T$ to emphasize that the variable of integration is $\log S_T$. The equivalent assumption for S_t is that the density is of the form

$$\Phi_{t,T}(S_T/S_t) \frac{dS_T}{S_T}.$$

15.10.1 Homogeneity

Let $C(S_0, K, T)$ denote the price of a call option struck at K, with expiry T at time 0 if spot is S_0. Our first result is

Theorem 15.2 *In a log-type model, the function $C(S_0, K, T)$ is homogeneous of order 1 in (S_0, K). That is*

$$C(\lambda S_0, \lambda K, T) = \lambda C(S_0, K, T),$$

for any $\lambda > 0$.

Proof We have, dropping T, that

$$C(S_0, K) = \int (S - K)_+ \Phi\left(\frac{S}{S_0}\right) \frac{dS}{S} \tag{15.23}$$

$$C(\lambda S_0, \lambda K) = \int (S - \lambda K)_+ \Phi\left(\frac{S}{\lambda S_0}\right) \frac{dS}{S}$$

$$= \int (\lambda S' - \lambda K)_+ \Phi\left(\frac{\lambda S'}{\lambda S_0}\right) \frac{dS'}{S'}$$

$$= \lambda \int (S' - K)_+ \Phi\left(\frac{S'}{S_0}\right) \frac{dS'}{S'}$$

$$= \lambda C(S_0, K). \tag{15.24}$$

Here in the second equality we have performed a change of variables $S = \lambda S'$. □

Note that this proof will hold for any pay-off function which is a homogeneous function of spot and strike. We therefore have

Corollary 15.1 *If the derivative, $D(S, K, T)$, has pay-off which is a homogeneous of order one in spot and strike then the price is also homogeneous of order one in spot and strike for any log-type model.*

15.10.2 Convexity

We can also show that the Gamma of a call option or indeed any option with a convex pay-off is positive for any log-type model. Recall that a function is convex if the chord between any two points on the graph lies above the graph, and that convexity is equivalent to the fact that the second derivative is non-negative.

When such a function not twice-differentiable, we can interpret this result in a distributional sense. For example, the call option's pay-off has second derivative equal to $\delta(S - K)$. The distribution $\delta(S - K)$ is positive in that its integral against any positive function is positive: if $f(x) > 0$ for all x, we have

$$\int f(x)\delta(S - K)dS = f(K) > 0.$$

Theorem 15.3 *If the derivative D has pay-off f, at time T, which is a convex function of S_T then the Gamma of D in any log-type model is non-negative, provided we have that $f(S_T)\Phi(S_T/S_0)$ tends to zero as S_T tends to infinity.*

Note that the technical condition here is very mild.

Proof We have that the value of D at time zero is equal to

$$e^{-rT} \int f(S)\Phi\left(\frac{S}{S_0}\right)\frac{dS}{S}.$$

Differentiating with respect to S_0 we obtain

$$-e^{-rT} \int f(S)\Phi'\left(\frac{S}{S_0}\right)\frac{dS}{S_0^2}.$$

Integrating by parts and using the technical part of the hypotheses, this becomes

$$e^{-rT} \int f'(S)\Phi\left(\frac{S}{S_0}\right)\frac{dS}{S_0}.$$

We now change variables and let

$$\bar{S} = \frac{S}{S_0},$$

to get

$$e^{-rT} \int f'(\bar{S}S_0) \Phi(\bar{S}) d\bar{S}.$$

Differentiating with respect to S_0, we find

$$e^{-rT} \int f''(\bar{S}S_0) \bar{S} \Phi(\bar{S}) d\bar{S}.$$

This will be non-negative since f is convex and Φ is supported where $\bar{S}$ is non-negative. As the Gamma is non-negative the value is convex as a function of spot. □

15.10.3 Floating smiles

One nice consequence of the homogeneity of call prices is that the implied volatility smile *floats*. Thus if strike is K and spot is S then the implied volatility function, $\hat{\sigma}(S, K)$, that is, the implied volatility of a call option struck at K given that spot is S satisfies

$$\hat{\sigma}(S, K) = g\left(\frac{K}{S}\right), \qquad (15.25)$$

for some function g. To see this observe that if

$$C(S, K, T) = \text{BS}(S, K, \sigma, T),$$

then it also true for any $\lambda > 0$ that

$$C(\lambda S, \lambda K, T) = \text{BS}(\lambda S, \lambda K, \sigma, T),$$

as the λ passes through everything. This means that $\hat{\sigma}(\lambda S, \lambda K, T)$ is independent of λ, which is equivalent to saying that it is a function of K/S.

We call K/S the *moneyness* as it expresses how much the option is in or out of the money as a ratio. The implied volatility is only a function of moneyness i.e. the smile will always look the same qualitatively. For example, if the smile is smile-shaped with a minimum at the money then this will remain true whatever spot does.

15.11 Key points

We have covered a lot of ground in this chapter. The main thing to take away is that whilst the incompleteness of the market means that option prices are not unique, many other aspects of the theory can be extended to jump-diffusion models.

- Jump-diffusion models encapsulate the idea that the stock price can jump with no possibility of rehedging during the move.

- A closed-form formula as an infinite sum can be developed for the price of a call or put option in a jump-diffusion model.
- The market consisting of a stock evolving to a jump-diffusion model and a riskless bond is incomplete.
- In an incomplete market an option does not have a unique price.
- When changing measure in a jump-diffusion world, we can change the drift, the intensity of the jumps and the jump distribution but we cannot change the volatility of the underlying.
- Increasing jump-intensity always increases the price of a European option, which has a convex final payoff.
- For a digital option increasing jump-intensity can either increase or decrease the price of an option.
- In an incomplete market it is the market which chooses the measure.
- It is possible to hedge in a jump-diffusion model using options provided we assume that the market does not change its choice of measure. That is provided the market is not *fickle*.
- Even if two models give identical prices to vanilla options, they can give quite different prices to exotic options.
- Jump-diffusion models give rise to deterministic future smiles so weak static replication can be used to price exotic options.
- Exotic options can be priced by Monte Carlo in jump-diffusion models by stepping between the look-at dates of the option.
- Jump-diffusion smiles are very sharp for small maturities and shallow for long maturities.

15.12 Further reading

The original paper on jump-diffusion models was written by Merton, [95], in 1976. This paper was written before the concept of an equivalent martingale measure had been introduced and is based on a partial-integro-differential equation approach. The issue of market incompleteness was avoided by arguing that the risk of a jump was diversifiable and therefore did not attract a risk-premium. The paper was reproduced in the highly recommended collection of Merton's papers, [94].

For some of the issues about choosing martingale measures and more highbrow proofs of option price monotonicity see [13], [31] and [103].

A jump-diffusion model with a different jump-distribution is given in [80].

Rebonato makes the case for jump-diffusion models at length in [106].

One approach to hedging to obtain minimal variance under the jump-diffusion model is given in [118]. Another approach under the assumption that the real-world measure is a martingale is given in [81].

The pricing of path-dependent exotic derivatives is discussed in [120]. The use of replication methods for pricing under jump-diffusion models is discussed in [74].

Some aspects of smiles under jump-diffusion models are discussed in [5]

The sub- and super-replication of exotic prices, whilst assuming the ability to trade in vanilla options today but not at future times, is studied in [62] and [28]. Note that prices developed in this fashion will be robust and not affected by market fickleness.

The original papers on models with volatility of the form $\sigma(S, t)$ are [44] and [41]. A recent paper on the practical difficulties involved in implementation is [85]. Another approach to finding to the fitting using PDEs is given in [106].

A quite different approach to developing a pricing formula using Fourier transform techniques is given in [84].

15.13 Exercises

Exercise 15.1 Suppose we have two assets with the same diffusive volatility and real-world drift. The first asset, S, jumps downwards according to a Poisson process and the second asset, T, never jumps. Show that the real-world expected payoff of a call option on S is always less than one on T. What does Theorem 15.1 say about the relative prices of call options on these assets?

Exercise 15.2 Show that an arbitrage exists if the two assets in Exercise 15.1 are driven by the same Brownian motion.

Exercise 15.3 Suppose the log S_t follows a Brownian motion over the period [0, 1] except at time 0.5 where it jumps by x. What are the first and second variations of log S_t over the period [0, 1].

Exercise 15.4 Develop a formula for the price of a digital call option in the context of a jump-diffusion model.

Exercise 15.5 Develop a formula for the risk-neutral density of a stock in a jump-diffusion world.

Exercise 15.6 Suppose a stock follows a process in the risk-neutral world which involves time-dependent parameters. For the pricing of a call option, which parameters can be replaced by the appropriate average without changing the price of the option?

Exercise 15.7 Show that the pay-off of a put option is a homogeneous function of spot and strike.

15.13 Exercises

Exercise 15.8 Show that the pay-off of a digital option is a homogeneous function of order zero in spot and strike.

Exercise 15.9 Find a formula for the price of a call option on a dividend-paying stock with constant dividend rate d under a jump-diffusion model.

Exercise 15.10 Suppose we wish to price an Asian option by Monte Carlo using a jump-diffusion model with log-normal jumps. If the option has N look-at dates, how many (pseudo-)random numbers will be needed per path? What if the jumps are not log-normal?

Exercise 15.11 Suppose a stock S_t follows a jump-diffusion process such that jumps can only occurs in the time period from 0 to t_1. An option pays $f(S_{t_2}/S_{t_1})$ for some reasonable function f. Show that there is a unique arbitrage-free price for f and write down an expression for it.

16

Stochastic volatility

16.1 Introduction

It is an observed fact in the market that the implied volatilities of traded options vary from day to day. We have seen that in an incomplete model, such as jump-diffusion, this can be caused by the changing risk preferences of the market's participants. An alternative and straightforward explanation is that the instantaneous volatility of a stock is itself a stochastic quantity. Thus during certain periods the stock is more heavily traded and more information arrives, causing the stock to wobble more rapidly. During such a period the total amount of vibration expected during the option's life will be greater, and one therefore expects the option's cost to be higher. Thus options' implied volatilities will become stochastic. It is important to realize that whilst making instantaneous volatility stochastic makes implied volatility stochastic, the relationship between the two is not straightforward. Indeed in the presence of stochastic volatility, it is necessary to redevelop the pricing formula to take account of the added uncertainty, and one then needs to plug the instantaneous volatility into the new formula and invert the Black–Scholes formula to get the new implied volatility.

Having decided to make the instantaneous volatility stochastic, it is necessary to decide what sort of process it follows. Volatility is generally chosen to follow a diffusive process though in more sophisticated models it can be allowed to jump, and indeed some models mix jump-diffusion and stochastic volatility to reflect the greater volatility in the market after a crash. This is achieved by giving a jump component to the volatility process which is correlated with the jump process for the underlying. We do not address jumps in volatility here but restrict ourselves to diffusive volatilities. We therefore take a process of the form

$$\frac{dS}{S} = \mu dt + V^{1/2} dW^{(1)}, \tag{16.1}$$

$$dV = \mu_V dt + \sigma_V V^{\alpha} dW^{(2)}, \tag{16.2}$$

with α a positive real number. We work with the instantaneous variance, V, which is the square of the instantaneous volatility σ. The Brownian motions $W^{(1)}$ and $W^{(2)}$ may be correlated or uncorrelated as we choose.

Many of the issues which arise with stochastic-volatility models are similar to those involved with jump-diffusion models. Both models imply an incomplete market and hence an infinity of prices. Pricing formulas can be developed but they are complicated. Pricing by Monte Carlo is possible but not very rapid. Hedging is possible provided one allows the use of an option and assumes that the market is not fickle, that is, the market does not change its choice of measure. We explore all these issues in this chapter.

16.2 Risk-neutral pricing with stochastic-volatility models

Given that the spot and the instantaneous variance evolve according to (16.2), what are the risk-neutral measures? Invoking the multi-dimensional version of Girsanov's theorem, changing the measure will allow us to change the drifts of the two stochastic processes but will allow nothing else. If we are working in a deterministic interest-rate world with constant continuous compounding rate r then the riskless bond at time t will be worth e^{rt}. Taking the riskless bond as numeraire, we want $e^{-rt}S$ to be a martingale. In this context, ignoring technical difficulties, being a martingale will be equivalent to the price process being driftless. The drift of the variance process is not relevant to being a martingale, as its size will only magnify up and down moves, not change the mean of the stock value.

A quick application of Ito's lemma shows that, as in the Black–Scholes world, for a risk-neutral measure the drift of S must be rS. Invoking Girsanov's theorem, we conclude that all risk-neutral measures are associated to processes of the form

$$\frac{dS}{S} = rdt + V^{1/2}dW^{(1)}, \tag{16.3}$$

$$dV = \tilde{\mu}_V(S, V, t)dt + \sigma_V V^\alpha dW^{(2)}, \tag{16.4}$$

The crucial point here is that $\tilde{\mu}_V$ is arbitrary: it need bear no relation to μ_V and can be any reasonable (i.e. measurable) function of (S, V, t) one likes.

The arbitrariness of $\tilde{\mu}_V$ reflects the fact that volatility is not a tradable quantity, and hence our market has two sources of uncertainty but only one underlying and so is incomplete. How do we choose $\tilde{\mu}_V$? A popular choice is to make the volatility process *mean-reverting*. We then have that

$$\tilde{\mu}_V(S, V, t) = \lambda(V_r - V). \tag{16.5}$$

This means that the instantaneous variance reverts to the level V_r at rate λ. A mean-reverting variance process is appealing because volatility tends to have a natural

370   *Stochastic volatility*

level which is occasionally perturbed. In particular, a turbulent period will eventually subside, and the volatility will fall back to the background level. However, one should be careful to distinguish between risk-neutral volatility and real-world volatility. The fact that the real world volatility is mean-reverting does not force us to use a mean-reverting risk-neutral volatility process. Of course, as in the jump-diffusion world, it is the market that chooses the measure, which in this case means that the market chooses the form of the drift function. A certain circularity now comes into play. If we attempt to fit the market, and a mean-reverting function fits the prices well, what does it signify? It may well be that other banks are using mean-reverting stochastic volatility models to price: the model could be driving the derivatives market.

Once the drift has been fixed we want to price. The hard thing is to actually use (16.4) for pricing. Of course, the price of a derivative is just given by its risk-neutral expectation but how do we evaluate that expectation?

16.3 Monte Carlo and related approaches to stochastic volatility pricing

When working with a Black–Scholes world, the evolution of the spot was greatly simplified by the fact that we could solve the stochastic differential equation which described the risk-neutral evolution. This allowed us to evolve the spot to the expiry time of a vanilla option in a single jump whilst ignoring the values in between. For stochastic volatility models, the stochastic differential equation is generally not solvable and Monte Carlo simulation is therefore much more cumbersome.

Any Ito process can be simulated by using small step increments, so we can certainly use Monte Carlo: it will simply be much slower. We choose a small step size, Δt, and put

$$\log S_{j\Delta t} = \log S_{(j-1)\Delta t} + \left(r - \frac{1}{2}V_{(j-1)\Delta t}\right)\Delta t + V_{(j-1)\Delta t}^{\frac{1}{2}}\sqrt{\Delta t}W_1, \quad (16.6)$$

$$V_{j\Delta t} = V_{(j-1)\Delta t} + \mu\left(S_{(j-1)\Delta t}, V_{(j-1)\Delta t}, (j-1)\Delta t\right)\Delta t$$
$$+ \sigma_V V_{(j-1)\Delta t}^{\alpha}\sqrt{\Delta t}W_2, \quad (16.7)$$

where W_1 and W_2 are correlated normal draws with correlation equal to the instantaneous correlation between V and S. We can certainly use this approach to price any derivative. However, the pricing will be rather slow! Monte Carlo is not particularly fast at the best of times, and if we divide the time segment into say one hundred steps then the time to run one path will increase a hundred-fold. We can use longer steps if we increase the stability of V. One way to increase the stability of (16.7) is to change variables so that the coefficient of W_2 is independent of V. In other words, we compute the SDE for V^β instead of V and choose β in such a way that the volatility term is a constant. This will make the drift term more complicated but since the volatility term dominates over short steps, this is a price well worth paying. The computation of the reduction is in Section 5.5.

We can look for simplifications which allow us to speed up the simulation. Suppose the volatility and spot are uncorrelated. Suppose in addition that the drift of the volatility process is independent of the value of spot. In other words, the volatility process does not see the value of spot in any way: information flows from the volatility to the spot but not the other way round. The density of the volatility paths can therefore be developed without reference to S.

Of course, the distribution of S still depends on the choice of volatility path. Suppose we are valuing a European option, and therefore want the risk-neutral distribution of S at time T. We first draw a volatility path and then use it to simulate S. Drawing a volatility path is really drawing a time-dependent function $V(t) = \sigma(t)^2$ for instantaneous variance. We are then drawing a path for S under the stochastic process

$$\frac{dS}{S} = r\,dt + \sigma(t)dW. \tag{16.8}$$

We know that the terminal distribution for S under such a process is given by

$$S_T = S_0 e^{(r - \frac{1}{2}\bar\sigma^2)T + \bar\sigma\sqrt{T}N(0,1)}, \tag{16.9}$$

where $\bar\sigma$ is the root-mean-square value of σ. This means we do not need to small-step S, which will greatly speed up the simulation.

Since the process for σ is independent of that for S, we can integrate over S for each value of $\bar\sigma$. What does this give us? We get

$$e^{-rT} \int f(S_T)\Phi(S_T, \bar\sigma)dS_T,$$

where $\Phi(S_T, \bar\sigma)$ is the log-normal density for S_T starting at S_0 with drift r and volatility $\bar\sigma$. However, we know the value of this integral; it is the Black–Scholes price for an option paying $f(S_T)$ at time T with volatility $\bar\sigma$.

How can we interpret this? We draw a volatility function $\sigma(t)$; this tells us the total amount of volatility experienced during the life of the option. The hedging costs for the option are then determined by the total amount of volatility, and the actual path for the underlying is not relevant. In other words, there are lucky and unlucky paths for the volatility but not for the underlying. Contrast this with the Black–Scholes world, where all paths were equivalent and luck was not an issue. This is also different from jump-diffusion where lucky and unlucky paths existed, but it was the movements in spot that were important.

To return to the issue of practical pricing, we can now price more effectively by drawing a path for the volatility, and then plugging the root-mean-square volatility for that path into the Black–Scholes formula. We then simply average these prices over many paths. This procedure is highly effective. As most of the work is in generating the volatility path, several options of different strikes but the same maturity can be priced simultaneously, without much extra computational effort. If one wished to do several different maturities simultaneously one could also stop the volatility path at various points in order to get the volatility up to each maturity. In practise, we cannot actually draw the entire volatility path, but instead we use short steps and draw the values of the volatility at a discrete set of points. We then integrate the variance along each path by assuming that the instantaneous variance is linear between sampling points. We could have taken the variance to be constant between times but this would result in requiring more steps for convergence since the approximation is cruder. It is the difference between using the trapezium rule and piecewise constant integration.

Our pricing of a European option only depends upon one aspect of the volatility path: the root-mean-square of the volatility, $\bar{\sigma}$, along the path. We therefore do not really need to simulate the path, all we need to do is to compute the distribution of $\bar{\sigma}$ and simulate it. The problem is that analytically computing the distribution of $\bar{\sigma}$ is not tractable. The root-mean-square volatility is, of course, the square root of the mean-square volatility. So if we can get the distribution of the latter that is enough. However, it is not particularly tractable either, but in special cases, we can compute its moments. Once we have computed the moments of the mean-square volatility we can then moment match using our favourite distribution. If g is the fitted density for the mean-square volatility, $\overline{V}$, and $\mathrm{BS}(\overline{V})$ is the option price for $\overline{V}$, then our implied price is

$$\int g(\overline{V}) \mathrm{BS}(\overline{V}) d\overline{V}.$$

This approach was pioneered by Hull & White and was one of the earliest approaches to implementing stochastic-volatility models.

When can we compute the moments? If the volatility process is log-normal and the risk-neutral drift is constant then the process for the square of the volatility is log-normal. It is then a question of computing the moments of the integral of a log-normal process. This is possible analytically.

16.4 Hedging issues

One reason why stochastic-volatility models tend to be more popular than jump-diffusion models is that they allow the illusion of hedgeability. There are two sources of uncertainty, the movement of the spot and the movement of volatility. This means that if we allow ourselves a second hedging instrument, we ought to be able to hedge. In particular, this means we can Vega-hedge the volatility.

How does this work in practice? Provided the risk-neutral measure is constant then we can assume for each of our options, O_1 and O_2, that it is a function of

(S, V, t). We then have from Ito's lemma that

$$dO_j = \frac{\partial O_j}{\partial t}dt + \frac{\partial O_j}{\partial S}dS + \frac{\partial O_j}{\partial V}dV + \frac{1}{2}\frac{\partial^2 O_j}{\partial V^2}dV^2 + \frac{\partial^2 O_j}{\partial S \partial V}dS\,dV$$
$$+ \frac{1}{2}\frac{\partial^2 O_j}{\partial S^2}dS^2, \quad \text{for} \quad j = 1, 2. \tag{16.10}$$

When hedging we are not interested in the drift, and, computing, we obtain

$$dO_j = \mu_j(S, V, t)dt + S\frac{\partial O_j}{\partial S}V^{1/2}dW^{(1)} + \frac{\partial O_j}{\partial V}\sigma_V V^\alpha dW^{(2)}, \tag{16.11}$$

with μ_j some unknown and unimportant function. Suppose we are short O_1. We can hedge the $W^{(2)}$ uncertainty in the option O_1 by holding

$$\left(\frac{\partial O_2}{\partial V}\right)^{-1}(S, V, t)\frac{\partial O_1}{\partial V}(S, V, t)$$

units of O_2. We are then left with a portfolio with non-zero Delta; however this can be hedged as usual by holding stock equal to the Delta.

Note that this argument works for any derivative O_1 – it could be any exotic option. For the hedging option O_2 the only important property is that it has non-zero Vega. This implies that, for example, a ten-year out-of-the-money call option could be hedged by a one-week in-the-money put option. Typically, we would, however, hedge with as similar an option as possible. For example, an at-the-money option might be used to hedge an out-of-the-money option with the same maturity. This is because it is really the implied volatilities that are moving around rather than the instantaneous. They are being driven as much by changing risk-preferences and expectations of future volatility as by the change in the instantaneous volatility. For example, something may happen that suggests the arrival time of some information in the future but does not particularly affect today's price or volatility. For example, if an election date is set we can expect a lot of volatility around the time of the election. We can expect a company share's price to react strongly to the publication of the annual report.

The argument for perfect hedging depends on the market's choice of risk-neutral measure not changing – the market must not be fickle. Thus whilst only one option is required for hedging in stochastic-volatility models, hedgeability really depends upon the assumption that the market does not change its risk-preferences, as we saw for jump-diffusion models.

16.5 PDE pricing and transform methods

Another reason for the popularity of stochastic-volatility models is that it is possible to produce a PDE for the price, and in certain cases to solve this PDE for the

Fourier transform of the price. This means that price evaluations can be reduced to computing an inverse Fourier transform at one point.

If the correlation between the two Brownian motions is ρ then an application of Ito's lemma using (16.4) yields that the drift of an option $O(S, V, t)$ in the risk-neutral measure is

$$\frac{\partial O}{\partial t} + rS\frac{\partial O}{\partial S} + \mu\frac{\partial O}{\partial V} + \frac{1}{2}VS^2\frac{\partial^2 O}{\partial S^2} + \rho V^{1/2}\sigma_V V^\alpha S\frac{\partial^2 O}{\partial S \partial V} + \frac{1}{2}\sigma_V^2 V^{2\alpha}\frac{\partial^2 O}{\partial V^2}.$$

In the risk-neutral measure, we must have that $e^{-rt}O(S, V, t)$ is a martingale. This is equivalent to saying that the drift of O must be rO. We therefore have the partial differential equation

$$\frac{\partial O}{\partial t} + rS\frac{\partial O}{\partial S} + \mu\frac{\partial O}{\partial V} + \frac{1}{2}VS^2\frac{\partial^2 O}{\partial S^2} + \rho V^{1/2}\sigma_V V^\alpha S\frac{\partial^2 O}{\partial S \partial V} + \frac{1}{2}\sigma_V^2 V^{2\alpha}\frac{\partial^2 O}{\partial V^2} = rO. \quad (16.12)$$

Note that the risk-preferences have entered this equation through μ, the choice of drift in the risk-neutral measure.

We now want to solve this equation. Whilst this equation can be solved in quite a few cases, we restrict ourselves to the case where $\alpha = 0.5$, $\mu = 0$ and $\rho = 0$. That is, the instantaneous variance follows a square root process which is uncorrelated with the stock process and has no drift.

$$dV = \sigma_V V^{1/2} dW^{(2)}. \quad (16.13)$$

The PDE now takes the form

$$\frac{\partial O}{\partial t} + rS\frac{\partial O}{\partial S} + \frac{1}{2}VS^2\frac{\partial^2 O}{\partial S^2} + \frac{1}{2}\sigma_V^2 V\frac{\partial^2 O}{\partial V^2} = rO. \quad (16.14)$$

Our first simplification is to use log coordinates for the spot. Letting $x = \log S$, we obtain

$$\frac{\partial O}{\partial t} + \left(r - \frac{1}{2}V\right)\frac{\partial O}{\partial x} + \frac{1}{2}V\frac{\partial^2 O}{\partial x^2} + \frac{1}{2}\sigma_V^2 V\frac{\partial^2 O}{\partial V^2} = rO. \quad (16.15)$$

We will solve this PDE by applying Fourier transform methods: they seem to be the only viable approach.

Recall that the Fourier transform of a function f is given by

$$\hat{f}(\xi) = \int e^{ix \cdot \xi} f(x) dx. \quad (16.16)$$

(I am following the physicists' convention on the sign of the exponent here as it seems to be prevalent in finance – probably because there are more physicists than pure mathematicians in mathematical finance.) The function f can be retrieved from $\hat{f}(\xi)$ via

$$f(x) = \frac{1}{2\pi} \int e^{-ix.\xi} \hat{f}(\xi) d\xi. \qquad (16.17)$$

For the integral (16.16) to exist, it is enough for the integral of f and $|f|$ to exist. When they do not, the Fourier transform can still be defined but it is more tricky. There are two basic approaches to taking Fourier transforms of functions whose integrals do not exist because of growth at infinity. The first approach is to use distribution theory to define the Fourier transform via duality. This works provided the function is polynomially bounded but really requires more theory than we can develop here. A second approach is to use complex values of ξ. Thus if $\xi = a + ib$, we have

$$\hat{f}(a+ib) = \int e^{iax - bx} f(x) dx. \qquad (16.18)$$

If f is zero for x large and negative, and is exponentially bounded for x large and positive, then this integral will converge for b sufficiently large. Similarly if f is zero for x large positive and exponentially bounded for x large negative then it will converge for b sufficiently large negative. Note that $\hat{f}(a+ib)$ is really the Fourier transform of $e^{-bx} f(x)$ which means that we can recover $e^{-bx} f(x)$ using (16.17), and hence $f(x)$ is recoverable from the knowledge of $\hat{f}(a+ib)$ for all a and one b.

Back to the problem at hand, our boundary condition for a call option is

$$C(S_T, V, T) = \max(S_T - K, 0). \qquad (16.19)$$

After transforming to log space, the boundary condition becomes

$$C(x, V, T) = \max(e^x - K, 0). \qquad (16.20)$$

The Fourier transform will therefore exist for $b > 1$. The put option is much more benign, the pay-off being

$$P(x, V, T) = \max(K - e^x, 0), \qquad (16.21)$$

which is bounded by 0 and K. The integral does not exist but the Fourier transform will exist for any $b < 0$. (Indeed the Fourier transform of the put option price exists as a distribution whereas that of the call option does not.) This makes the put option more tractable than the call option. Recall that by put-call parity, the price of the put option immediately determines the price of the call option. We therefore focus henceforth on the put option from here on.

A simple integration shows that

$$\hat{P}(\xi, V, T) = -\frac{K^{i\xi+1}}{\xi^2 - i\xi}, \qquad (16.22)$$

for $\operatorname{Im}\xi < 0$.

The crucial property of the Fourier transform in solving PDEs is that it transforms differentiation into multiplication by $-i\xi$. We have

$$\widehat{\frac{\partial f}{\partial x}}(\xi) = \int e^{ix.\xi} \frac{\partial f}{\partial x}(x)dx = -\int \frac{\partial}{\partial x} e^{ix.\xi} f(x)dx = -i\xi \int e^{ix.\xi} f(x)dx$$
$$= (-i\xi)\hat{f}(\xi), \quad (16.23)$$

Fourier transforming (16.15) in x, we obtain a PDE in t and V but not x. We have therefore reduced the dimensionality of the problem. Our Fourier transformed PDE is therefore

$$\frac{\partial \hat{O}}{\partial t} + \left(r - \frac{1}{2}V\right)(-i\xi)\hat{O} + \frac{1}{2}V(-i\xi)^2 \hat{O} + \frac{1}{2}\sigma_V^2 V \frac{\partial^2 \hat{O}}{\partial V^2} = r\hat{O}. \quad (16.24)$$

We can rewrite this as

$$\frac{\partial \hat{O}}{\partial t} + \frac{1}{2}\sigma_V^2 V \frac{\partial^2 \hat{O}}{\partial V^2} = \left(r + \left(r - \frac{1}{2}V\right)i\xi + \frac{1}{2}V\xi^2\right)\hat{O}. \quad (16.25)$$

If we let $\tau = T - t$, we can as usual transform a backwards equation into a forward one. We further simplify by letting $c(\xi) = (\xi^2 - i\xi)/2$, and putting

$$\hat{O} = e^{-r\tau - i\xi r\tau} \hat{Q}.$$

We then obtain

$$\frac{\partial \hat{Q}}{\partial \tau} = \frac{1}{2}\sigma_V^2 V \frac{\partial^2 \hat{Q}}{\partial V^2} - c(\xi) V \hat{Q}. \quad (16.26)$$

Our initial condition is that Q at $\tau = 0$ is given by $\hat{P}$ in (16.22).

Fortunately, it is possible to write down a solution to (16.26). In particular, if the boundary condition is 1 the solution is

$$\hat{Q}(\xi, V, \tau) = e^{-\frac{V}{\sigma_V}\sqrt{2c(\xi)} \tanh\left(\sqrt{\frac{c(\xi)}{2}} \sigma_V \tau\right)}. \quad (16.27)$$

In fact, since (16.26) does not involve any differentiations in ξ, if we want the boundary condition to be an arbitrary function $f(\xi)$, for example $\hat{P}$, then we just multiply $\hat{Q}$ by f. The function $\hat{Q}f$ has the correct boundary value by construction and since multiplication by $f(\xi)$ commutes with differentiation in the other variables, (16.26) is satisfied by $\hat{Q}f$.

We can now price any option for which we know the fundamental transform. We simply numerically invert the Fourier transform at the appropriate value of T and obtain a price.

16.6 Stochastic volatility smiles

Since the possibility of stochastic volatility getting large increases the probability of large movements in the underlying stock, stochastic-volatility models lead to fatter tails for the distribution of the final stock price. This leads to implied-volatility smiles which pick up out-of-the-money; that is, smile-shaped smiles!

If we allow correlation between the underlying and the volatility then a skew is introduced. Roughly, if the volatility and the underlying are negatively correlated then as the stock price falls, it becomes more volatile and so out-of-the-money put options require more hedging and thus are more valuable which leads to increased implied volatilities. On the other hand, increasing the stock price leads to lower volatility, and hence lower prices and implied volatilities.

The marked difference between stochastic-volatility smiles and jump-diffusion smiles is in their time decay. The amount of stochasticity in the volatility increases over time and this leads to long-maturity smiles not decaying. Jump diffusion models have the opposite property: the chance of a large move in a short time is much greater in a jump world than a diffusive world, but the relative impact of a large move in a long time is much smaller.

To a certain extent, the time behaviour can be controlled by the mean-reversion parameter. The faster the mean-reversion, the flatter long-time smiles will be, as the mean-reversion will stop volatility's effect from piling up. See Chapter 18 for graphs of various cases.

Note that we should be careful in using stochastic-volatility models. Just because a negative correlation can be used to produce a skewed smile does not mean that that is the financial mechanism which does produce it. We return to this point in Chapter 18.

16.7 Pricing exotic options

Pricing exotic options using stochastic-volatility models is tricky. The replication methods we have presented cannot be used. For a stochastic-volatility model, the future prices of options for a given time and value of spot are not determined today. Instead, they depend upon the then prevailing value of instantaneous volatility. Whilst this is a quite appealing feature of the model in that it reflects well experience in the markets, where the prices prevailing in the future are rarely predictable, it violates a fundamental requirement of pricing by replication.

We can, of course, price multi-look options by Monte Carlo. However, as the stochastic differential equation is not solvable, this requires short-stepping which means that the time to run one path is a lot slower than for other models. This, of course, implies that the time required to price an option is much greater and often too long to be useful.

Another approach is to use trees. However, we then have to use a two-dimensional tree to reflect the fact that we have two state variables. We also have that the size of spot movements are state-dependent which means that the trees will not naturally recombine.

As we have a PDE for stochastic volatility prices, we can apply PDE techniques but we will of course have to add on an extra dimension to cope with volatility. That is we have to solve a PDE with three state variables instead of two. If we wish to price path-dependent exotics, we may need to add in an extra auxiliary variable which increases the dimensionality again.

16.8 Key points

Stochastic-volatility models are currently quite popular. They provide a simple mechanism for allowing implied volatilities of options in the market to vary from day to day. A rapid pricing formula can be developed. They have the appealing property that it is possible to hedge using one option. They can also be used to produce convincing market smiles with appropriate parameters.

On the other hand, it is difficult to price exotic options, and the hedging is really too good to be true.

- Stochastic-volatility models introduces smiles by letting volatility be a stochastic quantity.
- Real-world volatility is mean-reverting.
- Any drift can be chosen for the volatility in the risk-neutral measure but in practice a mean-reverting volatility is used.
- In a stochastic-volatility model, the instantaneous volatility and the implied volatility are quite different things.
- Prices can be developed by Monte Carlo, transform methods and PDE solutions.
- If volatility and spot are uncorrelated then the spot can be long-stepped and the price of a vanilla option can be written as an integral over Black–Scholes prices.
- Stochastic-volatility smiles tend to be shallow relative to jump-diffusion smiles for short maturities and relatively steep for long maturities.

16.9 Further reading

The transform approach developed here is based on that in *Option Valuation Under Stochastic Volatility* by A. Lewis, [83] where much more general stochastic-volatility models are studied and solved. If you want to implement transform-based solutions to stochastic volatility models this is the book to buy.

The transform approach to stochastic-volatility pricing was started by Heston, [61]. A quite general jump-diffusion and stochastic-volatility model, which probably pushes the transform technique as far as it will go in this direction, has been developed by Duffie, Pan & Singleton, [43]. In [84], the transform technique is extended to cover a large class of models.

An alternate approach to stochastic volatility models using ideas from ergodic theory has been developed by Fouque, Papanicolaou & Sircar, [47]. Their model relies on the volatility having a very fast mean reversion which means that the only effective state variable is spot. The book is interesting, readable and accessible.

One of the first papers on stochastic volatility was by Hull & White, [64], where they developed a price in the uncorrelated case by moment-matching the density of the total variance along a path.

One appealing idea is to make the implied volatilities stochastic instead of the instantaneous volatilities. To do so in a way that avoids arbitrage is, however, tricky. One approach is to make the instantaneous volatilities stochastic but let the implied volatility drive the process. Such an approach has been developed by Schönbucher, [113].

16.10 Exercises

Exercise 16.1 Show that if spot and volatility are uncorrelated then the risk-neutral density of spot can be written as an integral over log-normal densities. How would you use Monte Carlo to estimate this density?

Exercise 16.2 Which of the replication techniques of Chapter 10 can be applied when using a stochastic-volatility model?

Exercise 16.3 Which PDE will an option on a dividend-paying asset following a stochastic-volatility process with constant dividend rate, d, satisfy?

Exercise 16.4 Suppose a stock moves according to a stochastic-volatility model and we Delta-hedge using a constant volatility; what will happen?

17

Variance Gamma models

17.1 The Variance Gamma process

If one examines the movements of stocks on small time scales, one finds that they do not look particularly similar to a Brownian motion. They move in little jumps rather than continuously and the total amount of up and down moves is finite rather than infinite. All of the models we have so far considered look very similar on small time scales. Jump-diffusion and the Black–Scholes model are identical except at crash times, and letting volatility be stochastic makes little qualitative difference at small scales. They all give continuous paths (except at jump events) and have an infinite first variation, that is an infinite amount of up and down moves.

The Variance Gamma model of stock evolution attempts to address these problems by letting 'experienced time' be a random process itself. The idea is that the volatility should be a measure of a stock's sensitivity to information as it arrives, but the amount of information arriving is random also, and needs to be described by a random process itself. One can think of trading volume as a proxy for information arrival, and there is some statistical evidence that stock price returns are more log-normal when rescaled to use trading volume for the time parameter instead of calendar time.

A mathematical motivation for using random times lies in the classification of martingales. If one has a continuous martingale then it is a diffusion, so moving within the space of continuous martingales will not buy us much. A second classification theorem says that a general martingale is a random-time-changed Brownian motion. This means that in mathematical terms, the introduction of random times is inevitable if we wish to study new processes.

The random process modelling information arrival should have certain properties. The market will not forget information so the amount of information can only increase. Our random process should therefore be monotone increasing. If we require that the speed of arrival of information should not be affected by the amount of information that has already arrived then the distribution of the amount

17.1 The Variance Gamma process

of information over any period should be independent of the total amount of information at the start of the period. We can also require that the distribution of the information arriving in a given period should only depend on the length of that period.

Note that this is quite different from stochastic-volatility models where an increase in volatility persists and keeps the stock more volatile until the volatility returns randomly or mean-reverts to its previous level.

If we denote our process for information by $\Gamma(t)$ then our requirements mean that $\Gamma(t) - \Gamma(s)$ is a positive random variable with density depending only on $t-s$. Note that we cannot use a Brownian motion for $\Gamma(t)$ because it will not result in an increasing Γ. We therefore look for a family of positive random variables Y_h such that $Y_h + Y_r$ has the same distribution as Y_{h+r}. We then want $\Gamma(t) - \Gamma(s)$ to be distributed as Y_{t-s} for t larger than s.

Such a family of random variables are the Gamma distributions. We have two parameters: μ, the mean and, ν, the variance. Having fixed these the density function of Y_h is

$$p_h(x) = H(x) \left(\frac{\mu}{\nu}\right)^{\frac{\mu^2 h}{\nu}} \frac{x^{\frac{\mu^2 h}{\nu}-1} \exp\left(-\frac{\mu}{\nu}x\right)}{\Gamma\left(\frac{\mu^2 h}{\nu}\right)}, \qquad (17.1)$$

where $H(x) = 0$ for $x \leq 0$ and Γ is the Gamma function. Recall that $\Gamma(x)$ is a generalization of the factorial function defined for non-integers with the properties

$$\Gamma(x+1) = x\Gamma(x)$$

and

$$\Gamma(n) = (n-1)!$$

for integers n. We recall that

$$\Gamma(x) = \int_0^\infty s^{x-1} e^{-s} ds. \qquad (17.2)$$

The random variable Y_h has mean μh and variance νh.

We now define a Gamma process to be a sequence of random variables, T_t, such that $T_t - T_s$ is distributed as Y_{t-s}. The group property, that $Y_h + Y_r$ has the same distribution as Y_{h+r}, guarantees that we obtain the same distribution for T_t regardless of how many stops we make before time t. That is if we take times

$$t_1 < t_2 < \cdots < t_n = t,$$

and sample T_{t_1} as Y_{t_1}, and T_{t_j} as $T_{t_{j-1}} + Y_{t_j - t_{j-1}}$, we get the same distribution for T_t regardless of the values of t_j and n. The parameter μ is said to be the *mean rate* of

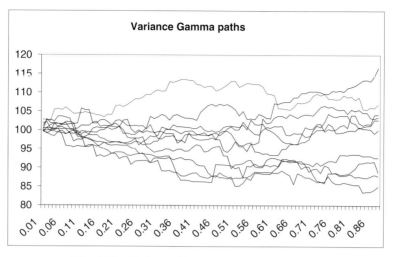

Fig. 17.1. Variance Gamma paths with small v.

T and we shall typically take it to be 1. The parameter v is said to be the *variance rate* of T. As T_t and Y_t have the same distribution, we have that T_t has mean $t\mu$ and variance tv.

Having defined the Gamma process, we can now define a Variance Gamma process. Let $b(t; \theta, \sigma)$ be a Brownian motion, with drift θ and volatility σ, that is,

$$db = \theta dt + \sigma dW, \qquad (17.3)$$

and let

$$X(t; \sigma, v, \theta) = b(T_t; \theta, \sigma), \qquad (17.4)$$

where T_t is the value of a Gamma process at time t with $\mu = 1$ and variance rate v. We then say that X is a *Variance Gamma process*. We illustrate this process in Figures 17.1 and 17.2.

Thus to take a random draw from X we first take a random draw from a Gamma distribution to get T, and then we take a second random draw, Z, from a standard Gaussian distribution to get

$$X = \theta T_t + \sigma T_t^{1/2} Z. \qquad (17.5)$$

Our model for stock evolution is now to let the log of the stock follow a Variance Gamma process with an additional drift, α, based on the actual calendar time rather than the randomly drawn time. Thus we take

$$S(t) = S(0)e^{\alpha t + X(t; \sigma, v, \theta)}. \qquad (17.6)$$

We now proceed to examine to price options using this model.

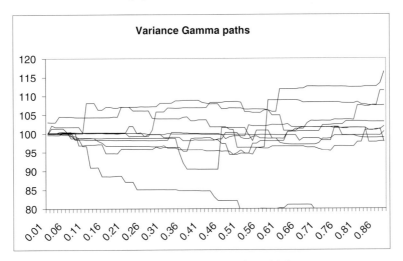

Fig. 17.2. Variance Gamma paths with large ν.

17.2 Pricing options with Variance Gamma models

If we plot the paths of a Variance Gamma process, we discover that they consist of a large number of very small jumps. We have repeatedly seen that perfect hedging in the presence of jumps is impossible. We can therefore expect that a Variance Gamma model for stock movements will lead to an incomplete market and therefore the existence of many equivalent martingale measures.

We will show in Section 17.4 that

$$S(0)e^{X(t;\sigma,\nu,\theta)+\omega t}$$

is a martingale if

$$\omega = \frac{1}{\nu} \log\left(1 - \theta\nu - \frac{\sigma^2 \nu}{2}\right). \qquad (17.7)$$

Note that as ν tends to zero, this ω converges to $-\theta - \frac{1}{2}\sigma^2$ which is the correction term required to make geometric Brownian motion a martingale.

This means that if we work with a continuously-compounding interest rate, r, then for the discounted stock price to be a martingale, then the stock process must be of the form

$$S(0)e^{rt+X(t;\sigma,\nu,\theta)+\omega t}.$$

It turns out that when taking an equivalent measure we can change the parameters $\alpha, \sigma, \nu,$ and θ to whatever we want. Proving this is however way beyond our scope. The essential idea is that the process is made up of a large number small jumps of varying sizes; if we think in terms of each jump size occurring according to a Poisson process then we can let each Poisson process have any intensity we like. Note the contrast to the Black–Scholes world where only the drift could be changed. This means that we can obtain an equivalent martingale measure by setting σ, ν, θ to anything we choose and we are then constrained to put

$$\alpha = r + \omega. \tag{17.8}$$

Once we have chosen an equivalent martingale measure, that is once we have chosen σ, ν and θ, we can price options as expectations in the usual fashion. For a vanilla European option, C, with payoff function $f(S_T)$ at time T, we have that the value at time zero is

$$C(0) = e^{-rT} \mathbb{E}\big(f\big(S(0)e^{rT + X(T;\sigma,\nu,\theta) + \omega T}\big)\big). \tag{17.9}$$

This is easily evaluated by a Monte Carlo simulation. We simply take a random draw, S, from a Gamma distribution with mean T and variance νT, and then let

$$X = \theta S + \sigma \sqrt{S} W, \tag{17.10}$$

where W is a draw from an $N(0, 1)$ distribution. We then just plug X into (17.9); repeating many times we get a Monte Carlo estimate for the price. Note the big advantage here over stochastic-volatility models – we never need to substep. We always just take two draws: one Gamma draw for the variance and one normal draw for the Brownian motion, and the distribution is precisely simulated. One subtlety here is that we need to be able to draw quickly from the Gamma distribution. One method of rapidly computing the incomplete Gamma function, that is the integral of the density up to a point x, is given in [104]. Its inverse can then be computed via Newton–Raphson search. In practice, we might want to develop a table before running the Monte Carlo simulation depending on how many paths we are doing.

For vanilla call and put options, we can develop the price as an integral over Black–Scholes prices. If we fix a random time S and integrate $N(0, 1)$ draws, we are taking an integral

$$C(0) = e^{-rT} \int \gamma(R) \int f\big(S(0)e^{rT + \theta R + \omega T + \sqrt{R}\sigma W}\big) g(W) dW dR, \tag{17.11}$$

where $\gamma(R)$ is the density of the relevant Gamma distribution and g is the density of a standard normal. We can rewrite (17.11) as

$$C(0) = \int \gamma(R)$$
$$e^{-rT} \int f\big(S(0)e^{\theta R + \omega T + \frac{1}{2}\sigma^2 R} e^{rT - \frac{1}{2}\sigma^2 R + \sqrt{T}\sqrt{R/T}\sigma W}\big) g(W) dW$$
$$dR \tag{17.12}$$

17.2 Pricing options with Variance Gamma models

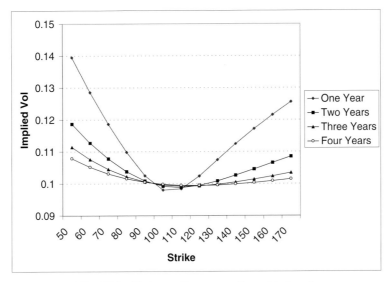

Fig. 17.3. Variance Gamma smiles with $\theta = 0$.

The inner integral is now just the Black–Scholes price of an option with payoff f with spot equal to $S(0)e^{\theta R + \omega T + \frac{1}{2}\sigma^2 R}$ and volatility $\sigma\sqrt{R/T}$. We can therefore substitute the Black–Scholes price for the inner integral and then do the outer integral numerically; this pricing will be very fast.

Having developed the pricing method, we can examine the qualitative shape of Variance Gamma smiles. There are two effects depending on the values of ν and θ. If θ is zero then the smile is symmetric and goes down in the middle and up at either side, i.e. the 'smile' is smile-shaped. This reflects the fact that increasing ν increases the probability that the spot (in the risk-neutral world) will end up far out-of-the-money, making far out-of-the-money calls and puts more valuable; the distribution has fatter tails than the normal distribution. When ν is non-zero, the θ parameter determines the skewness of the smile. A positive θ will yield an upwards sloping smile, whilst a negative one will give a downwards slope. See Figures 17.3 and 17.4.

As well as examining the smile at one time horizon, it is interesting to see how the smile looks across many maturities. In common with jump-diffusion models, we see that it is much sharper at short times and becomes flat at long time horizons. This qualitative behaviour is often found in market smiles and therefore is a good argument in favour of the model. However, as with jump-diffusion models there is a tendency for the smile to flatten too quickly. This contrasts with stochastic volatility models, where the total amount of variation of volatility increases over time and for certain parameter sets, smiles flatten quite slowly.

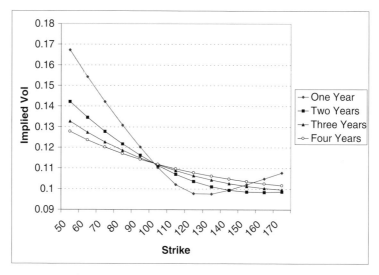

Fig. 17.4. Variance Gamma smiles with $\theta < 0$.

If we take constant parameters then the observed smiles will be a constant function of time-to-expiry and moneyness (i.e. strike divided by spot.) To see this, observe that in both Variance Gamma and Black–Scholes, the process for the log is given by a process with increments that are distributed independently of current level and time. Our smiles therefore float. This property is shared with jump-diffusion models.

To what extent is the Variance Gamma model correct? It gives an accurate model for stock price movements in the small scale. It does not include the mechanisms for crashes which are an important qualitative feature of equity markets. However, one could easily develop a combination of jump-diffusion and Variance Gamma which includes crashes.

17.3 Pricing exotic options with Variance Gamma models

Two of the methods we have presented for pricing exotic options easily go over to the Variance Gamma model. These are Monte Carlo simulation and replication. Having chosen the risk-neutral measure, we need as usual to evaluate the discounted expectation of the derivative's payoff.

To price a multi-look option paying $f(S_{t_1}, \ldots S_{t_k})$ at time t_k in a constant-interest-rate world, we simply simulate the values of S_{t_j} one after another. Thus we draw the random time T_{t_1}, and use this to draw S_{t_1} with a normal draw. We then draw the increment $T_{t_2} - T_{t_1}$ which is independent of the value of T_{t_1}, and use this to determine S_{t_2}/S_{t_1} which is, of course, independent of S_{t_1} and so on. To approximate the expectation

$$e^{-rt_k}\mathbb{E}(f(S_{t_1},\ldots,S_{t_k})),$$

we therefore just plug in drawn values of S_{t_j} into f, discount and average over a large number of draws.

For the use of replication methods, the important point to note about the Variance Gamma model is that the state is only dependent on the current value of spot – this is the same as jump-diffusion models and the Black–Scholes model but unlike stochastic-volatility models. The price of an option at some future time for a given value of spot is therefore determined within the model today. This means that the replication methods presented in Sections 10.4 and 10.3 go over verbatim.

17.4 Deriving the properties

In this section, we derive some of the easier properties of Variance Gamma processes which were needed earlier in the chapter. As p_h is zero for $x \leq 0$, Y_h, will, by construction, be a positive random variable. The easiest way to see that it has the other requisite properties is to use its *characteristic function*. Let A be a random variable. Recall that the characteristic function, $\phi_A(u)$, is the expectation of e^{iAu}. It is essentially the Fourier transform of the density function and as Fourier transforms are invertible, knowing the characteristic function is equivalent to knowing the density function. It is immediate from the definition of the characteristic function that A and B are independent random variables if and only if

$$\phi_{A+B}(u) = \phi_A(u)\phi_B(u). \tag{17.13}$$

This means that to prove that $Y_s + Y_t = Y_{s+t}$ in distribution, we need only show that

$$\phi_{Y_{s+t}}(u) = \phi_{Y_s}(u)\phi_{Y_t}(u). \tag{17.14}$$

In fact, the characteristic function of Y_t is

$$\phi_{Y_t}(u) = \left(\frac{1}{1 - iu\frac{\nu}{\mu}}\right)^{\frac{\mu^2 t}{\nu}}, \tag{17.15}$$

where μ, ν are the mean and variance rates respectively, which immediately yields (17.14). The energetic reader can derive the characteristic function using contour integration.

To get an expression for the density of $X(t; \sigma, \nu, \theta)$, we use the fact that, if we know the density, $p_{A|B}(x, b)$, of A given B and the density, $p_B(b)$, of B, then the density of A is

$$p_A(x) = \int p_{A|B}(x,b) p_B(b) db. \tag{17.16}$$

Applying this to the Variance Gamma density, where A corresponds to a normal distribution of which the variance is determined by the second draw and B is the random time drawn from the Gamma distribution, we deduce that the density of $X(t)$ is given by

$$f_{X(t)}(x) = \int \frac{1}{\sigma\sqrt{2\pi g}} e^{-\frac{(x-\theta g)^2}{2\sigma^2 g}} \frac{g^{\frac{t}{\nu}-1} e^{-\frac{g}{\nu}}}{\nu^{\frac{t}{\nu}} \Gamma\left(\frac{t}{\nu}\right)} dg. \tag{17.17}$$

The characteristic function can be evaluated as an integral over characteristic functions of Gaussians and is equal to

$$\phi_{X(t)}(u) = \left(\frac{1}{1 - i\theta\nu u + (\sigma^2\nu/2)u^2}\right)^{t/\nu}.$$

If we evaluate the characteristic function at $u = -i$, we obtain the expected value of $e^{X(t)}$ and this is equal to $e^{-\omega t}$, which is why we have defined ω in the way that we did. In particular, it follows that $e^{X(t)+\omega t}$ is a martingale.

Another fact that we can easily deduce from the characteristic function is that Variance Gamma paths are of finite first variation. This is very different from Brownian motion where the second variation is finite and non-zero, and the first variation is always infinite. We prove that the first variation is finite by showing that the Variance Gamma process is a difference of two increasing processes. An increasing process is necessarily of finite first variation since the variation is then just the total amount of up moves which is just the final value minus the initial value.

To see that it is the difference of two increasing functions, we simply rewrite the characteristic function as

$$\left(\frac{1}{1 - i(\nu_1/\mu_1)u}\right)^{\left(\frac{\mu_1^2}{\nu_1}\right)t} \left(\frac{1}{1 + i(\nu_2/\mu_2)u}\right)^{\left(\frac{\mu_2^2}{\nu_2}\right)t},$$

where μ_j, ν_j satisfy

$$\nu_j^{-1}\mu_j^2 = \nu^{-1}, \quad \frac{\nu_1\nu_2}{\mu_1\mu_2} = \frac{\sigma^2\nu}{2}, \quad \frac{\nu_1}{\mu_1} - \frac{\nu_2}{\mu_2} = \theta\nu. \tag{17.18}$$

The first term is the characteristic function of a Gamma process with parameters μ_1, ν_1, and the second is the characteristic function of the negative of a Gamma process with parameters μ_2, ν_2. Recalling that Gamma processes are increasing, we have the desired result. Note that if $\theta = 0$, the equations for μ_j and ν_j are symmetric in j and so the Variance Gamma process can be written as the difference of two identically distributed Gamma processes.

17.5 Key points

The Variance Gamma process has some nice properties. It is possible to develop Monte Carlo and numeric integral prices for vanilla options which allow rapid calibration to the market. The paths seems to mimic market prices well. The process leads to an incomplete market with a great deal of choice for risk-neutral parameters – perhaps too much. It is not however clear how to hedge since all movements are jumps. For pricing exotic options, the Variance Gamma model is well-adapted to both Monte Carlo and replication techniques.

- The Variance Gamma is a model based on the notion of random time.
- Random time increments have the properties of being proportional in mean and variance to the length of calendar time, and of being independent of previously elapsed time.
- Variance Gamma paths consist of many small jumps.
- Variance Gamma paths are of finite first variation whereas Brownian motion paths are of infinite first variation.
- In passing to an equivalent measure, there are no constraints on the changes in parameter values unlike in the diffusion setting.
- Variance Gamma smiles are a fixed function of time-to-expiry and moneyness.
- Variance Gamma smiles become flatter as time-to-expiry increases.
- Vanilla options can be priced using Variance Gamma models as an integral over Black–Scholes prices or by Monte Carlo.
- Exotic options can be priced using Variance Gamma models by Monte Carlo or replication.

17.6 Further reading

Variance Gamma models were introduced by Dilip Madan and various collaborators. The fundamental papers are:

- 'The Variance Gamma model for share market returns,' by Dilip Madan & Eugene Seneta, [88]. This paper introduces the Variance Gamma process and discusses how good a model it is for stock market returns.
- 'Option Pricing with VG martingale components' by Dilip Madan & Frank Milne, [89]. This paper introduces option pricing with Variance Gamma and produces the formula for pricing vanilla options.
- 'The Variance Gamma process and option pricing' by Dilip Madan, Peter Carr & Eric C. Chang, [90]. This paper extends the theory and deduces formulas for the Variance Gamma density in terms of Bessel functions.

A quite different approach to developing a pricing formula using Fourier transform techniques is given in [84].

17.7 Exercises

Exercise 17.1 Which of the replication techniques of Chapter 10 can be applied when pricing with a Variance Gamma model?

Exercise 17.2 Suppose we with to price an Asian option with N look-at dates with a Variance Gamma model using Monte Carlo. How many (pseudo-)random numbers will we need per path?

18

Smile dynamics and the pricing of exotic options

18.1 Introduction

We have examined in varying depths a range of models for stock price evolution and developed various methods of pricing using them. This yields differing methods of pricing exotic options which will lead to varying prices. The topic of this chapter is how to choose between them, and what the choice means. Our objective is more to make the reader aware of the questions than to answer them. Indeed, the questions we raise are still the subject of much ongoing research.

Our starting point will always be that there is a liquid market in vanilla options. We want to price the exotic options in a manner compatible with this market. A simple and fundamental constraint is that the price of an exotic must not be arbitrageable by using a static portfolio of vanilla options.

Recall that in Chapter 6 we showed that the specification of a process for the underlying was equivalent to specifying a measure on the space of paths. We also saw that the price of a derivative contract could be obtained by taking an expectation in a risk-neutral measure associated with the process. What are we doing when we take this expectation? Typically, our derivative contract depends upon the value of the underlying at a finite set of times. The choice of a risk-neutral measure then specifies a joint probability distribution of the underlying for this set of times. Suppose our look-at times are

$$t_0 < t_1 < \cdots < t_n.$$

Let S_j denote the value of the underlying at time t_j.

Let $\Phi(S_1, \ldots, S_n)$ denote the joint probability density function implied the by risk-neutral pricing measure. Unfortunately, Φ is not determined by the vanilla option prices. However, they do imply certain constraints on Φ. We showed in Section 6.3 that the risk-neutral density of the underlying at time t_j is determined by the call option prices.

This means that we can observe $\Phi_j(S_j)$, the integral of Φ over all coordinates except S_j. Risk-neutrality implies that the expectation of S_j under Φ_j must be the forward of S_0, the spot at time zero. In fact, the martingale conditions also means that the future implied densities must always be risk-neutral: that is, the conditional density of S_l given the values of $S_1, \ldots, S_k$ for some $k < l$, must have expectation equal to the forward of S_k.

The upshot of all this is that there are many possible choices of Φ, and the way in which they will differ is in the future densities implied by S_j for $j \le k$ taking given values. A choice of density for a single time horizon is equivalent to specifying the call option prices, and that in turn is equivalent to specifying the implied volatility smiles. Our choice of Φ is therefore a choice of smile dynamics. In other words, how we choose Φ is a statement about how we believe the smile evolves over time.

Our purpose is this chapter is therefore to examine some of the possible ways smiles can evolve, and to relate these to the models we have developed. The moral is that we should price the exotic option with a model that produces believable smile dynamics.

18.2 Smile dynamics in the market

Before we start examining what smile dynamics our models imply, we need to examine the various sorts of smile dynamics that can be observed in the market. There are basically two things we need to think about: how the smile changes with spot, and how the smile changes with time.

18.2.1 Sticky or floating

The implied volatility smile is a function of strike. The crucial question is how does that function change when spot is moved. Two fairly obvious functional forms are

$$\hat{\sigma}_1(K) \quad \text{and} \quad \hat{\sigma}_2(K/S),$$

where K is the strike and S is spot. In the first case, the smile does not change as S changes. Such a smile is said to be *sticky*. In the second case, the smile *floats* with spot and the smile is said to be *floating*. It is sometimes called a *sticky-delta* model.

For example, a foreign exchange smile is typically a 'U' shape with the bottom of the 'U' at-the-money. If we modelled our smile by a sticky smile model then we would be saying that at-the-money would be up the sides of the 'U' if spot changed. Clearly this would be a poor model for market behaviour. We therefore conclude that for foreign exchange we need a floating smile model.

On the other hand, in interest-rate markets the caplet smile tends to be a downward-sloping curve whose shape is attached to the level of strike rather than relative to at-the-money. We conclude that the smile is therefore *sticky*.

We can expect equity smiles to float. Remember that the principal cause of the equity smile is the risk-aversion of investors against rapid downward movements. This is always manifested in terms of movements relative to the current value of spot. An investor is always worried about losing the value he currently has, not what he might have in a year, or what he did have a year ago. There is however some evidence that implied volatilities increase when spot decreases which suggests that the smile has a sticky component.

It is important to realize that the decision between sticky and floating can have consequences for hedging when working with a Black–Scholes type model. Suppose we have sold a call option, C, on an underlying S, and we are Delta-hedging it. We should hold $\frac{\partial C}{\partial S}$ units of S. If we use a Black–Scholes formula with the implied volatility smile, $\hat{\sigma}$, taken into account, then this implies that we have to hold

$$\frac{\partial}{\partial S}(C(S, \hat{\sigma}(S, K))) = \frac{\partial C}{\partial S}(S, \hat{\sigma}(S, K)) + \frac{\partial C}{\partial \hat{\sigma}}\frac{\partial \hat{\sigma}}{\partial S}(S, K) \qquad (18.1)$$

units of S. For a sticky smile, the second term is zero but for a floating smile it is not. How we hedge is therefore affected by our belief about smile movements. If our belief is wrong, we will end up with a non-delta neutral position, and be left with undesirable extra risk arising from the hedging error.

18.2.2 Time dependence

There are really two types of time dependence of the smile. The first is the time dependence as seen from today: we can observe various maturities of vanilla options in the market and for each maturity we get a smile. There is nothing forcing these smiles to be the same. The second sort of time dependence comes from the why the smile evolves over time. Do we expect the smile to have the same qualitative properties in the future as it has today? Should these qualitative properties be relative to current time or should they be associated to fixed calendar dates?

Typically in the equities market one sees a steeply downwardly sloping smile with sometimes a slight up-kick for out-of-the-money calls. This smile is much sharper for short maturity options than for long-dated ones. Short-dated here means about one to three months. This behaviour has persisted over time and has been generally the same qualitatively since the 1987 crash. Before then smiles were much shallower but the crash seems to have triggered a much greater use of options in hedging against jump-risk which increased the smile's slope.

The fact that the equities smile has persisted over time means that we can expect the future equities smile to have a similar shape to today's.

The foreign exchange smile is more constant as a function of maturity. It displays a 'U' shape for each maturity but these 'U' shapes are roughly similar for each maturity. Again, these smiles have persisted over a long period of time, and we therefore expect the smile to be the same shape qualitatively in the future.

The interest-rate smile really has two components. The first is a downward slope which has been around for quite a long time and which first appeared in the mid 1990s. The second component is a slight upward curve for out-of-the-money caplets which appeared in the aftermath of the Asian crisis in 1998. This component reflects risk-aversion to large market moves. This second component is fairly homogeneous across maturities existing even for very long-dated options, for example ten-year caplets.

18.3 Dynamics implied by models

We have studied a number of alternative models in varying detail. We recall them here

 (i) jump-diffusion,
 (ii) stochastic-volatility,
 (iii) Variance Gamma,
 (iv) displaced diffusion, that is a Black–Scholes type model in which the underlying plus a constant is log-normal instead of the underlying,
 (v) a Derman–Kani or Dupire type model where the underlying follows a process

$$dS = S\mu dt + \sigma(S, t) S dW. \qquad (18.2)$$

What sort of smile dynamics do these models give rise to?

18.3.1 Jump-diffusion smiles

If we use a log-normal jump-diffusion model with constant parameters, then everything is defined relative to the current value of spot and the current time. As we saw in Section 15.10 this leads to a smile which is a constant function of moneyness. This means that if we write the implied volatility as a function of strike, K, spot, S, current time, t, and expiry time, T, we obtain a function

$$\hat{\sigma}(S, K, t, T) = \hat{\sigma}(K/S, T - t). \qquad (18.3)$$

If we make the mean jump size (ratio) less than 1 then we obtain a downwards-sloping smile. However, this smile will be much sharper for small values of $T - t$ than large ones. Over long time periods, the smile becomes more horizontal as the diffusive component of the model wins out. See Figures 18.1 and 18.2.

18.3 Dynamics implied by models

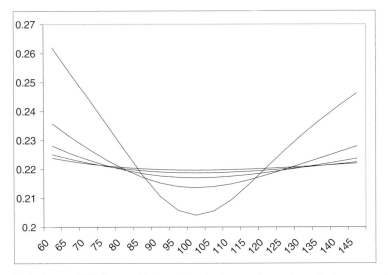

Fig. 18.1. Jump-diffusion smiles for time horizons of one through five years. The sharpest smile is one year, and the shallowest is five years. Spot is 100 and jumps are symmetric.

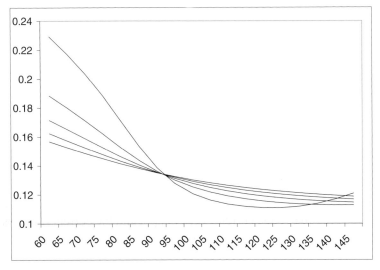

Fig. 18.2. Jump-diffusion smiles for time horizons of one through five years. The sharpest smile is one year, and the shallowest is five years. Spot is 100 and jumps are asymmetric with mean ratio equal to 0.8.

18.3.2 Stochastic-volatility smiles

If we use a stochastic-volatility model with constant parameters the model is of log-type and again everything is defined relative to the current value of spot and time,

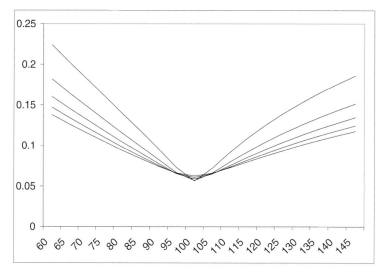

Fig. 18.3. Stochastic-volatility smiles (Heston model) for time horizons of one through five years. The sharpest smile is one year, and the shallowest is five years. Spot is 100 and volatility is uncorrelated with spot. The reversion speed is 1 and the volatility of variance is 1. Initial volatility is 10%.

and we obtain a functional dependence for the implied volatility of the form

$$\hat{\sigma}(S, K, t, T) = \hat{\sigma}(K/S, T - t). \tag{18.4}$$

The principal difference between stochastic-volatility and jump-diffusion smiles is that there is an implicit assumption that the volatility has not changed in (18.4). A big difference between jump-diffusion and stochastic volatility is therefore that we expect the smile's shape and level to evolve, even if risk-preferences do not change, in stochastic-volatility models.

Another big difference with stochastic volatility is that for long maturities, the stochasticity of volatility has more time to affect the moves of the underlying. The smiles therefore do not decay so rapidly at long maturities. The relative sharpness of long-dated smiles can be controlled by the speed of mean reversion of the volatility. If a very strong mean reversion is used then the volatility cannot get too far away from the mean for any length of time, and so there will be less opportunity for the effects of the stochastic volatility to build up. See Figures 18.3, 18.4, 18.5, 18.6, 18.7 and 18.8 for examples of these effects.

18.3.3 Variance Gamma smiles

If we use a Variance Gamma model with constant parameters, then everything is defined relative to the current value of spot and the current time. Once again we

18.3 Dynamics implied by models 397

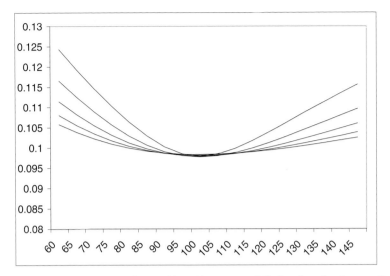

Fig. 18.4. Stochastic-volatility smiles (Heston model) for time horizons of one through five years. The sharpest smile is one year, and the shallowest is five years. Spot is 100 and volatility is uncorrelated with spot. The reversion speed is 1 and the volatility of variance is 0.1. Initial volatility is 10%.

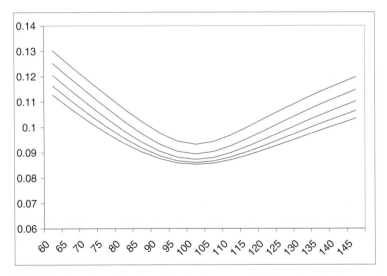

Fig. 18.5. Stochastic-volatility smiles (Heston model) for time horizons of two through ten years. The highest smile is two years, and the bottom is ten years. Spot is 100 and volatility is uncorrelated with spot. The reversion speed is 0.1 and the volatility of variance is 1. Initial volatility is 10%.

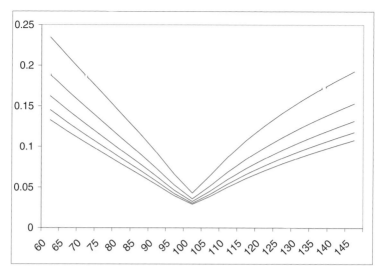

Fig. 18.6. Stochastic-volatility smiles (Heston model) for time horizons of one through five years. The sharpest smile is one year, and the shallowest is five years. Spot is 100 and volatility is uncorrelated with spot. The reversion speed is 0.1 and the volatility of variance is 1. Initial volatility is 10%.

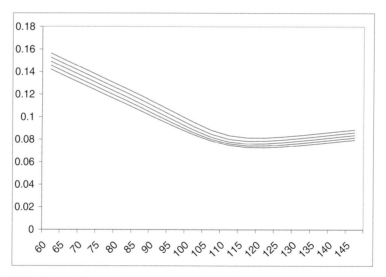

Fig. 18.7. Stochastic volatility smiles (Heston model) for time horizons of one through five years. The highest smile is one year, and the lowest is five years. Spot is 100 and volatility is negatively correlated (-0.6) with spot. The reversion speed is 0.1 and the volatility of variance is 0.1. Initial volatility is 10%.

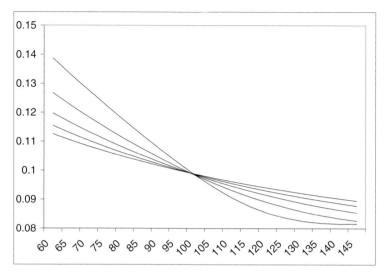

Fig. 18.8. Stochastic-volatility smiles (Heston model) for time horizons of one through five years. The steepest smile is one year, and the shallowest is five years. Spot is 100 and volatility is negatively correlated (-0.6) with spot. The reversion speed is 2 and the volatility of variance is 0.1. Initial volatility is 10%.

obtain an implied volatility function of the form

$$\hat{\sigma}(S, K, t, T) = \hat{\sigma}(K/S, T - t). \quad (18.5)$$

The smile is symmetric, unlike a jump-diffusion smile. However, skewness can be introduced using the θ parameter. For a single fixed maturity, Variance Gamma and stochastic-volatility smiles look very similar. As with jump-diffusion models, this smile will be much sharper for small values of $T - t$ than large ones. Over long time periods, the smile becomes more horizontal, since the model becomes more and more similar to a purely diffusive model.

18.3.4 Displaced-diffusion smiles

Displaced diffusion will give us a downward sloping smile. It is a quite sticky smile in that the shape is quite insensitive to the value of spot. The overall level will change a little but if one rescales volatility so that the smile is at the same level then one obtains an almost identical smile.

The shape of the displaced-diffusion smile is highly insensitive to maturity, see Figure 18.9.

Note that as none of the parameters involve the current time, the smile implied will necessarily be purely a function of $T - t$, not t and T individually.

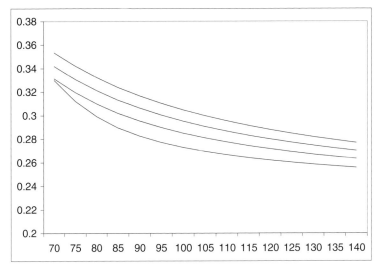

Fig. 18.9. The displaced-diffusion implied volatility smile for maturities 1, 3, 5 and 9 years. The highest graph is 9 years and the lowest is 1.

18.3.5 Dupire/Derman–Kani smiles

If we calibrate a Dupire/Derman–Kani model to the market then we obtain a function $\sigma(S, t)$ to use in the model. Typically the function $\sigma(S, t)$ is fairly constant for large t. This means that if we use the model to predict the smiles that will be observable a year from now, then we find that there will not be any. That is the model predicts that smiles are destined to disappear. As smiles have persisted for a long period of time, this is an undesirable feature.

The other aspect of the model is that the volatility function σ is highly dependent on spot. This means that as spot changes, the predicted smile will change greatly, in possibly strange ways. This again is less than ideal.

18.4 Matching the smile to the model

In the previous two sections, we looked at the smile dynamics in various markets, and which ones are implied by various models. In this section, we put the two together.

18.4.1 Equity smiles

The equity smile is highly skewed, much sharper for short maturities and is much flatter for long maturities than for short maturities. These behaviours persist over

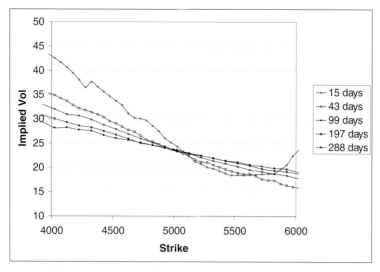

Fig. 18.10. Implied volatilities of short-dated options on the FTSE.

time. The obvious match is therefore the jump-diffusion model as it naturally gives all these properties. See Figure 18.10.

However, if we try to fit the smile with constant parameters we make a surprising discovery: the market-implied parameters will not be a constant function of time. The market skewness may decay even faster than implied by the model or at times it may decay more slowly. We can interpret this fact in a couple of ways. One obvious way is simply to decide that jump-diffusion is not the whole story and that we need a more sophisticated model; however it is not so clear how to define the requisite more sophisticated model that gives a time-homogeneous fit.

A second way is to remember that the market chooses the martingale measure, which in this case corresponds to choosing the jump-intensity. If jump-intensity is not constant, we can then conclude that the market is choosing to price with greater jump-intensity for short-dated jumps. How could this come about? Ultimately, the market's *choice* is determined by supply and demand. Thus if there is a great deal of demand for out-of-the-money put options from fund managers trying to protect their portfolios then the short-term jump-intensity will be driven up, and increase skewness in the short-dated smiles.

A second problem with jump-diffusion models is that the at-the-money implied volatility is much greater than the diffusive volatility since a large component of the price comes from jump risk. However, if one measures the diffusive volatility of a major index such as the S&P and compares it with the implied volatility of

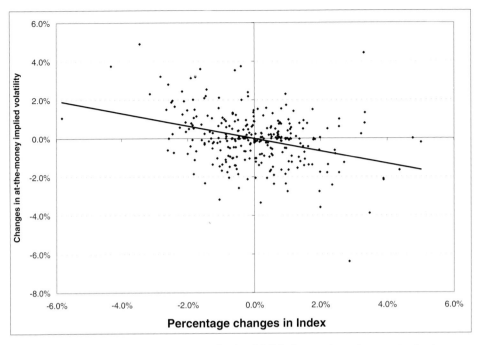

Fig. 18.11. Scatter plot of changes in the S&P index against changes in the implied volatilities of at-the-money call options from February 2000 to May 2001.

at-the-money options they are not particularly different. This is an argument against using jump-diffusion models. (See Figure 18.12.)

A third issue is that jump-diffusion models imply deterministic future smiles that float perfectly. Future smiles are not deterministic so we are failing to capture an important aspect of smile evolution. There is also some evidence that at-the-money implied volatilities increase when spot decreases which the model also fails to capture. (See Figure 18.11.) We could capture this second feature by using a displacement in combination with a jump-diffusion model. To get indeterminacy in future smiles we need to make some parameter stochastic. This parameter could be the instantaneous volatility but could equally be the jump-intensity or jump-mean.

18.4.2 FX smiles

The FX smile between major currencies is fairly symmetric and time-constant. These properties can be obtained by the use of a stochastic-volatility model with rapid mean-reversion. As there is no skew, we can use a model with uncorrelated volatility. Occasionally, jumps do occur in exchange rates. For example, the dollar-yen rate once moved 20% in one day.

18.5 Hedging

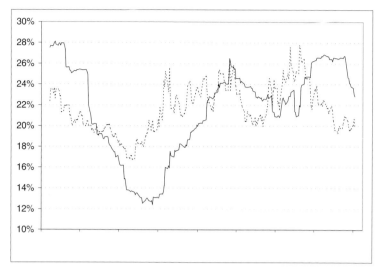

Fig. 18.12. Three-month historic volatilities (solid line) and three-month implied volatilities for the S&P index from February 2000 to June 2001.

18.4.3 Caplet smiles

The caplet smile really has two components. One is the skew which is sticky and time-constant. We can naturally achieve this part via displaced-diffusion. The second component is the slight increase in volatilities for high strikes. This is also fairly time-constant. It is less clear whether it is sticky or floating. We can achieve a good match to it using an uncorrelated mean-reverting stochastic-volatility process and maintain time-homogeneity.

18.5 Hedging

The pricing of exotic options is not just about finding prices that are compatible with market dynamics in the sense of being non-arbitrageable; it is equally about realizing those prices via hedging. Thus if a model is to be useful to a trader it must tell him how to hedge and that the hedges must work.

Typically, the way a trader will hedge is to fit the model to the market and then to hedge each of the parameters by using simple options. Typically, an exotic option will be hedged using calls and puts of various maturities. Therefore after fitting the model to the vanilla market, the trader measures the derivative of the price with respect to each parameter of the model, possibly breaking up time into pieces to do this if he is using a variable-parameter model so as to get the exposure of the model to changes in the parameter over the various time slices. The trader then buys a portfolio of vanilla options so as to cancel all these exposures and uses the underlying at the end to remove any residual Delta.

The trader returns a day later and repeats the process. The market will have changed a little in the meantime. If the market fit has also changed only a little, his hedge has been successful and the value of his portfolio has only changed a small amount. He need then only make small adjustments to his hedging portfolio to keep himself hedged.

However, suppose he runs his fitting routine and it outputs a vastly different parameter-set. He then has a problem: although the market has not changed much his fitter is telling him to totally dissolve his original hedge and set up a new one. In addition, the new fit will probably give a wildly different price for the exotic option. The trader will be very unhappy at this and probably throw the model away (and shoot the quant!)

This means that an important criterion for trading off a model is that it should fit the market stably. That is, if one changes the market slightly, the fit should also change slightly. This is also related to uniqueness of fits. If a model has many parameters – for example a jump-diffusion stochastic-volatility model with all parameters time-dependent, then it is possible to get similar qualities of fits with vastly different parameter sets. This implies instability in that small changes in the market-observed prices to be fit will lead to great jumps in the parameter-set.

The ultimate test of a model is therefore how does it perform at hedging? This can be tested by taking historical market data and then running historical simulations of hedging over various periods. In [10], an empirical study of the performance of various option pricing models at hedging vanilla options on the S&P is carried out. The authors find that sophisticated models, particularly stochastic-volatility models do a lot better than Black–Scholes.

18.6 Matching the model to the product

We have discussed matching the model to the market in terms of deciding whether the model produces dynamics for the underlying and the smile which replicate well those observed in the market. A second equally important aspect of pricing derivatives is that we must consider what quantity an option is most exposed to, and we must make sure our model prices that quantity correctly.

We consider some examples. A *cliquet* is an option on the ratio of the value of spot at two different times. If the strike is A, and the two times are T_1 and T_2, then at T_2 the holder receives the sum

$$\left(\frac{S_{T_2}}{S_{T_1}} - A\right)_+.$$

This is a *call cliquet*, one could easily define a *put cliquet* analogously.

18.6 Matching the model to the product

We can regard the cliquet as a forward option, in particular we can rewrite the payoff as

$$\frac{\left(S_{T_2} - A S_{T_1}\right)_+}{S_{T_1}}.$$

In other words, at time T_1 it becomes a call option with strike $A S_{T_1}$ and notional $1/S_{T_1}$. A quick examination of the Black–Scholes formula shows that the Black–Scholes value of this call option is independent of the value of S_{T_1} for fixed volatility.

For simplicity, suppose that there are no interest rates. Suppose we take $A = 1$; what will the value of this option be at time T_1? We have the approximation to the Black–Scholes formula that an at-the-money call option is worth

$$0.4 S \sigma \sqrt{T}.$$

This means that the value of our cliquet at time T_1 is

$$0.4 \sigma \sqrt{T_2 - T_1},$$

where σ is the implied volatility observable in the market at time T_1 for options expiring at time T_2. Thus when we buy and sell cliquets, we are really trading the forward value of implied volatility. For a cliquet, as the value is linear in implied volatility the product is reasonably benign, and for our valuation all that matters is that the mean volatility is captured correctly. Thus if we use a model that reproduces today's smile in the future we can be reasonably confident of our pricing.

Suppose we now add a twist to our product. The holder has to pay an additional fee at time T_1 in order to receive the payoff at time T_2. In other words, we have a compound option. We denote the additional fee by K. Thus we have an option to buy a call option struck at S_{T_1} with notional $S_{T_1}^{-1}$ and expiry T_2, and the strike price of the option is K. We will call this option an *optional cliquet*.

Suppose we use a deterministic smile model such as Black–Scholes, jump-diffusion or Variance Gamma. Then the implied volatility prevailing at time T_1 is already known at time T_0 and a unique price, $C(1)$, at time T_1 for the cliquet is known. The value of the optional cliquet is then

$$\max(C(1) - K, 0).$$

This is very neat but also very dangerous. If $C(1) \leq K$, then we are saying that the option has zero value. Suppose $C(1)$ is equal to K. If you ever get the opportunity do this trade, and sell the option for zero or almost zero, you will almost certainly get the sack immediately. Why? Implied volatilities do change. There is a roughly

50% chance that the option will be worth something at time T_1 and by selling the option for zero, you have given away money. By using a deterministic-volatility model, you have failed to capture the essential feature of this option, namely that it is an option on volatility. To price this option correctly, you need a model which captures the variation in volatility well. The obvious candidate for such a model is a stochastic-volatility model. Of course, by this we mean a stochastic *instantaneous* volatility model not a stochastic *implied* volatility model, so the connection is not direct as the name suggests.

Another possibility is to use a jump-diffusion model with stochastic parameters. For example, if we believe that changing risk-aversions to jumps are an important component of changes in volatility levels we could let the jump-intensity be a stochastic parameter, and then price by Monte Carlo.

A general moral to be drawn from this product is that when compound optionality is involved it is very important to take account of the stochastic nature of implied volatility.

Consider another related cliquet product. Suppose we trade two cliquets on the same underlying with different strikes. Suppose one has strike ratio 1.1 and is a call, whilst the other has strike ratio $1/1.1$ and is a put. If we consider the portfolio consisting of the difference of the two options, what are we trading? If the two options are trading at the same implied volatility at time T_1, then their values will cancel as they have the same moneyness, and the trade will be of zero value. Our value at time T_1 is therefore a function of the skewness of the smile. If we use a model which can never produce changes in skewness, such as a deterministic smile model, or a stochastic-volatility model with uncorrelated spot and vol, we are failing to capture the salient features. Once again, this failure will be exacerbated if we introduce optionality at time T_1. To price accurately, we will need a model that accurately reproduces random changes in skew. We could again use a model with stochastic jump-intensity or a stochastic-volatility model in which correlation between spot and volatility is stochastic.

These cliquet-related products exemplify the need to capture the smile dynamics well when pricing certain products. Other products depend more on the ability to capture spot moves well. Suppose we consider a crash option. The crash option pays one after a year if spot drops by a fixed ratio on any one day during the year. If the ratio is say 20%, then the holder receives a payment if and only if the index loses at least 20% of its value in one day. If we price this product with a purely diffusive model, we will get a very small number as the probability of a move of a certain size in a time interval of length t behaves likes $e^{-1/t}$. One the other hand, if we price with a jump model, the probability of a move of a certain size behaves like λt which is much, much bigger. When selling this option, we would definitely use a jump-type model even in markets which do not generally display much jumpiness, such as major FX markets.

In conclusion, when pricing an exotic option we should take especial care to analyze what features of the market the exotic option is particularly sensitive to, and to make sure that our model captures those features accurately.

18.7 Key points

- A smile can either *float* or be *sticky* according to whether it behaves as a function of strike or of strike divided by spot.
- FX smiles tend to float.
- Equity smiles tend to be downward-sloping and display a mix of floating and sticky behaviour.
- Interest-rate smiles are partially sticky.
- Different markets display differing term structures for smiles. Equity smiles display a decrease in skew with time. FX and interest-rate smiles are more time-constant.
- One method of evaluating a model is whether or not it predicts the future will be different from the present.
- An important criterion for selecting a model is its performance at hedging.
- Jump-diffusion, stochastic-volatility and Variance Gamma models predict floating smiles.
- Displaced-diffusion predicts a stickier smile.
- When pricing an exotic option we should be careful to examine what features of the model it is particularly sensitive to.

18.8 Further reading

This chapter has been highly influenced by many conversations with Riccardo Rebonato and by his book *Volatility and Correlation*, [106], which is highly recommended.

Emanuel Derman carried out an analysis of how changes in spot related to changes in skew for the S&P 500 and identified differing regimes over time, [40]. Other papers applying the same methodologies in different contexts are [1] and [99]. See also [11], [72] and [73] for further discussion of how option price movements are affected by spot movements in the real world.

Eric Reiner examined many of the different possible smile dynamics in [109].

Alexander discusses methodologies for modelling real-world market processes in [2]. For a perspective more driven by economics see [29].

Whilst complicated models have the upside of producing more realistic smile dynamics, they have the downside of non-perfect reproduction of market smiles. One compromise is therefore to overlay a Dupire/Derman–Kani style model on top of a jump-diffusion or stochastic-volatility model. This is done in [6] and [23].

In [10], an empirical study of the performance of various option-pricing models at hedging vanilla options on the S&P is carried out. The authors find that sophisticated models, particularly stochastic-volatility models perform better than the Black–Scholes model.

Appendix A

Financial and mathematical jargon

Finance is full of arbitrary terms that appear to make little sense. In this appendix, we provide definitions of the more commonly used terms in finance and mathematical finance for general reference.

Accreting notional An instrument has an accreting notional if the notional increases during its life. Typically used in interest-rate derivatives such as swaps and Bermudan swaptions.

American option An American option is an option that can be exercised at any time before expiry. See also *European option* and *Bermudan option*.

Amortising notional An instrument has an amortising notional if the notional decreases during its life. Typically used in interest-rate derivatives such as swaps and Bermudan swaptions.

Arbitrage An arbitrage is a trading strategy which results in a risk-free profit. In other words, an opportunity to make money for nothing.

Asian option An Asian option pays off according to the average value of an asset over a number of dates.

Auto cap A cap which is limited so that only the first k caplets which are in-the-money pay off for some prespecified k.

Barrier option A barrier option is an option that only pays off if the underlying has either passed or not passed some prespecified barrier level. See *knock-out* and *knock-in*.

Basis point 0.01%.

Basket option An option that allows the holder to buy or sell a basket of securities.

Binary option Another name for a *digital* option.

Bermudan option A Bermudan option is an option that can be exercised on any one of a finite number of times before expiry. See also *American option* and *European option*.

BGM or BGM/J BGM stands for Brace, Gatarek & Musiela; and J stands for Jamishidian. The BGM model is a model based on letting forward rates have their own log-normal processes. It is also known as the LIBOR market model. The Jamishidian model is based on letting swap rates have their own log-normal processes. Such models are examples of market models.

Black–Scholes model A model consisting of an asset following geometric Brownian and a riskless bond allowing frictionless trading.

Bond A unit of debt issued by a company or country that involves periodic payment of an interest payment called the coupon and return of its face value at the time of maturity.

Brownian motion A random process in which the distribution of increments between time t and time s is independent of behaviour up to time s, and is distributed as a normal with mean zero and variance $t - s$.

Call option A contract that carries the right but not the obligation to buy an asset for a predetermined price. See also *put option*.

Cap A series of *caplets*.

Caplet The right but not the obligation to enter into a forward-rate agreement at a pre-agreed strike. So called because it caps the cost of borrowing. See also *cap* and *floorlet*.

Caption An option on a cap.

Cash bond Another name for the *continuously compounding money-market account*.

Cliquet An option that pays off according to the ratio of the underlying's value across two different dates.

Complete market A market in which every contingent claim can be replicated by trading in the underlying asset or assets.

Consol A bond that pays a regular coupon but has no maturity date and therefore goes on forever.

Contingent claim A contract whose payoff depends on the price behaviour of another asset.

Continuously compounding money-market account The riskless money-market market account in which interest is continuously accumulated.

Convertible bond A bond that can be exchanged for a stock if the holder so desires.

Coupon A regular payment made to the holder of a bond.

Credit rating A rating assigned to debt that assesses the probability that the obligor will pay back the debt.

Delta The derivative of the price of an option with respect to spot.

Derivative An instrument that pays off according to the price of another asset.

Digital option An instrument that pays either a fixed amount or zero according to the value of some reference rate.

Discount curve The theoretical prices of zero-coupon bonds of all maturities.

Diversifiable risk Risk that can hedged away by judicious holdings of other assets.

Dividend A sum paid to the owner of a stock by a company out of its profits at the discretion of the board.

European option A European option is an option that can only be exercised at one fixed time. See also *American option* and *Bermudan option*.

Expectation The expected value of a random variable. Mathematically defined as the integral of its density function, f, against x:

$$\mathbb{E}(X) = \int x f(x) dx.$$

Fat tails A distribution has fat tails if its *kurtosis* is higher than that of a normal distribution.

Fixed rate A rate for lending or deposit that is fixed across the lifetime of a contract.

Floating rate A rate that changes during a contract according to market conditions.

Floor A series of *floorlets*.

Floorlet The right but not the obligation to enter into a forward-rate agreement at a pre-agreed strike. So called because it puts a floor on the interest received for putting money on deposit. See also *caplet* and *floor*.

Floortion An option on a floor.

Forward contract A contract that carries the obligation to buy an asset at a pre-determined price on a fixed date.

Forward-rate agreement A contract to put some money on deposit for a fixed period in the future at a pre-agreed interest rate. The interest is paid at the end of the contract. The interest rate is called the *strike* of the contract. Also known as a FRA.

FRA short for *Forward-rate agreement*.

Gamma The second derivative of the price of an instrument with respect to spot.

Girsanov's theorem states that changing to an equivalent measure changes the drift of a Brownian motion but nothing else.

Greek The derivative of the price of an instrument with respect to any parameter or variable.

Hedging Holding an asset in order to reduce the risk exposure due to some other asset.

Incomplete market A market in which is not *complete*.

Knock-in option A derivative that pays off only if some reference level is passed.

Knock-out option A derivative that pays off only if some reference level is not passed.

Kurtosis The fourth moment of a random variable minus its mean divided by the variance squared:

$$\frac{\mathbb{E}((X - \mathbb{E}(X))^4)}{\text{Var}(X)^2}.$$

LIBID London Interbank Bid Rate. The rate at which the bank can deposit short-term money in the interbank market. See also *LIBOR*.

LIBOR London Interbank Offer Rate. The rate at which the bank can borrow short-term money in the interbank market. See also *LIBID*.

Long A long position is a positive holding of an asset. Opposite to a *short* position.

Market model A model for interest rates in which the movement of some market-observable rates are modelled directly. See also *BGM, BGM/J*.

Martingale A random variable whose value is always equal to its expected future value.

Moment The kth moment of a random variable is the expectation of its kth power.

Parisian option A barrier option which requires the barrier to be breached for some prespecified period of time.

Payer's swap A *swap* in which the holder pays the *fixed rate* and receives the *floating rate*.

Payer's swaption An option on a *payer's swap*.

Put option A contract that carries the right but not the obligation to sell an asset for a predetermined price. See also *call option*.

Receiver's swap A *swap* in which the holder receives the fixed rate and pays the floating rate.

Receiver's swaption An option on a *receiver's swap*.

Rho The derivative of the price of an instrument with respect to r, the continuously compounding interest rate.

Risk-neutral measure A probability measure is risk-neutral if all assets grow at the same rate as a riskless bond.

Risk premium The additional return expected on an asset in order to compensate for the riskiness in its future value.

Share A fraction of the ownership of a public limited company which carries the right to receive dividends and voting rights. It does not carry any obligations. Stock is an equivalent term.

Short To go short is to sell something you do not own. Thus one effectively has a negative holding in the asset. See also *long*.

Short rate The theoretical interest rate available for depositing money for very short periods of time.

Skew A normalization of the third moment of a random variable:
$$\frac{\mathbb{E}((X - \mathbb{E}(X))^3)}{\text{Var}(X)^{3/2}}.$$

Stochastic A fancy word for random.

Stock See *share*.

Strike The price that an options allows an asset to be ought or sold for.

Swap A contract to swap a fixed stream of interest rate payments for a floating stream of interest rate payments. The fixed rate is called the *strike* of the swap.

Swap rate The rate such that a swap with that strike has zero value.

Swaption The option but not the obligation to enter into a swap.

Theta The derivative of the price of an instrument with respect to time.

Trigger option An option that requires the holder to buy or sell an asset at a fixed price according to the level of some reference rate.

Value at risk (or VAR) The amount that a portfolio can lose over some period of time with a given probability. For example, the amount the bank can lose in one day with 5% probability.

VAR Short for *value at risk*.

Variance Variance is defined as
$$\text{Var}(X) = \mathbb{E}((X - \mathbb{E}(X))^2).$$

Vanna The derivative of the Vega with respect to the underlying.

Vega The derivative of the price of an instrument with respect to volatility.

Yield The effective interest rate receivable by purchasing a bond. (There are lots of different sorts of yields.)

Yield curve Another name for a discount curve.

Zero-coupon bond A bond which pays no *coupons*.

Appendix B

Computer projects

B.1 Introduction

In this appendix, we look at some basic methods of simulating financially important mathematical functions, and then list a number of projects the reader is encouraged to try for himself. Ultimately, quantitative analysis is about the implementation of financial models not the theory, and the reader will not have truly learnt the topic until he, or she, has programmed a few models. For the reader who is familiar with object-oriented programming, we include a few pointers on how to make use of O.O. techniques when programming financial models. We refer all readers to this book's parallel text: "C++ Design Patterns and Derivatives Pricing" for detailed discussion of how to implement financial models in object-oriented C++. It also contains code for a variety of purposes including random number generation, the cumulative normal function and the inverse cumulative normal function.

One point I wish to stress is that any model should be implemented at least twice. Once in an efficient fast manner for pricing use, and a second time in a robust, straightforward manner for checking the first method. Often the second method is Monte Carlo since it is generally easy to implement but slow to converge, which makes it ideal for testing purposes where speed is generally not an issue but accuracy is. In a bank, there will generally be a model-validation team whose job is to carry out the second implementation completely separately and compare the results with the first implementation. Fortunes have been lost by the incorrect implementation of models, so being lax on testing is an unaffordable risk.

One extremely good and affordable resource, which any serious quant should have access to, is *Numerical Recipes in C++*, [104], which comes in book and CD form. It lists implementations of a large number of mathematical techniques and algorithms in C++, and one should make use of these techniques and programs whenever possible!

This leads on to an important point which is that programs should always be written with re-use in mind. If one can use part of an old program or a library routine one should always do so. It saves a lot of time not just in code writing, but also more importantly in debugging – if the code has been used several times without problems it is much more likely to be robust.

B.2 Two important functions

The cumulative normal function and its inverse are the two most important functions in mathematical finance. We therefore need to be able to evaluate them quickly and accurately.

Recall that the cumulative normal function is defined by

$$N(x) = \frac{1}{\sqrt{2\pi}} \int_{-\infty}^{x} e^{-\frac{s^2}{2}} ds. \tag{B.1}$$

Its importance lies in the Black–Scholes formula, (3.25).

The inverse cumulative normal function, $N^{-1}(x)$, is simply the inverse of $N(x)$. It is useful in the simulation of normal random variables. A computer random number generator typically generates a integer, m, between zero and some fixed large number RANDMAX. We can create a uniform random variable X on the unit interval by taking just

$$m/\text{RANDMAX}.$$

(If this gives you zero then you are probably accidentally using integer arithmetic!) We then have

$$\mathbb{P}(X \leq x) = x, \quad \text{for } x \in [0, 1]. \tag{B.2}$$

We want to convert X into a standard normal variable. We can do this by taking $N^{-1}(X)$. To see this, observe that

$$\mathbb{P}(N^{-1}(X) \leq x) = \mathbb{P}(X \leq N(x)), \tag{B.3}$$

since N is increasing. The right-hand side is just the definition of a normal random variable.

We give a method due to Moro, [93], which is accurate to within $1E - 12$. We remark that there are other methods of generating normal random variables using draws from a uniform distribution which do not rely on the inverse cumulative normal function. We avoid these other methods here as they typically require more than one uniform variate to generate a normal variate. This causes problems when one shifts from the use of random numbers to the use of low-discrepancy numbers, as the special structure is destroyed by the mixing around. Thus if the reader wishes to plug in a low-discrepancy number generator at a later time then no additional problems will be introduced if he has adopted the inverse cumulative normal distribution approach from the start.

B.2.1 The inverse cumulative normal function

We give the Moro algorithm for computing the inverse cumulative normal function. The function is defined for x between 0 and 1.

Let

$$a_0 = 2.50662823884, \quad \text{(B.4)}$$
$$a_1 = -18.61500062529, \quad \text{(B.5)}$$
$$a_2 = 41.39119773534, \quad \text{(B.6)}$$
$$a_3 = -25.44106049637, \quad \text{(B.7)}$$
$$b_0 = -8.47351093090, \quad \text{(B.8)}$$
$$b_1 = 23.08336743743, \quad \text{(B.9)}$$
$$b_2 = -21.06224101826, \quad \text{(B.10)}$$
$$b_3 = 3.13082909833 \quad \text{(B.11)}$$

and

$$c_0 = 0.3374754822726147, \quad \text{(B.12)}$$
$$c_1 = 0.9761690190917186, \quad \text{(B.13)}$$
$$c_2 = 0.1607979714918209, \quad \text{(B.14)}$$
$$c_3 = 0.0276438810333863, \quad \text{(B.15)}$$
$$c_4 = 0.0038405729373609, \quad \text{(B.16)}$$
$$c_5 = 0.0003951896511919, \quad \text{(B.17)}$$
$$c_6 = 0.0000321767881768, \quad \text{(B.18)}$$
$$c_7 = 0.0000002888167364, \quad \text{(B.19)}$$
$$c_8 = 0.0000003960315187 \quad \text{(B.20)}$$

Let y equal $x - 0.5$; then if, $|y| < 0.42$ let r equal y^2, then the required value is

$$\frac{y \sum_{j=0}^{3} a_j r^j}{\sum_{j=0}^{3} b_j r^{j+1} + 1.0}$$

If $|y| \geq 0.42$, let r equal x if y negative, and $1-x$ otherwise. Let s equal

$$\log(-\log(r))$$

and let t equal

$$\sum_{j=0}^{8} c_j s^j.$$

If $x > 0.5$ then the required value is t otherwise it is $-t$.

B.2.2 Cumulative normal function

The cumulative normal function is defined by

$$N(x) = \frac{1}{\sqrt{2\pi}} \int_{-\infty}^{x} e^{-\frac{s^2}{2}} ds.$$

If $x > 0$, let

$$k = 1/(1 + 0.2316419x);$$

then $N(x)$ is equal to

$$1 - \frac{1}{\sqrt{2\pi}} e^{-\frac{x^2}{2}} k(0.319381530 + k(-0.356563782 + k(1.781477937 \\ + k(-1.821255978 + 1.330274429k))))).$$

If $x < 0$, simply evaluate $1 - N(-x)$. We have written the function in a slightly odd way as it will be faster to evaluate in a computer when written in that fashion.

B.3 Project 1: Vanilla options in a Black–Scholes world

The purpose of this project is to implement the pricing of some vanilla options with the Black–Scholes model by multiple methods. Some of the functions implemented here will be extremely useful for other projects.

Formulas

The first thing to do is to implement the Black–Scholes formulas for various options:

- Implement the price of a forward as a function of time-to-maturity, T, continuously compounding rate, r, dividend rate, d, strike, K, spot, S, and volatility, σ.
- Ditto for a call option.

- Ditto for a put option.
- Ditto for a digital-call option.
- Ditto for a digital-put option.
- Ditto for a zero-coupon bond.

Consistency

We need to be sure that the formulas have been implemented correctly so run the following consistency checks.

(i) We should have put-call parity: the price of a call minus the price of a put equals the value of a forward.
(ii) The price of a call option should be monotone decreasing with strike.
(iii) A call option price should be between S and $S - Ke^{-rT}$, for all inputs.
(iv) A call option price should be monotone increasing in volatility.
(v) If $d = 0$, the call option price should be increasing with T.
(vi) The call option price should be a convex function of strike.
(vii) The price of a call-spread should approximate the price of a digital-call option.
(viii) The price of a digital-call option plus a digital-put option is equal to the price of a zero-coupon bond.

Validation via Monte Carlo

We want to test the prices against a Monte Carlo simulation. It is worthwhile writing code to work using a random number generator class which can then be changed later on. This will allow you to check whether the random number generator is biased, and also to easily plug in a low-discrepancy generator at a later time. Each path in any Monte Carlo simulation will require a certain number of random draws. The maximum number needed is the *dimensionality*. It is best to set-up the class to draw a vector of this size from the random number generator at the start of each run, and make sure the random number generator is actually capable of that dimensionality. Several methods of generating random numbers are given in [104]. Once you have done this, you should

(i) Implement an engine which randomly evolves a stock price from time 0 to time T according to a geometric Brownian motion with drift $r - d$, and volatility σ. Use the formula

$$S_T = S_0 e^{(r-d)T - \frac{1}{2}\sigma^2 T + \sigma \sqrt{T} W}, \qquad (B.21)$$

where W is a standard normal random variable.

(ii) Use the engine to write Monte Carlo pricers for all the products mentioned above. The engine generates a final stock value. The option's pay-off for that final value is then evaluated and discounted. These values are then averaged over a large number of paths. Get it to return the price for successive powers of two for the number of paths so you can see the convergence. Also get it to return the variance of the samples and standard error.

When implementing the pricer and the engine try to do it in an orthogonal way so that the engine just takes in an option object which states its pay-off and expiry. If done correctly, the engine should then need no modifications when a new option type, such as a straddle, is added in. Also the option objects can then be reused when doing Monte Carlo simulations based on different engines.

We can now run some tests.

(i) Compute Monte Carlo and formula prices for a large range of inputs for each of the options above. They should all agree up to the degree of convergence of the Monte Carlo.

If the above tests worked, we can be reasonably confident in both our Black–Scholes functions and in our Monte Carlo engine.

Investigations

We can now use these routines for implementing tougher projects and doing some investigations.

(i) How does the Black–Scholes price of a call option vary as a function of volatility? What happens when volatility is zero, or volatility is very large?
(ii) What about a digital call option?
(iii) For various at-the-money call options, how does the price vary with volatility? Plot the ratio of price to volatility.
(iv) For various put options plot the price and intrinsic value on the same graph. Find at least one example where the two graphs cross.

Stepping methods

One further thing to implement is an alternative engine based on Euler stepping. Divide the time, T, into a large number of steps, N. Let

$$\Delta t = T/N. \tag{B.22}$$

Evolve the stock price across each step by

$$S_{(j+1)\Delta T} = S_{j\Delta T} + r S_{j\Delta T} \Delta T + S_{j\Delta T} \sigma \sqrt{\Delta T} W_j, \tag{B.23}$$

where the W_j are independent normal variables. Running up to the last step this gives an alternate way of generating the final stock value. Use this to develop pricers for the basic options above. The engine will need as inputs the number of steps and the number of paths.

(i) Plot the final price as a function of the number of steps to see how many steps are required for convergence.
(ii) Compare the number of paths required for the two Monte Carlo methods to get a given degree of convergence.
(iii) Make sure the two methods give the same prices.
(iv) Compare the times required to get a given level of accuracy.

B.4 Project 2: Vanilla Greeks

The purpose of this project is to implement formulas for the Greeks and then use them to test various methods of doing Monte Carlo Greeks. The functions written in the first section will be used in later projects.

Implementing the formulas

Implement formulas for

(i) the Delta (spot derivative) of a call option,
(ii) the Gamma (2nd spot derivative) of a call option,
(iii) the Vega (volatility derivative) of a call option,
(iv) the Rho (r derivative) of a call option,
(v) the Theta of a call option, this is equal to minus the T derivative,

The formulas are deducible by simply differentiating the Black–Scholes prices.

Testing

Test them all by comparing with the finite differencing price. That is let ϵ be a small number, and compute the price change for bumping the parameter for ϵ and divide by ϵ. To approximate the delta, for example, take

$$\frac{1}{\epsilon} (\mathrm{BS}(S + \epsilon, T, \sigma, r, d) - \mathrm{BS}(S, T, \sigma, r, d)).$$

The Gamma can be approximated by finite differencing the Delta, or by taking the formula

$$\frac{1}{\epsilon^2} (\mathrm{BS}(S + \epsilon, T, \sigma, r, d) - 2\mathrm{BS}(S, T, \sigma, r, d) + \mathrm{BS}(S - \epsilon, T, \sigma, r, d)).$$

(Why does this work?)

Graphs

Once you have all the formulas working and tested. Plot the following graphs and try to interpret them.

(i) The Delta of a call option as a function of spot.
(ii) The Delta of a call option as a function of time for in-the-money, out-of-the-money and at-the-money options.
(iii) The Gamma of a call option as a function of spot,
(iv) The Vega of a call option as a function of volatility, as a function of spot and as a function of time.

Monte Carlo Greeks

We can now try various Monte Carlo methods for computing the Greeks of vanilla options and compare them to the analytical formulas. Implement the following methods and compare to the formulas. How do the convergence speeds compare?

(i) Run the Monte Carlo twice. The second time with the parameter slightly bumped, and finite difference to get the Greek. Use different random numbers for the two simulations.
(ii) Do the same again but using the same random numbers for the two simulations. (Depending upon the language you are using, it will either default to different random numbers or default to the same ones. Setting the random number seed is the way to achieve either.)
(iii) Implement the pathwise method for the Delta.
(iv) Implement the likelihood ratio method for the Delta.

B.5 Project 3: Hedging

The essence of the Black–Scholes approach to derivatives pricing is that the uncertainty in the final pay-off can be removed by trading in the underlying so try it out in the context of hedging a vanilla call option.

The perfect Black–Scholes world

Implement an engine which evolves a stock under a geometric Brownian motion with drift μ, volatility σ, in N steps using the solution for the stochastic differential equation. Write a hedging simulator that accounts for the profits and losses of a hedging strategy against an option payoff, if the interest rate is r. Implement the Black–Scholes hedging strategy for a call option: hold a Delta amount of stock across each time step. Obviously, you will need to have already done part of the Vanilla Greeks project for this.

Note here we could do the three things quite separately. For the reader au fait with object-oriented programming, one class could handle generation of stock paths, a second could define hedging strategies and the third could take a path generation object and a hedging strategy object and actually carry out the simulation. The best way to approach this would be to use abstract base classes for the path generation and the hedging strategy from which the specific classes could then be inherited. If the simulator then takes in objects from the abstract base classes, new strategies and generators can easily be plugged in later.

Assessing step-size dependence

Use the hedging simulator to compute the variance of the Delta hedging strategy for various input parameters. Plot the variance against the time-step size. How is the variance affected by changing μ and σ?

The stop-loss strategy

Implement the stop-loss hedging strategy: across a time step hold one unit of stock if spot is greater than or equal to the call option strike and none otherwise. How does the variance change with step size? How are the mean and variance of the final portfolio affected by changing μ and σ?

Gamma hedging

Extend the hedging simulator to allow hedging with options. Implement the Gamma hedging of a far-out-of-the-money option with spot and another option. How does the variance change with time-step size?

Time-dependent volatility

We now try hedging our option when volatility is time-dependent but deterministic. We therefore have

$$\frac{dS}{S} = \mu dt + \sigma(t)dW_t. \tag{B.24}$$

To simulate perfectly across a time-step, we take the root-mean-square volatility for that step and put

$$S_{t+\Delta t} = S_t e^{(\mu - \frac{1}{2}\bar{\sigma}^2)\Delta t + \bar{\sigma}\sqrt{\Delta t}N(0,1)}, \tag{B.25}$$

with $\bar{\sigma}$ the root-mean-square value of $\sigma(t)$ across $[t, t + \Delta t]$, and $N(0, 1)$ a normal draw.

Implement the following hedging methods:

(i) Delta hedge using the current value of $\sigma(t)$ in the formula for the Black–Scholes Delta;
(ii) Delta hedge using the root-mean-square value of $\sigma(t)$ across $[0, T]$ at all times;
(iii) Delta hedge using the root-mean-square value of $\sigma(t)$ across $[s, T]$ at time s.

For each method, plot the graph of the variance of final portfolio value against time-step size. Extrapolate to get the variance for instantaneous hedging. Which one works perfectly, and why is it the only one that does?

B.6 Project 4: Recombining trees

Recombining trees are a standard method for pricing options. In this project, we try them out for some simple options. The crucial point when implementing a recombining tree is to make use of the fact that an up-move followed by a down-move is the same as a down-move followed by an up-move. This keeps the total number of nodes tractable, otherwise the number of nodes grows exponentially. A tree should be implemented purely in the risk-neutral world, the real-world tree is useful for justifying risk-neutral pricing but not for actually doing the pricing. We also work with the log as the geometry is simple.

We wish to price an option under geometric Brownian motion so we discretize

$$d(\log S) = \left(r - \frac{1}{2}\sigma^2\right) dt + \sigma dW. \tag{B.26}$$

We divide time into N steps of length Δt. For a binomial tree at each step $\log S$, goes either up or down by $\sigma\sqrt{\Delta t}$ and increases by $(r - \frac{1}{2}\sigma^2)\Delta t$. One can therefore easily construct the set of all possible nodes across N steps, and work out how they relate to each other.

Pricing rules

A tree is well-suited to pricing an option whose value can be written as a function of the current spot and the option's expected value in the future. In practical terms, we can price options whose value can be specified at a node as a function of spot at the current node, and the discounted average of the values at its two daughter nodes. Note the discounted average will be

$$e^{-r\Delta t}\frac{1}{2}(\text{Value-up} + \text{Value-down}).$$

Work out what the rule for computing the price at a node is for each of the following derivatives

(i) a vanilla call option,
(ii) a forward,
(iii) a put option,
(iv) a down-and-out call or put option,
(v) an American put option,
(vi) a digital option.

Pricing on the tree

Now implement the binomial tree and apply it to the pricing of each of the above. If you are using an objected-oriented language keep the definitions of the tree and of the option rule as separate as possible so you can plug in various different rules without recoding the tree. The option will be specified by the rule for its value at a node, and by its final payoff.

Plot the value of each option as a function of the number of steps for reasonable parameters. (e.g. spot 100, strike 100, $r = 0.05$, $\sigma = 0.1$, and $T = 2$.) Check the answers you get against analytic formulas or another pricing method.

For which options does the price oscillate? Compare the rate of convergence with that obtained by taking the sequence obtained by averaging the N-step price with the $(N+1)$-step price.

Price an option with a strike that puts the option far out-of-the-money. How does the speed of convergence change?

Trinomial trees

Repeat everything with a trinomial tree. Compare the rate of convergence both as a function of the number of steps and as a function of the number of nodes. See if you can get improved convergence rates by adapting the position of the nodes to specific properties of the derivative product.

B.7 Project 5: Exotic options by Monte Carlo

Monte Carlo is the most straightforward way to price path-dependent exotic options, and therefore should be implemented early on as a benchmark for testing other implementations.

The pricing engine

We wish to be able to price arithmetic Asian options and discrete barrier options in a Black–Scholes world. To do this we need three things: the first is a path-generator

B.7 Project 5: Exotic options by Monte Carlo

which, given the parameters, S_0, r, d, and σ, and a set of times, $t_1, t_2, \ldots, t_n$, generates a random path $S_{t_1}, S_{t_2}, \ldots, S_{t_n}$, for the option price at those times. For future flexibility, allow the possibility that the parameters are step functions which are constant on the intervals $(t_j, t_{j+1}]$ but need not be constant everywhere. The second thing is a product specification which converts the values of the set of times into a cashflow, that is, a sum of money at a given time. For the Asian option, this would be

$$\left(\frac{1}{n} \sum_{j=1}^{n} S_{t_j} - K \right)_+$$

at time t_n. The third thing is a control engine which calls the path-generator, calls the product definition, discounts the cashflow back to time zero and averages the results. We wish to return not just the final value, but also the intervening averages so that we can see the convergence trend; return, for example, the results for all the powers of two. Try to implement the random number generator as an object which plugs into the path generator and returns n independent Gaussian draws on request – this will make life easier when doing Project 6.

Pricing Asian options

Having implemented the engine, price the following options with $S_0 = 100$, $\sigma = 0.1$, $r = 0.05$, $d = 0.03$, and strike 103.

(i) an Asian call option with maturity in one year and monthly setting dates;
(ii) an Asian call option with maturity in one year and three-monthly setting dates;
(iii) an Asian call option with maturity in one year and weekly setting dates.

How do the prices compare? How do they compare with a vanilla option? How does the speed of convergence vary?

Pricing discrete barrier options

Price some discrete barrier options, all with maturity one year and struck at 103.

(i) a down-and-out call with barrier at 80 and monthly barrier dates;
(ii) a down-and-in call with barrier at 80 and monthly barrier dates;
(iii) a down-and-out put with barrier at 80 and monthly barrier dates;
(iv) a down-and-out put with barrier at 120 and barrier dates at 0.05, 0.15, $\ldots$, 0.95.

Compare prices and speed of convergence. Also compare prices with the vanilla option.

Speeding up the simulation

At time t_{n-1}, the discrete barrier and Asian options become vanilla options as all the path-dependence has been used. Now change the product definitions so that instead of using the value of S_n, they return a cashflow at time t_{n-1} which is the Black–Scholes value of the option at that time. Reprice all the options, make sure the final prices are the same and compare convergence speeds.

B.8 Project 6: Using low-discrepancy numbers

For this project, you will need a good low-discrepancy number generator. There is some discussion and code for the implementation of Sobol numbers in [104].

Straight-forward implementation

First adapt the code in Project 5 to use the low-discrepancy numbers. Rerun all the option pricing exercises from that project and compare the rates of convergence.

Bridging

You should have seen some improvement but not necessarily a huge amount. The reason is that the power of low-discrepancy numbers is much better in low dimensions that in high ones. We therefore wish to put as much dependence on the low dimensions as possible. There are two ways to do this: the Brownian bridge and spectral decomposition. Our objective is to produce draw a path, W_t, from a Brownian motion and pass back the increments $W_{t_j} - W_{t_{j-1}}$. When we use the intuitive method of incremental path-generation, we are effectively drawing n Gaussian variables, Z_j, and letting

$$W_{t_j} = W_{t_{j-1}} + Z_j, \tag{B.27}$$

with the net effect that we pass back the vector (Z_j).

However, all that is important is that we synthesize n normal variates W_j which have covariance matrix

$$(\min(i, j)). \tag{B.28}$$

As we observed in Chapter 9, setting

$$W = AZ, \quad \text{where} \quad AA^T = C, \tag{B.29}$$

is a necessary and sufficient condition. We now wish to try out various choices of A. The incremental method is the Cholesky decomposition, that is, the unique choice of A which is lower triangular. Try the following methods and compare convergences for Asian options,

(i) Use spectral decomposition to write $C = PDP^T$ with D diagonal and with the diagonal elements decreasing. Try $A = PD^{1/2}$ and $A = PD^{1/2}P^T$. See [104] for code to carry out diagonalization. Here P is of course the matrix of orthonormal eigenvectors.
(ii) Use Cholesky decomposition but first reverse the order.
(iii) Use Cholesky decomposition but reorder so that the first index is the last one, and each successive one is in the middle of the longest set of indices not chosen (not unique). Thus if there are ten indices the ordering would be say 10, 5, 2, 7, 3, 8, 1, 4, 6, 9.

B.9 Project 7: Replication models for continuous barrier options

In Chapter 10, we looked at a number of methods of replicating exotic options using vanilla options. As well as being useful for hedging, these methods can actually be applied to pricing.

Convergence

Implement the method of Section 10.2 for the pricing of continuous barrier options. Write your implementation in such a way that you can plug in any pricing function for the vanillas as a function of spot, strike, current time and the expiry time of the option. Also allow as arbitrary input the final pay-off of the option.

In a Black–Scholes world with spot equal to 100, r equal to 0.05, d equal to 0, and $\sigma = 0.1$, use the engine to price an up-and-out call option with strike 100 and barrier at 120, and a down-and-out call option with the same strike and a barrier at 80. Take both options to have expiry time of one year. How do the rate of convergences as a function of the number of steps compare?

Varying the final pay-off

We do not need to replicate the final pay-off behind the barrier as the replicating portfolio is dissolved on first touching the barrier. We can therefore change the pay-off behind the barrier and see if that helps. Try pricing the up-and-out call with the following final pay-offs and compare the convergence rates:

(i) The pay-off is constant above the barrier;
(ii) A tight call spread of width ϵ is used to bring the value at the barrier down to 0 and then the pay-off is zero above $120 + \epsilon$. Do multiple values of ϵ;
(iii) A tight call spread of width 2ϵ starting at 120 is used to bring the value just above the barrier down to -20 and then the pay-off goes back up to zero with gradient one. Once it reaches zero it stays zero;

(iv) A tight call spread of width 2ϵ starting at $120-\epsilon$ is used to bring the value just above the barrier down to -20 and then the payoff goes back up to zero with gradient one. Once it reaches zero it stays zero. (Note that the pay-off below the barrier is not exactly correct for this one.)

Try to explain your results.

Time-step size

Try varying the time-step sizes so that the steps are shorter close to expiry. What difference does this make?

Time-dependent volatility

Now reprice the options using a variety of time-dependent functions for σ. Choose the functions so that the root-mean-square value of σ over the year is always the same. Make sure the variety of functions include:

(i) a function which is rapidly decreasing;
(ii) a function which is rapidly increasing;
(iii) a function with a bump in the middle.

Do the tests with zero interest rates; with small interest rates and with large interest rates.

B.10 Project 8: Multi-asset options

We want to price multi-asset options depending upon the evolution of several underlyings. As usual, we want to price the options in multiple fashions in order to check our models are correct.

Quantos

Implement an analytic formula for a quanto option and a Monte Carlo pricer. The Monte Carlo pricer will need to simulate correlated random draws for the foreign exchange and the stock. Check the two pricers give the same value. How much impact does correlation have on the price? Do the Monte Carlo both in a single step and in several steps.

Margrabe

Implement a pricer for a Margrabe option. Check that it gives the correct price by Monte Carlo. The Monte Carlo pricer will have an input for interest rates since the stocks will grow at the riskless rate. How much does correlation affect the price?

Is the pricer injective in correlation? i.e. can one deduce the correlation given the other inputs and the price?

B.11 Project 9: Simple interest-rate derivative pricing

Swap-rate formulas

Let $0 \leq t_0 < t_1 < \cdots < t_n$. Let P_j be the zero-coupon bond expiring at time t_j. Write a function which computes the swap-rate for the times t_j in terms of P_j. Write one that computes the annuity of the swap also.

Discounts from swap-rates

If SR_i is the swap-rate for the times $t_i, t_{i+1}, \ldots, t_n$, write a routine which compute P_j for $j > 0$ given all the rates SR_i and P_0. Check that the two functions are inverse to each other. Such a collection of swap-rates is said to be co-terminal.

Do the same problem for co-initial swap-rates. i.e. take the swap rates starting at time t_0 and finishing at time t_i for each i.

Black formulas

Implement the Black formulas for each of the following:

(i) a payers or receivers swaption as a function of annuity, swap-rate, strike, volatility and expiry;
(ii) a receiver's swaption;
(iii) a caplet;
(iv) a floorlet.

The easiest way to do these is to use the Black–Scholes formula with zero interest rates and multiply by the annuity.

B.12 Project 10: LIBOR-in-arrears

This project is a precursor to the BGM project. Some of the issues that arise there appear here without being so fiddly.

We have multiple ways to price the LIBOR-in-arrears forward rate agreement (henceforth the arrears FRA) and the LIBOR-in-arrears caplet.

Analytic formula

If f runs from t_0 to t_1 and the strike is K, the arrears FRA pays $(f - K)(t_1 - t_0)$ at time t_0 instead of at time t_1. This is equivalent to paying

$$(f - K)(1 + f(t_1 - t_0))(t_1 - t_0)$$

at time t_1. Use this fact to derive an analytic formula for the price of the arrears FRA using the zero-coupon bond expiring at time t_1 as numeraire; assume that f is log-normal.

Pricers

Implement the following pricers:

(i) an analytic pricer for the arrears FRA;
(ii) a numeric integration pricer for the arrears FRA;
(iii) a numeric integration pricer for the arrears caplet;
(iv) a Monte Carlo pricer for the FRA and caplet using the t_1 bond as numeraire.

Use all the methods to price a FRA and caplet starting in ten years with current forward rate 6%, strike 7%, volatility 20%, price of the zero-coupon bond expiring in ten years is 0.5 and they are both of length 0.5.

Change of numeraire

Suppose we now use the t_0 bond as numeraire. We have to price by Monte Carlo as we do not know the density explicitly. Implement a Monte Carlo pricer for the arrears FRA and caplet. Do it twice. The first time use an Euler integration method. The second time use the predictor-corrector type method outlined in Section 14.6. In both cases, divide time into a number of steps and plot the converged price as a function of the number of steps.

B.13 Project 11: BGM

The purpose of this project is to implement an engine for pricing products using BGM. This is a large task and you should not embark on it unless you have plenty of time to do it. On the other hand, if you can successfully do it, you are well on the way to being a quant.

To do a project like this well, one really has to use object-oriented techniques. I am therefore setting the project in terms of writing various classes of objects rather than pricing things. Whilst I tend to state objects using the terminology of C++, any object-oriented language could be used. However, if you want to be a quant implement it in C++ as that's what the bank will want.

We will want to simulate the movements of n forward rates, f_j, associated to times

$$t_0 < t_1 < \cdots < t_n,$$

such that f_j runs from t_j to t_{j+1}.

The forward volatility structure

We will repeatedly need the covariance matrix of our forward rates across arbitrary time steps. Assume that the forward f_j has volatility

$$K_j\big((a + b(t_j - t))e^{-c(t_j - t)} + d\big)$$

for $t < t_j$, and zero otherwise. Assume that the instantaneous correlation between f_i and f_j is $e^{-\beta|t_i - t_j|}$.

Write a class that stores all the necessary information and has a method that returns the covariance between f_i and f_j over any time step. (Use an analytic integration which is a bitch to compute in itself.)

For speed, it will be better to write a method that computes the entire covariance matrix for the time-step and stores it in a matrix that has been passed by reference.

At some stage, you may want to use a different functional form for your forward rates. One easy way to have this functionality is to set up an abstract base class which has the method of computing covariances, and then inherit this particular volatility structure from it. One alternative volatility structure to implement would be flat volatilities: every forward rate has some constant volatility up to its setting time and then has zero volatility.

The products

We want to implement our engine and the products separately so we can plug new products into the engine easily, and also if we invent a new model then we can still plug the products in without having to rewrite them all which would be an unnecessary pain.

We start with an abstract base class BGMProduct. This class should have abstract methods implemented in the inherited class which defines the product as follows.

- GetUnderlyingTimes – this returns an array of times between which the forward rate to be used in the simulation will run.
- GetEvolutionTimes – this states at what times the product needs to know the forward rates,
- Reset – this resets the object for a new path of the simulation
- DoNextStep – this passes in the current forward rates, and should pass back whether to continue to the next step and the values and timings of any cash flows generated.

For example, for a swaption associated to a set of times $t_0, t_1, \ldots, t_n$, with strike K,

- GetUnderlyingTimes – return the times $t_0, t_1, \ldots, t_n$.
- GetEvolutionTimes – return t_0

- Reset – does not do anything
- DoNextStep – returns a signal to terminate and computes the value of the swap-rate, SR, and annuity, A. It returns the intrinsic value of the swaption i.e. $(SR - K)_+ A$.

For a swap-based trigger swap which at each stage knocks-out if the swap-rate for the remaining times is above a reference rate, R, and otherwise is a swap at strike K:

- GetUnderlyingTimes – returns the times $t_0, t_1, \ldots, t_n$.
- GetEvolutionTimes – returns $t_0, t_1, \ldots, t_{n-1}$.
- Reset – sets a variable I to zero to indicate we are the beginning.
- DoNextStep – returns a signal to terminate if I is $n-1$ or if the remaining swap-rate is above R. Generates a cash flow of value $(f_I - K)(t_{I+1} - t_I)$ at time t_{I+1}. Increases I by 1.

Work out how to do a FRA-based trigger swap with only one evolution time t_{n-1}.

The engine

The engine is the class that does all the work. As inputs we need:

- the product;
- the covariance structure (which must cover the forward rates used in the product);
- the number of paths to be used in the simulation;
- a number generator – do this as an input to give some flexibility;
- the amount of substepping to be carried out – whilst we only need to know the forward rates at certain times, the Monte Carlo will be more accurate if you put in intervening times because of the state-dependent drift;
- the method of approximating the drift, e.g. the Euler stepping method or the predictor-corrector method;
- the initial values of the forward rates;
- what numeraire to use and its initial value.

The engine should precompute as much as possible. For example, the covariance matrices over each time-step will be the same every time; this means that they can be computed once and for all in the engine's constructor. We need to know the pseudo-square root of each covariance matrix, and we can precompute them too. The initial drifts for the first step can be precomputed, but all other drifts will have to be computed on the fly.

For each path, the engine will have to evolve the forward rates up to each evolution time specified by the product, possibly using multiple steps to get there. First the Reset method of the product should be called. At each evolution time, it will then call the product method DoNextStep to decide whether to terminate and discover if any cashflows have been generated. The generated cashflows will have to be accounted for by storing the ratio of their value to the current value of the numeraire. Note that the value of the cashflow and the numeraire will have have to be computed by discounting using the current forward rates.

The engine stores the final ratio of product value to numeraire for each path, and then averages over all paths.

We then multiply the final value of this average by the initial value of the numeraire to get the price. In practice, we would want to output the average for smaller numbers of paths also, in order to see how converged the Monte Carlo is.

Testing the engine

Having written the engine, we want to be sure that it works. One important property it should have is that changing the numeraire should not change the price.

Our first test is therefore to price a caplet with the *wrong* numeraire. We take

$$t_j = 10 + j/2. \tag{B.30}$$

Let P_j denote the bond expiring at time t_j. If the initial curve is flat with compounding continuous rate 5%, the value of a bond expiring at time t will be $\exp(0.05t)$. Price a caplet on the first forward rate with strike 6% by using the engine with each different P_j. Compare the price with that obtained from the Black formula. To define the forward volatility structure, take for example

$$a = 0.05, \tag{B.31}$$
$$b = 0.09, \tag{B.32}$$
$$c = 0.44, \tag{B.33}$$
$$d = 0.11, \tag{B.34}$$

with all the K factors equal to 1. Take $\beta = 0.1$. You may need multiple steps if you use an Euler approximation type method for the drift.

Once you have got the caplet price to be numeraire invariant, implement the swaption and trigger swap, and test that they are numeraire invariant also.

The approximation formula

In Section 14.7, we developed a formula for pricing swaptions in a BGM model instantaneously. Implement this formula and compare the prices obtained with those implied by the BGM engine.

Sensitivity to shape

We want to see how much changing the shape of the instantaneous volatility curves affects the price of an exotic option. Work out the effective constant volatility that gives each forward rate the same total volatility, and thus gives the same price to all the caplets.

With the same value of β, now price the trigger swap with flat volatilities and with the a, b, c, d volatility structure.

Compare the changes in value with the Vega of the option, that is, the change in value obtained by bumping all the volatilities up by 1%. (Typically, we know volatilities within about 1% so a change in price which is large compared to the Vega is important, whilst one small is unimportant.)

Also test the price sensitivity to changing β. How different is the $\beta = 0$ flat volatilities price from the $\beta = 0.1$ variable volatilities price?

Log-normality of swap-rates

Forward rates and swap-rates cannot be simultaneously log-normal; to see this just derive the SDE for a swap rate from the log-normal SDE for forward rates.

However, it does not really matter whether the rates are perfectly log-normal. If the swap-rates are almost log-normal then the failure of perfect log-normality does not really matter. How can we test the rates log-normality? If we price swaptions using our BGM engine then the deviation of the swap-rates from log-normality will be displayed in the shape of the swaption smile.

Price options on 1-, 5- and 10-year swaps starting in 1, 5, and 10 years with a variety of strikes. Plot the implied volatility smile of the swaptions in each case. (See Project 12 for some discussion of how to implement an implied volatility function.) What can we conclude about log-normality?

B.14 Project 12: Jump-diffusion models

In this project, we investigate how to implement a pricer for a jump-diffusion model, see what sort of smiles are implied and look at pricing variations for exotic options.

Vanilla options

Implement a pricer for vanilla options for a jump-diffusion model with log-normal jumps. Implement a Monte Carlo pricer also and check they give the same answers.

Implied volatility

Implement an implied volatility function – this is a function which inverts the Black–Scholes price function to get the unique volatility which gives the correct price for the option. There is no analytic formula so you will have to use Newton–Raphson or repeated bisection to invert the map

$$\sigma \mapsto \text{Black–Scholes Price}.$$

Smiles

With $r = 0.05$, $d = 0$, $\sigma = 0.1$, $\lambda = 0.2$, $m = 0.9$ and $\nu = 0.1$, plot the implied volatility as a function of strike for options with maturity 0.25, 0.5, 1, 2, 4, and 8 years. How does the shape change with maturity?

Repeat the exercise but now take $m = 1$.

Varying intensity

Plot the price of a vanilla call as a function of λ. Do the same for a variety of digital options.

Exotic options

Write a pricer for Asian options using a Monte Carlo implementation of jump-diffusion. With parameters as in the previous setting, spot equal to 100 and strike equal to 100, price a one-year Asian call option with monthly resets.

Comparison with Black–Scholes

Fit a Black–Scholes model with time-dependent volatility so that it gives the same implied volatility to at-the-money call options at the monthly reset times. Reprice the Asian option with this model.

Now do a discrete barrier call option with the same reset times and strike, with an up-and-out barrier at 105. Compare with the Black–Scholes prices obtained by calibrating to the at-the-money and at-the-barrier prices.

B.15 Project 13: Stochastic volatility

In this project, we investigate how to implement a pricer for a stochastic-volatility model, see what sort of smiles are implied and look at pricing variations for exotic options.

Vanilla options

Implement a pricer for vanilla options for a stochastic-volatility model with uncorrelated volatility and spot. Implement a Monte Carlo pricer also and check they give the same answers. Do the Monte Carlo pricer in the following ways:

 (i) short-step the volatility and the spot;
 (ii) short-step the volatility, compute the root-mean-square volatility for the path and long-step the spot;
 (iii) short-step the volatility, compute the root-mean-square volatility for the path and plug it into the Black–Scholes formula.

With $r = 0.05$, $d = 0$, $\sigma_0 = 0.1$, $\alpha = 0.5$, zero drift of volatility, and volatility of volatility 0.2, plot the implied volatility as a function of strike for options with maturity 0.25, 0.5, 1, 2, 4, and 8 years. (See Project 12 for discussion of how to do implied volatilities.) How does the shape change with maturity? Compare with jump-diffusion smiles.

Exotic options

Write a pricer for Asian options using a Monte Carlo implementation of stochastic volatility. With parameters as in the previous setting, spot equal to 100 and strike equal to 100, price a one-year Asian call option with monthly resets.

Comparison with Black–Scholes

Fit a Black–Scholes model with time-dependent volatility so that it gives the same implied volatility to at-the-money call options at the monthly reset times. Reprice the Asian option with this model.

Now do a discrete barrier call option with the same reset times and strike, with an up-and-out barrier at 105. Compare with the Black–Scholes prices obtained by calibrating to the at-the-money and at-the-barrier prices.

B.16 Project 14: Variance Gamma

In this project, we investigate how to implement a pricer for the Variance Gamma model, see what sort of smiles are implied and look at pricing variations for exotic options.

Vanilla options

Implement a pricer for the Variance Gamma model as an integral over Black–Scholes prices. Implement a Monte Carlo pricer also and check they give the same answers.

With $r = 0.05$, $d = 0$, $\sigma_0 = 0.1$, $\theta = 0$, and $v = 0.2$, plot the implied volatility as a function of strike for options with maturity $0.25, 0.5, 1, 2, 4,$ and 8 years. (See project 12 for discussion of how to do implied volatilities.) How does the shape change with maturity? Compare with jump-diffusion smiles and stochastic volatility smiles. Repeat trying varying values of θ.

Exotic options

Write a pricer for Asian options using a Monte Carlo implementation of variance gamma. With parameters as in the previous setting, spot equal to 100 and strike equal to 100, price a one-year Asian call option with monthly resets.

Comparison with Black–Scholes

Fit a Black–Scholes model with time-dependent volatility so that it gives the same implied volatility to at-the-money call options at the monthly reset times. Reprice the Asian option with this model.

Now do a discrete barrier call option with the same reset times and strike, with an up-and-out barrier at 105. Compare with the Black–Scholes prices obtained by calibrating to the at-the-money and at-the-barrier prices.

Appendix C

Elements of probability theory

C.1 Definitions

Our objective in this appendix is to recall some of the basic definitions and some elementary results from probability theory. We do not attempt to teach the reader who knows no probability theory but instead wish to fix notation, remind the reader of some basic results, and perhaps shift his point of view a little. We refer the reader who has never studied probability theory to Grimmett & Stirzaker, [53].

To define probabilities we need three things. The first is a *sample space* generally denoted Ω. The sample space can be viewed as a set which encapsulates the notion of a state-space. For example, it could be just the set $\{0, 1\}$, or it could be the real numbers, or it could be the set of continuous functions from $[0, 1]$ to $\mathbb{R}$. For us, the sample space will often be the set of continuous paths.

The second thing we need is a collection of events. In probability theory, we wish to assign numbers between 0 and 1 to subsets of Ω to represent the likelihood of that subset containing the random element drawn. These subsets are called events. We require the set of events to be closed under certain simple operations.

Definition C.1 A collection of subsets $\mathcal{F}$ of Ω is called a σ-field if

(i) the empty set and Ω are in $\mathcal{F}$;
(ii) $\mathcal{F}$ is closed under countable unions and intersections;
(iii) $A \in \mathcal{F}$ if and only if $A^c \in \mathcal{F}$.

The third thing we need is a probability measure. That is, we need to be able to assign to every event a probability. As events are subsets of Ω, the probability measure is a map from a set of subsets of Ω to $[0, 1]$. For technical reasons we generally require the probability measure only to be defined on a σ-field, $\mathcal{F}$, rather than everywhere; this is essentially because the set of all subsets is too big a set in general. We require the probability measure to have certain consistency properties:

C.1 Definitions

Definition C.2 A probability measure $\mathbb{P}$ on a sample space, Ω, and a σ-field $\mathcal{F}$ of Ω is a function $\mathbb{P}$ from $\mathcal{F}$ to $[0, 1]$ such that

(i) $\mathbb{P}(\emptyset) = 0$;
(ii) $\mathbb{P}(\Omega) = 1$;
(iii) *if* $A_1, A_2, A_3, \ldots$ *are pairwise disjoint elements of* $\mathcal{F}$ *then*

$$\mathbb{P}\left(\bigcup_j A_j\right) = \sum_{j=1}^{\infty} \mathbb{P}(A_j).$$

An immediate consequence of the definition is that

$$\mathbb{P}(A) + \mathbb{P}(A^c) = 1, \tag{C.1}$$

for any A in $\mathcal{F}$.

We therefore define a probability space to be a triple $(\Omega, \mathcal{F}, \mathbb{P})$, where Ω is the sample space, $\mathcal{F}$ is the σ-field, and $\mathbb{P}$ is a probability measure on $\mathcal{F}$.

Often when we are considering simple random variables this definition is unnecessarily complicated; however when studying probability events on spaces of paths it becomes necessary.

Definition C.3 A random variable on a triple $(\Omega, \mathcal{F}, \mathbb{P})$ is a map, X, from the sample space to the real numbers such that for any $x \in \mathbb{R}$, we have

$$\{\omega \in \Omega : X(\omega) \leq x\} \in \mathcal{F}.$$

Thus the event $\{X \leq x\}$ is in the σ-field, $\mathcal{F}$, for any x. This means that we can define a probability to the event $\{X \leq x\}$. Thus we can write

$$\mathbb{P}(X \leq x) = \mathbb{P}(\{\omega \in \Omega : X(\omega) \leq x\}).$$

Much of the time, we do not need to think of the random variable X and the sample space as being different things. In particular, in simple cases we can take Ω to be the real numbers and X to be the identity map.

Example C.1 We can model the toss of a coin by taking Ω to be the set $\{0, 1\}$, the σ-field to be the sets

$$\emptyset, \{0\}, \{1\}, \{0, 1\},$$

and the probability of these sets to be

$$0, 0.5, 0.5, \text{ and } 1,$$

respectively. In this case, our random variable is simply the identity map taking 0 to 0 and 1 to 1. ◇

It is when studying multiple random variables simultaneously that the distinction between sample space and random variable becomes important.

Example C.2 Suppose we wish to simulate two random coin tosses. We take Ω to be the set
$$\{0, 1\} \times \{0, 1\}.$$
We assign to each event a probability 0.25 times the number of points in it. If we let X_1 denote the (projection onto the) first coordinate and X_2 the second coordinate then each of X_1 and X_2 define a random variable. The probability that X_1 takes the value zero is equal to the probability of the event
$$\{0\} \times \{0, 1\}$$
which is equal to twice 0.25 as it has 2 elements. Thus X_1 defines the same random variable as in our previous example. Similarly, for X_2. ◇

Consider a related but slightly different example.

Example C.3 Suppose we again wish to simulate two random coin tosses. We again take Ω to be the set
$$\{0, 1\} \times \{0, 1\}.$$
We assign probabilities as follows: if the event contains $\{0, 0\}$ or $\{1, 1\}$ it has probability 1/2; if it contains both it has probability 1. In all other cases it has probability 0.

Let Y_1 denote the (projection onto the) first coordinate and Y_2 the second coordinate, then each of Y_1, and Y_2 defines a random variable. The probability that Y_1 takes the value 0 is equal to the probability of the event
$$\{0\} \times \{0, 1\}$$
which is equal to 0.5. Thus Y_1 defines the same random variable as in our previous example. Similarly, for Y_2. However, the probability of the event $Y_1 = Y_2$ is now equal to 1. Whereas the probability of the event $X_1 = X_2$ was 0.5. ◇

The moral of this example is that the nature of the sample space is important when trying to understand the interaction between different random variables.

When studying multiple random variables, we also will need conditional probabilities: the probability of an event given that another event has occurred. Given events A and B, we define the *conditional probability*
$$\mathbb{P}(A|B) = \frac{\mathbb{P}(A \cap B)}{\mathbb{P}(B)}. \tag{C.2}$$

This captures the notion that we should only consider the part of the event A that is in B if we know that B has occurred.

With two random variables, knowledge of one of the variables can affect the value of the other. For example, in the example above if we know Y_1 then we also know Y_2. This implies that the probability that Y_2 takes a given value is affected by the value of Y_1. In this case, the two variables are said to be *dependent*. Variables are said to be *independent* if the value of one does not affect the value of the other. In other words, X_1 and X_2 are independent if

$$\mathbb{P}(X_1 \in A) = \mathbb{P}(X_1 \in A | X_2 \in B) \tag{C.3}$$

for any sets A and B.

When studying random variables, we can specify their behaviour by using the *cumulative distribution function*. If X is a random variable then this is defined by

$$F_X(x) = \mathbb{P}(X \leq x). \tag{C.4}$$

Since increasing x increases the probability that X is less than x, we have that F_X must be an increasing function of x and will range between 0 and 1.

Note that F_X need not be continuous. For example if F_X equals 0 for $x < 0$, 0.5 for $0 \leq x < 1$, and 1 for $x \geq 1$, then X takes the values 0 and 1 with equal probability 0.5.

When F_X is continuous it can generally be written in the form

$$F_X(x) = \int_{-\infty}^{x} f_X(s) ds. \tag{C.5}$$

We then say that f_X is the *probability density function* of X.

For us the two most important examples of random variables are the uniform distribution and the normal distribution.

Definition C.4 A random variable X is said to be uniformly distributed if it takes values between 0 and 1, and the probability of $X \in J$ is equal to the length of J for any J subinterval of $[0, 1]$.

A uniform random variable, U, has probability density function f_U, equal to 0 outside $[0, 1]$ and 1 inside.

Definition C.5 A random variable X is said to have a standard normal distribution if it has probability density function equal to

$$\frac{1}{\sqrt{2\pi}} e^{-\frac{s^2}{2}}.$$

It is then written as $N(0, 1)$. More generally we say a random variable is $N(\mu, \sigma^2)$ if it can be written as $\sigma N(0, 1) + \mu$.

To justify this notation, we need to recall the concept of expectation.

C.2 Expectations and moments

The expectation of a random variable encapsulates the intuitive notion of the average value of a random variable. If X has continuous density function f_X then the expectation is equal to

$$\mathbb{E}(X) = \int_{-\infty}^{\infty} f_X(x) x \, dx. \tag{C.6}$$

We have the simple relations

$$\mathbb{E}(aX + bY) = a\mathbb{E}(X) + b\mathbb{E}(Y), \tag{C.7}$$

for any $a, b \in \mathbb{R}$ and any random variables X and Y.

For the uniform distribution, U, we have

$$\mathbb{E}(U) = \int_0^1 s \, ds = \frac{1}{2}. \tag{C.8}$$

For the standard normal, $N(0, 1)$, we have

$$\mathbb{E}(N(0, 1)) = \frac{1}{\sqrt{2\pi}} \int s e^{-\frac{s^2}{2}} ds, \tag{C.9}$$

which is equal to zero since the integral is odd.

The *law of large numbers* tells us that the expectation does indeed capture the notion of a long-term average:

Theorem C.1 *If X_j are identically distributed independent random variables then*

$$\lim_{N \to \infty} \frac{1}{N} \sum_{j=1}^{N} X_j = \mathbb{E}(X),$$

with probability 1.

This theorem is very important in mathematical finance since it gives us a method of evaluating the expectation. Rather than computing the integral analytically or via numerical integration, we simply repeatedly draw random variables and take the long run average to approximate the integral. This method is called *Monte Carlo simulation*.

C.2 Expectations and moments

As well as needing the notion of the average of a random variable, we also need to understand how likely it is to stray from the average. This notion is captured by the *variance*, defined to be equal to

$$\text{Var}(X) = \mathbb{E}((X - \mathbb{E}(X))^2). \tag{C.10}$$

As the expectation of a positive quantity, the variance is always positive. Its square root is called the *standard deviation*.

We have trivially that

$$\text{Var}(X) = \mathbb{E}(X^2) - \mathbb{E}(X)^2, \tag{C.11}$$

and

$$\text{Var}(\alpha X) = \alpha^2 \text{Var}(X). \tag{C.12}$$

We also have that if X and Y are independent then

$$\text{Var}(X + Y) = \text{Var}(X) + \text{Var}(Y). \tag{C.13}$$

When evaluating expectations of functions of X we have the handy result, sometimes called the *law of the unconscious statistician*, that

$$\mathbb{E}(g(X)) = \int f_X(s) g(s) ds \tag{C.14}$$

for any function g.

If we take $\sigma N(0, 1) + \mu$ we have

$$\mathbb{E}(\sigma N(0, 1) + \mu) = \mu + \sigma \mathbb{E}(N(0, 1)) = \mu, \tag{C.15}$$

and

$$\text{Var}(\mu + \sigma N(0, 1)) = \text{Var}(\sigma N(0, 1)) = \sigma^2 \text{Var}(N(0, 1)). \tag{C.16}$$

The variance of $N(0, 1)$ is equal to

$$\frac{1}{\sqrt{2\pi}} \int s^2 e^{-\frac{s^2}{2}} ds = 1.$$

So an $N(\mu, \sigma^2)$ distribution has mean μ and variance σ^2.

The normal distribution is important for two reasons; the first is that it underlies the definition of Brownian motion which will be crucial to us in modelling stock price movements. The second is that it is, in a certain sense, the distribution one obtains by averaging a large number of random variables. In particular, if we add together a sequence of independent random variables, whilst rescaling in such a way as to keep the mean and variance fixed, then we obtain a normal distribution. This is the *Central Limit theorem*:

Theorem C.2 *Let Y_j be a sequence of identically distributed random variables with mean μ and variance σ^2; if we let*

$$Z_j = \frac{\sum_{i=1}^{j} Y_j - j\mu}{\sigma \sqrt{j}},\qquad(C.17)$$

then, as $j \to \infty$, Z_j converges to a standard normal distribution.

Note that Z_j has been defined so that it has mean 0 and variance 1.

For example, suppose we let Y_j be a sequence of random variables defined by taking 1 and -1 with probability 0.5. Then Y_j has mean 0 and variance 1. We set

$$Z_k = \frac{1}{\sqrt{k}} \sum_{j=1}^{k} Y_j. \qquad(C.18)$$

We then have that Z_k converges to a standard normal random variable as k tends to infinity. This means that we can approximate a normal distribution as a sum of binary random variables by taking k large but finite.

As well as studying the mean and variance, the higher-order moments can often tell us important things about a distribution. Let μ denote the mean of X and σ^2 the variance. The *skew* is defined to be equal to

$$\frac{\mathbb{E}((X-\mu)^3)}{\sigma^3}.$$

Similarly, we define the *kurtosis* to be equal to

$$\frac{\mathbb{E}((X-\mu)^4)}{\sigma^4}.$$

Note that the skew will not be affected by adding a positive constant to X or by multiplying by a constant. The skew of a normal random variable is 0 and the kurtosis is 3. (Sometimes 3 is subtracted from the definition in order to make the kurtosis of a normal random variable equal to 0.) When a random variable has kurtosis bigger than the normal, it is said to have *fat tails*, expressing the notion that the distribution does not decay to infinity as quickly as a normal with the same variance. The skew expresses the idea that a distribution may be asymmetric and therefore tilted to one side or the other.

C.3 Joint density and distribution functions

If we have a collection of random variables then as well as knowing the density or distribution function of each one, we will also want to know their interdependencies. This means that we need to the joint probability that each will take given

C.3 Joint density and distribution functions

values or lie in given sets. The random variables could be the values of different stocks or different forward rates or simply the value of a given stock for a collection of times. Clearly, the value of a stock at time 2 will be affected by its value at time 1 and it will certainly not be independent of it.

We can express these dependencies by using joint distribution functions. Thus, given random variables $X_1, \ldots, X_n$, the joint distribution function is

$$F(x_1, x_2, \ldots, x_n) = \mathbb{P}(X_1 \leq x_1, X_2 \leq x_2, \ldots, X_n \leq x_n). \tag{C.19}$$

Similarly, the joint density function, if it exists, is the function f such that for any reasonable (i.e. measurable) set, $A \subset \mathbb{R}^n$,

$$\mathbb{P}((X_1, \ldots, X_n) \in A) = \int_A f(x_1, x_2, \ldots, x_n) dx_1 dx_2 \ldots dx_n. \tag{C.20}$$

We can recover the density function of any individual X_j by integrating out the other variables since the event that $X_j \in B$ is clearly equal to

$$\int_{x_j \in B} f(x_1, \ldots, x_n) dx_1 \ldots dx_n$$

$$= \int_{x_j \in B} \left(\int f(x_1, \ldots, x_n) dx_1 \ldots dx_{j-1} dx_{j+1} \ldots dx_n \right) dx_j.$$

From the other direction this means that if we know the density functions of the individual X_j, we do not know the joint density function but only its value when integrated against any $n - 1$ coordinates. We will call these individual distributions the *marginal* distributions.

When studying the pricing of path-dependent derivatives this fact is particularly pertinent since we will be able to deduce the probability densities of the stock's pricing measure across each single time slice from the value of vanilla options, but we will not be given the joint distribution functions. The purpose of our model will then be to determine the form of the joint density function from these marginals.

Note that when the individual random variables are independent, the joint density takes on a particularly simple form; it is just a product of one-dimensional functions

$$f_1(x_1) f_2(x_2) \ldots f_n(x_n).$$

Whilst this will never be the case for the value of the stock at various time horizons it is not so unreasonable for the value of the ratio of the stock between time horizons; thus we could put

$$X_j = \frac{S_{t_j}}{S_{t_{j-1}}},$$

and assume each X_j is independent. Fixing the distribution of the X_j would then be the same as fixing their joint distribution.

C.4 Covariances and correlations

We will often need methods of expressing the level of dependence for random variables which are not independent. It is here that the concepts of covariance and correlation are important. In certain limited but important circumstances they express the relationship between two normal random variables.

The *covariance* of two random variables X, Y is defined by

$$\text{Cov}(X, Y) = \mathbb{E}((X - \mathbb{E}(X))(Y - \mathbb{E}(Y))). \tag{C.21}$$

It is equal to

$$\mathbb{E}(XY) - \mathbb{E}(X)\mathbb{E}(Y).$$

Note that for this definition to make sense X and Y must be defined on the same sample space.

Clearly, we have

$$\text{Cov}(\alpha X, \beta Y) = \alpha\beta\text{Cov}(X, Y). \tag{C.22}$$

It is therefore sometimes useful to strip out the size of X and Y by dividing by their standard deviations to get the *correlation coefficient*

$$\rho(X, Y) = \frac{\text{Cov}(X, Y)}{\sqrt{\text{Var}(X)\text{Var}(Y)}}. \tag{C.23}$$

It can be shown that

$$|\mathbb{E}(XY)| \leq |\mathbb{E}(X)\mathbb{E}(Y)| \tag{C.24}$$

and thus we have that the correlation lies between -1 and $+1$. If $X = Y$ then

$$\rho(X, Y) = 1, \tag{C.25}$$

as the covariance is the variance of X, and if X is $-Y$ then we get -1. If X and Y are independent we get 0.

Given a vector X_j of random variables we can form a matrix

$$\text{Cov}(X_i, X_j)$$

called the *covariance matrix*, and similarly we can form the *correlation matrix*.

These matrices have some special properties. Note that as the correlation matrix is the covariance matrix of the random variables $\frac{X_j}{\sqrt{\text{Var}(X_j)}}$, any property of the former will equally well hold for the latter. Indeed we can distinguish correlation matrices amongst covariance matrices by the property of having 1s on the diagonal.

C.4 Covariances and correlations

The first obvious property is that the covariance matrix is symmetric. Let C be a covariance matrix and let C_{ij} denote the ij element. Since it is symmetric, we can use C to define a bilinear form via

$$C(v, w) = v^T C w = \sum_{i,j=1}^{n} v_i C_{ij} w_j. \tag{C.26}$$

If we take $C(v, v)$ we obtain

$$\sum_{i,j=1}^{n} v_i \text{Cov}(X_i, X_j) v_j = \text{Cov}\left(\sum_{i=1}^{n} v_i X_i, \sum_{j=1}^{n} v_j X_j\right),$$

which is the variance of $\sum_{i=1}^{n} v_i X_i$ and is therefore non-negative.

This says that the covariance matrix is a positive semi-definite matrix. Since it is symmetric, it can be diagonalized and has a complete basis of eigenvectors. If e_j is an eigenvector of eigenvalue λ_j then we have

$$0 \leq C(e_j, e_j) = e_j^T C e_j = \lambda_j e_j^T e_j = \lambda_j.$$

So all the eigenvectors are non-negative. One consequence of this is that the covariance matrix always has a square root, that is there exists a matrix A such that

$$C = A^2. \tag{C.27}$$

We can construct A via diagonalization. We can write $C = PDP^T$ where P is the matrix with columns equal to the eigenvectors of C and D is a diagonal matrix with elements equal to the eigenvalues (with the same ordering.) We then set

$$A = PD^{\frac{1}{2}} P^T,$$

where $D^{\frac{1}{2}}$ is found by taking the square root of the diagonal elements.

One important issue for us is to be able construct vectors of normal random variables with a given covariance matrix. The key to this is the fact that a sum of two independent normal variables is also normal with mean the sum of the means and variance the sum of the variances. We can write this as

$$N(\mu_1, \sigma_1^2) + N(\mu_2, \sigma_2^2) = N(\mu_1 + \mu_2, \sigma_1^2 + \sigma_2^2).$$

Thus if we take two independent normal variables X, Y with mean 0 and variance 1, and consider

$$Z = \rho X + \sqrt{1 - \rho^2} Y \tag{C.28}$$

we obtain a normal variable with mean 0 and variance 1. The covariance of X and Z is equal to ρ since X and Y are independent.

More generally, if we take a vector of independent $N(0, 1)$ random variables, X_j, we can form a vector of correlated random variables, by multiplying by a matrix A. Thus if we set

$$Y = AX, \tag{C.29}$$

the random variables Y_j are all normal variables and are correlated. In fact, we can compute

$$\begin{aligned} \text{Cov}(Y_i, Y_j) &= \text{Cov}\left(\sum_{k=1}^{n} a_{ik} X_k, \sum_{l=1}^{n} a_{il} X_l\right), \\ &= \sum_{k,l=1}^{n} a_{ik} a_{jl} \text{Cov}(X_k, X_l), \\ &= \sum_{k=1}^{n} a_{ik} a_{jk}. \end{aligned} \tag{C.30}$$

In matrix notation, we have that the covariance matrix of the vector Y is equal to AA^T.

This means that we can achieve any positive semi-definite covariance matrix, C, by taking a matrix A such that

$$C = AA^T. \tag{C.31}$$

The matrix A is then said to be a *pseudo-square* root of A. We saw above that a symmetric pseudo-square root always exists. There are in general many pseudo-square roots. We discuss pseudo-square rooting further in Section 9.4.

Appendix D

Hints and answers to exercises

Chapter 1

1. It would generally trade for less than 1/6.
2. The sum of the assets would trade for 1 and each asset would trade for 1/6. The risk is diversifiable.
3. They go down.
4. The corporate bond will be worth less.
5. The yield will go lower and the price higher.

Chapter 2

1. £1 is 120 × 1.4 yen.
2.
   (i) Zero and 100/110.
   (ii) 20 is upper and lower.
   (iii) Zero and 100/6.
   (iv) Zero and infinity.
3.
   (i) The sum is the price of a zero-coupon bond with the same expiry.
   (ii) The sum is less than the price of a zero-coupon bond.
   (iii) The sum is more than the price of a zero-coupon bond.
4. The forward price will go up. The value of a forward contract will increase also.
5. The bounds will be at least as wide since an arbitrage portfolio in a world with transaction costs will also be one in a world without transaction costs.
6. We have to use Theorem 2.10. The lower bound will be $-Z(T)$.

7. Construct a portfolio of these options which approximates taking the third derivative and then proceed as for proving call options are convex. The basic point is that the linear relation holds for the final time slice for each value of S and therefore will hold at previous times as well.

8. This follows from put-call parity and the fact that the forward with this strike has zero value.

9. The lower bounding portfolio must be below zero at infinity so must have zero or negative slope. That is, the number of stocks is non-positive. At zero the value of the stock is zero, as is the digital-call, so the number of bonds must be non-positive too. The most valuable lower bound portfolio is therefore obtained by letting both be zero.

10. Just go short the asset and use the money to buy bonds. When the asset drops in value, buy it back.

11. $A \leq \alpha Z + \beta S_0 + \gamma B$, where Z is the price of a zero-coupon bond with the same expiry.

12. A will be worth at least as much as B.

13. We have with the same hypotheses $KZ \geq P(K) \geq KZ - S$. We also have that $P(K)$ is an increasing function of K, is Lipshitz continuous and convex. We do not have that $P(K, T)$ is an increasing function of T.

14. The bounds will widen by Xe^{rT}.

15. To get S_{t_2} just buy the stock at time 0 at cost S_0. To get S_{t_1} at time t_2 we must buy $S_{t_1} e^{-r(t_2-t_1)}$ bonds at time t_1. We can achieve this by buying $e^{-r(t_2-t_1)}$ stocks today. The overall cost is therefore $S_{t_0} - S_{t_0} e^{-r(t_2-t_1)}$.

Chapter 3

1. They will both be worth 5.

2. 10, 20/3, 0.

3. It will be worth precisely 1.

4. It will be worth between 2.5 and 7.5. For the 110 option between zero and 5/3.

5. $\sum_{j \geq n/2}^{n} \binom{n}{j} (2j - n) 2^{-n}$.

6. The European is worth 13.06 and the American 13.38.

7. The down move has probability 2/3.

8. Let the probabilities be p_1 for 40, p_2 for 55 and p_3 for 70. We must have that

$$p_1 + p_2 + p_3 = 1$$

which determines p_3 from the other two. We also have that

$$4 = 6p_1 + 3p_2.$$

As probabilities must be between 0 and 1, we have that p_1 ranges from $1/6$ to $2/3$ and then the other probabilities are determined.

9. Replicate by holding α of A and β of B. Solve the system

$$110\alpha + 120\beta = 10, \qquad \text{(D.1)}$$
$$90\alpha + 80\beta = 0, \qquad \text{(D.2)}$$

to find α and β. The value is then $100(\alpha + \beta)$.

10. 6.805.

11. This follows from a backwards induction. In the final layer the values are the same. In each previous layer, we assume the result has already been proven for the next layer. At each node the discounted expectation of the next layer must then be at least as much for the American as for the European. Taking the max with the intrinsic value can only increase the value. So at each node the value is at least as much. The result now follows by inducting back to the initial point.

12. Use the same argument as the previous question.

13. This can be done either by a power series expansion or by direct integration. For direct integration we have

$$\frac{1}{\sqrt{2\pi}} \int_{-\infty}^{\infty} e^{-\frac{x^2}{2} + \sigma x} dx.$$

Write

$$\frac{x^2}{2} - \sigma x = \left(\frac{x}{\sqrt{2}} - \frac{\sigma}{\sqrt{2}} \right)^2 - \frac{1}{2}\sigma^2,$$

and put $y = x - \sigma$, then the answer is clear.

Chapter 4

1. For a single option the ratio of Vega to Gamma is a non-zero function depending only on spot and expiry. This remains true for a portfolio of vanilla options with the same expiry so one is 0 if and only if the other is. It is not true for multiple expiries as the function depends on expiry.

2. The Vega will increase with time for reasonable lengths of time but for very large expiries the Vega will fall away.

3. The Greek of a digital-call will be minus the Greek of the digital-put except for ρ and θ. To see this differentiate the fact that the sum is a zero-coupon bond. We can get a slightly more complicated relationship this way for the ρ and θ.

4. It goes down.

5. The concavity of the portfolio's value means that it will decrease in value as the maximum will be at today's value of spot.

6. Our approximation for the price is $0.4 S_0 \sigma \sqrt{T}$, so we just differentiate this.

7. Differentiate put-call parity.

Chapter 5

1.
$$dY_t = \left(k\mu + \frac{k(k-1)}{2}\sigma^2\right) Y_t dt + k\sigma Y_t dW_t.$$

2. The Brownian part of $d(f(X_t))$ will have coefficient $f'(X_t)\sigma(X_t)$. The solution of $f'(x) = V\sigma(x)^{-1}$ will do. The value of f can be changed by a linear multiple or adding a constant.

3.
$$dF_t = (\mu - r)F_t dt + \sigma F_t dW_t.$$

4. S_t^{-1} satisfies
$$\frac{dy_t}{y_t} = (-\mu + \sigma^2)dt - \sigma dW_t.$$

5. It will revert to the level μ. Increasing α will increase the reversion rate. The price of a call option is not affected by any drift terms.

6. It follows by direct differentiation. It must be a solution since it is the value of a derivative that pays S_T at some time T in the future.

7. It follows by direct differentiation. It must be a solution since it is the value of a derivative paying Ae^{rT} at time T in the future.

8. We have from our rational bounds theorem that the value of call option must lie between 0 and S so it must hold true for the Black–Scholes value.

9. Let derivative D pay $g - f$ at time T. It must then be of non-negative value at all previous times since it is non-negative at expiry. Its value at previous times is just $g - f$.

10. The SDE for the difference is

$$d(X_t^{(1)} - X_t^{(0)}) = (\mu_2 - \mu_1),$$

which is an ODE with solution $(\mu_2 - \mu_1)t$.

11. Replace σ by σ/S in the Black–Scholes equation.

12. It will satisfy the Black–Scholes equation. The easy way to compute the solution is to use the techniques of the next chapter. The hard way is to crank it through but it can be done.

Chapter 6

1. Write F_t in terms of S_t and compute. $dF_t = \sigma F_t dW_t$.
2. (i), (iii), (v), (vi) and (vii).
3. The first two.
4.
$$dS = rSdt + \sigma dW.$$

To develop the option prices, it is easier to work with the forward price for a forward contract with the same expiry. The forward price follows the process

$$dF = \sigma e^{r(T-t)} dW$$

and an option on it will pay the same as the option on the underlying since they will agree at expiry time. The value of an option with payoff g is then

$$\mathbb{E}(g(F))e^{-rT}.$$

This is easily computed since F is distributed as $S_0 e^{rT} + \bar{\sigma}\sqrt{T} N(0, 1)$, where $\bar{\sigma}$ is the root mean square vol.

5. The risk-neutral drift is decreased so the price of a call option will decrease.
6. The value is

$$e^{-rT}\mathbb{E}\big((S_t^2 - K)_+\big)$$

where

$$S_T = S_0 e^{(r-0.5\sigma^2)T + \sigma\sqrt{T}N(0,1)}.$$

7. Our hedge will be wrong and our final portfolio will have non-zero variance.
8. The value is equal to

$$e^{-rT}\mathbb{E}(\log S_0 + (r - 0.5\sigma^2)T + \sigma\sqrt{T}N(0, 1)) = e^{-rT}(\log S_0 + (r - 0.5\sigma^2)T).$$

9. It's the same as the Black–Scholes formula except that in d_j we replace the strike by the trigger level. It's most easily seen by dividing into two pieces with different numeraires, as for the Black–Scholes formula.

Chapter 7

1. If X is our discretized variable and our step is dt long, then we need

$$\mathbb{E}(X) = 0,$$
$$\mathbb{E}(X^2) = dt$$
$$\mathbb{E}(X^3) = 0$$
$$\mathbb{E}(X^4) = 3dt^2.$$

Since Brownian motion is symmetric, we guess that X is too. We assume it takes values $\alpha, 0$ or $-\alpha$. Let the probability of an up-move be p; then to get the expectation correct we must have the probability of a down-move is p too. The third moment is now correct. We are left with two equations in two unknowns. We solve to get $p = 1/6$ and $\alpha = \sqrt{3}$.

2. Since the integral is symmetric about the midpoint for each step, we have

$$0.5(x_j - x_{j-1})(f(x_j) + f(x_{j-1})),$$

so if we have $n+1$ points evenly spaced with $x_j - x_{j-1}$ we can rewrite the integral as

$$\delta x \left(0.5 f(x_0) + 0.5 f(x_n) + \sum_{j=1}^{n} f(x_j) \right).$$

3. Take $\tan(\pi(U - 0.5))$.

4. We have to compute

$$\frac{1}{2N} \left(\sum_{j=1}^{N} s_j^k + \sum_{j=1}^{N} (-s_j)^k \right)$$

which is equal to zero for k odd.

5. A put and a call struck at K.

6. A put struck at 95 minus 5 zero-coupon bonds.

7. The price of the digital-call with downwards-sloping smile will be worth more.

8. This is just the linearity of differentiation.

9. Take a Taylor expansion and note that in the symmetric case the first term cancels.

Chapter 8

1. If the value of the sum is higher do the vanilla minus the sum. Otherwise do the negative.

2. If interest-rates are zero, time-dependence of volatility is irrelevant. Otherwise the time-dependence matters since the stock will drift in the risk-neutral measure and whether the volatility occurs before or after the drifting will make a difference. Forward-rates are always driftless so we are back in the zero-interest-rate case.

3. If L is the level, and τ is the passage time, the crucial observation is that

$$\mathbb{P}(\tau \leq t) = \mathbb{P}(M_t \geq L)$$

so differentiate the right-hand side to get the density.

4. The stock will wobble about more so the probability of breaching the barrier is increased.

5. It will increase it. (Unless it has already breached the barrier in which case there is, of course, no effect.)

6. The graph will be similar to the Black–Scholes price of a put option but will have lower value for spot above the barrier. For spot below the barrier, it will be an ordinary put option.

7. Identically zero!

8. Just differentiate.

9. We have to compute

$$e^{-rT}\mathbb{P}(m_T^s \geq K).$$

We just plug the parameters of the risk-neutral process into Corollary 8.1.

10. The American put will cost more than the European digital put with the same strike and expiry.

11. The asset is driftless so when the American pays off there is precisely a 50% chance (in the risk-neutral measure) that the European option will pay off too so the European is worth half as much as the American.

Chapter 9

1. Vanilla greater than discrete greater than continuous.

2. In general, it will be lower since the averaging process reduces overall variance in the risk-neutral measure.

3. By PDE, use running maximum as an auxiliary variable and then proceed as for an Asian. For Monte Carlo, proceed as for Asian but take the maximum along each path.

4.
$$\mathbb{E}((X+a)^k) = \sum_{j=0}^{k} \binom{k}{j} a^{k-j} \mathbb{E}(X^j).$$

5. We take as auxiliary variable the value 1 if the barrier has not been breached and 0 if it has. The updating rule is then make it 0 if the barrier is breached at the new time, otherwise leave it alone. The final payoff is the auxiliary times the option payoff.

6. For auxiliary variable, take the running geometric mean. For Monte Carlo, compute the geometric mean instead of the ordinary one at the end of the path. The key to developing the formula is to observe that a geometric mean of log-normal random variables is also log-normal. The geometric option is worth less by the inequality of arithmetic and geometric means so for every path the pay-off of the geometric option is lower.

7. For the PDE method, we need two auxiliary variables, the running min and the running max. For Monte Carlo just compute the min and the max at the end of each path, and take the difference.

8. The implied rate is
$$\frac{\log P(t_j) - \log P(t_{j-1})}{t_{j+1} - t_j}$$
which is positive if $P(t_j) > P(t_{j+1})$.

Chapter 10

1. We can write the range accrual as a sum of double digitals with pay-off times moved to the end of accrual time. Each double digital corresponds to one day of accrual. We can replicate the double digitals in the usual way. We have to assume deterministic interest rates so that the change in discounting is known.

2. Sections 10.2, 10.5, 10.6

3. $n^2 N$, Nn

4. The first relation follows from the fact that all the parameters are time-independent so it is only residual time that matters. The second is by direct observation or can be proven using the fact that the process for the log is constant coefficient. With time-dependent volatility the second relation still holds but the first does not.

5. This can be done taking time-dependent volatility which is rapidly decaying. Or more extremely, take the F that is implied by having zero volatility after one year. Then one- and two-year options have the same value initially if $r = 0$, but after we

get to one year the two-year option is now a one-year option and is worth more than the one-year option which is at its pay-off time. So the portfolio of two-year option minus one-year option is an arbitrage portfolio.

Chapter 11

1. Price in pounds and then convert the premium into dollars at today's exchange rate. This will work because we can replicate the option by converting the premium into pounds and then carrying out the usual Black–Scholes replication argument.

2. To price the option with strike in dollars, we work out the process for the dollar price of the stock which will be affected by the correlation between exchange rate and the stock price. Once we know the process for the dollar price we can proceed in the standard fashion.

3. We must have that $(\mu_j - r)/\sigma_j$ is independent of j.

Chapter 12

1. Any exercise strategy for B is also a strategy for A so the price of A is a maximum over a larger set and therefore must be at least as large. The American and European calls on a non-dividend-paying stock is an example where they have the same value.

2. It will be more valuable than a forward but less so than the American call. The obvious way to price this is to use a tree. In the final layer, we take the value of the forward contract. In the preceding layers, we take the max of the discounted unexercised value and the exercised value until we get back to time t_1 and before that we do not do anything except discount.

3. No it does not. For example, if we have zero dividend rate but positive interest rate, the American put is worth more than the European put but the two calls are worth the same. This means that put-call parity cannot hold.

4. This can be done by backwards induction. In the final layer, they are equal. In preceding layers, each node is the max of the exercised value and the discounted values in the next layer, which are assumed to be at least as big, so the value at the node must be more.

5. We just take t_j so that $\int_{t_j}^{t_{j+1}} \sigma^2(t)dt$ is independent of j.

6. We can do this via PDE methods using an auxiliary variable.

7. Its value will be the stock price in both cases. In the first case, use the rational bounds theorem. In the second use the Black–Scholes formula.

8. C carries more rights than A since we can exercise the two options separately. But it will have the same value since the two pieces will have the same optimal exercise time.

9. Zero. If it is not worth zero after six months the person who it is against will cancel it.

Chapter 13

1. 4.88%
2. 7.84%
3. 3.88%
4. Let f run from t_0 to t_1. Take $P(t_1)$ as numeraire. The value is then

$$\mathbb{E}((f - K)(1 + f\tau)\tau) P(t_1),$$

with f log-normal and driftless. The expectation can be evaluated since it is a quadratic in f and f is log-normal.

5. Proceed by induction. We have $P(t_0) = 0$, and

$$P(t_1) = (1 + X_1(t_1 - t_0))^{-1}.$$

If we know P_{t_i} for $i < j$, we have

$$X_j = \frac{P_0 - P_j}{\sum_{i \leq j} P_i}.$$

All terms in this equation except P_j are known so we can solve for P_j.

6. This is similar to the previous question except that we work backwards from the end and initially find the ratios of the discount factors to the value of the final discount factor.

7. Compute the volatility of the swap-rate; it will be of the form

$$\sqrt{\sum_{i,j} \sigma_i \sigma_j f_i f_j \rho_{ij} \frac{\partial \mathrm{SR}}{\partial f_i} \frac{\partial \mathrm{SR}}{\partial f_i}},$$

with ρ_{ij} the instantaneous correlations between forward rates and σ_i the volatility of f_i. This is clearly not of the form swap-rate times a constant.

8. In the Black formula, we replace the strike inside the $N(d_j)$ terms by the trigger level.

Chapter 14

1. The key to this problem is to observe that f has two different roles. One is its part in the volatility term σf which must be replaced by $\sigma(f+\alpha)$ and the other is its role in the discounting, which is unchanged. So the displaced diffusion drift is the original drift with all σf terms replaced by $\sigma(f+\alpha)$ terms.

2. Express the swap-rate formula in terms of forward rates, and differentiate.

3. In the drift computation, we will have replaced each occurence of σf by σ. We evolve the rates instead of their log. Basically not much is different except that we must develop and define everything for the rate instead for its log.

4. See [52].

5. The inverse floater can be decomposed as a sum of floorlets and FRAs so we can price by Black. If we price by BGM, we need only one evolution time since it is only the terminal values of rates that matter (but if we evolve over too long we may get a drift error.)

6. This is very similar to taking the stock as numeraire. We get that f has drift σ^2. We have that fP is a mulitple of the difference of the bonds at start and end of the FRA, so fP is a tradable and can be taken to be the numeraire.

7. Just add α to trigger, strike and initial rate.

8. We need to compute

$$0.5 \int_0^1 (0.1 + e^{-2t})(0.2 + e^{-3t}) dt$$

$$= 0.5 \int_0^1 (0.02 + 0.2 e^{-2t} + 0.1 e^{-3t} + e^{-5t}) dt.$$

This is easily evaluated to give

$$0.5 \left(0.02 + 0.1(1 - e^{-2}) + 0.1 \frac{1 - e^{-3}}{3} + \frac{1 - e^{-5}}{5} \right).$$

Chapter 15

1. If we condition on no jumps then the expected pay-off is the same. If there are jumps, the expectation is lower since the asset has jumped down so the overall expectation must be lower. However, the price of the second call option will be higher – the difference is that the risk-neutral drift will compensate in the second case.

2. Without loss of generality, take S and T to have the same initial value. (Otherwise take a multiple of S.) Then consider the portfolio $S - T$.

3. The first variation is infinite. The second variation is equal to 1.25.

4. Just differentiate the call formula with respect to strike.

5. Twice differentiate the call formula with respect to strike.

6. The intensity by its average. The vol by the rms value. The others cannot. For example, if we average jump-means, we may end up with no jumps at all.

7.
$$(\lambda K - \lambda S)_+ = \lambda (K - S)_+.$$

8. Multiplying spot and strike by λ does not change the sign of $S - K$ so will not affect the value of $H(S - K)$.

9. We just have to adjust the risk-neutral drift down by d and everywhere it appears we must reduce it by d. So for the log-normal jumps case we reduce the r_n by d.

10. We need $2N$ as we can bundle all the jumps together with the diffusive moves. If the jumps are not log-normal but we know how to bundle them with each other, then we need $3N$. If we do not know how to bundle the jumps then we need a potentially infinite number as there is upper bound on the number of jumps per interval. In practice, we would neglect the possibility of more than, say, ten jumps occurring in a short time.

11. At time t_1, all the jumps are over. We are therefore in a Black–Scholes world and for each value of S_1 there is a unique price which can be replicated in the standard way. In fact, since the option payoff is homogeneous of order zero, this price will be independent of S_1. Let this price be C. We can hedge the option by buying $e^{-rt_1}C$ bonds at time 0, doing nothing until time t_1 and then carrying out the Black–Scholes hedging strategy. The value at time t_0 is therefore $e^{-rt_1}C$. If the rms vol of S from t_1 to t_2 is σ the value of $e^{-rt_1}C$ will be
$$e^{-rt_2}\mathbb{E}\left(e^{(r-0.5\sigma^2)\tau + \sigma\sqrt{\tau}N(0,1)}\right),$$
where $\tau = t_2 - t_1$.

Chapter 16

1. Just differentiate the price with respect to strike twice to get the expression for the density. To estimate the density by Monte Carlo we could short step the vol, integrate it, draw the final value of spot, and then store all the final values of spot. Dividing the spot values into bins and finding the fraction in each bin would then give an estimate of the density.

2. Most will not work. Strong static replication of digitals does work. The up and in put with barrier at strike method works. Put-call symmetry works when there are zero interest rates and the vol is uncorrelated with spot.

3. The $rS\frac{\partial O}{\partial S}$ term will change to $(r-d)S\frac{\partial O}{\partial S}$. See this by using delivery contracts just as in the deterministic vol case.

4. We will be imperfectly hedged and our final portfolio will have final variance which is non-zero. However, the variance will generally be quite small.

Chapter 17

1. All the techniques that do not depend on the stock price being continuous will work.

2. $2N$.

References

[1] C. Alexander, Principal component analysis of implied volatility and skews, ISMA Centre Discussion Paper in Finance, 2000–10.
[2] C. Alexander, *Market Models: a Guide to Financial Data Analysis*, Wiley, 2001.
[3] L. Anderson, A simple approach to the pricing of Bermudan swaptions in the multi-factor LIBOR market model, *Journal of Computational Finance* **3**(2), Winter 1999/2000, 5–32.
[4] L. Andersen, J. Andreasen, Volatility skews and extensions of the LIBOR market model, *Mathematical Finance* **7**, 2000, 1–32.
[5] L. Andersen, J. Andreasen, Jumping smiles, *Risk* **12**, November 1999, 65–8.
[6] L. Andersen, J. Andreasen, Jump diffusion processes: volatility smile fitting and numerical methods for pricing, Gen Re working paper, 1999.
[7] L. Andersen, J. Andreasen, Factor dependence of Bermudan swaptions: fact or fiction, Gen Re working paper, October 2000.
[8] L. Andersen, M. Broadie, A primal-dual simulation algorithm for pricing multi-dimensional American options, preprint 2001.
[9] G. Baker, G.D. Smith, *The New Financial Capitalists*, Cambridge University Press, 1998.
[10] G. Bakshi, C. Cao, Z. Chen, Empirical performance of alternative option pricing models, *Journal of Finance* **52**(5), 1997, 2003–49.
[11] G. Bakshi, C. Cao, Z. Chen, Do call prices and the underlying stock always move in the same direction *Review of Financial Studies* **13**(3), 2000, 549–84.
[12] M. Baxter, A. Rennie, *Financial Calculus*, Cambridge University Press, 1999.
[13] N. Bellamy, M. Jeanblanc, Incompleteness of markets driven by a mixed diffusion, *Finance and Stochastics* **4**(2), February 2000, 209–22.
[14] E. Benhamou, A. Duguet, Volatility and model risk for discrete Asian options, preprint 2001.
[15] Y.Z. Bergman, Pricing path contingent claims, *Research in Finance* **5**, 1985, 229–41.
[16] P.L. Bernstein, *Against the Gods: the Remarkable Story of Risk*, Wiley, 1998.
[17] T. Björk, *Arbitrage Theory in Continuous Time*, Oxford University Press, 1998.
[18] F. Black, The pricing of commodity contracts, *Journal of Financial Economics* **3** 1976, 167–79.
[19] F. Black, M. Scholes, The pricing of options and corporate liabilities, *Journal of Political Economy* **81**, 1973, 637–54.

[20] A. Brace, D. Gatarek, M. Musiela, The market model of interest-rate dynamics, *Mathematical Finance* **7**, 1997, 127–155.

[21] D. Breeden, R. Litzenberger, Prices of state-contingent claims implicit in option prices, *Journal of Business* **51**, 1978, 621–51.

[22] D. Brigo, F. Mercurio, *Interest Rate Models – Theory and Practice*, Springer Verlag, 2001.

[23] M. Britten-Jones, A. Neuberger, Option prices, implied price processes and stochastic volatility, *Journal of Finance* **55**(2), April 2000, 839–66.

[24] M. Broadie, P. Glasserman, Estimating security derivative prices by simulation, *Management Science* **42**, 1996, 269–85.

[25] M. Broadie, P. Glasserman, Pricing American-style securities using simulation, *Journal of Economic Dynamics and Control* **21**(8/9), 1997, 1323–52.

[26] M. Broadie, P. Glasserman, A stochastic mesh method for pricing high-dimensional American securities, working paper, Columbia University, 1997.

[27] O. Brockhaus, M. Farkas, A. Ferraris, D. Long, M. Overhaus, *Equity Derivatives and Market Risk Models*, Risk Books, 2000.

[28] H. Brown, D. Hobson, L.C.G. Rogers, Robust hedging of barrier options, *Mathematical Finance* **11**, 2001, 285–314.

[29] J.Y. Campbell, A.W. Lo, A.C. MacKinlay, *The Econometrics of Financial Markets*, Princeton University Press, 1997.

[30] P. Carr, K. Ellis, V. Gupta, Static hedging of exotic options, *Journal of Finance* **53**, 1998, 1165–91.

[31] T. Chan, Pricing contingent claims on stocks driven by Levy processes, *Annals of Applied Probability* **9**(2), 1999, 504–28.

[32] W. Cheney, D. Kincaid, *Numerical Analysis: Mathematics of Scientific Computing*, Brooks/Cole, 2001.

[33] L. Clewlow, C. Strickland, *Implementing Derivatives Models*, Wiley, 1998.

[34] L. Clewlow, C. Strickland, *Exotic Options: the State of the Art*, Thompson International Press, 1997.

[35] J.H. Cochrane, *Asset Pricing*, Princeton University Press, 2001.

[36] J.C. Cox, S. Ross, M. Rubinstein, Option Pricing: a simplified approach, *Journal of Financial Economics* **7**, 1979, 229–63.

[37] M. Curran, Beyond average intelligence, *Risk* **5**, 1992, 60.

[38] M. Curran, Valuing Asian and portfolio options by conditioning on the geometric mean price, *Management Science* **40**, 1994, 1705–11.

[39] F. Delbaen, W. Schachermayer, A general version of the fundamental theorem of asset pricing, *Mathematische Annalen* **300**, 1997, 463–520.

[40] E. Derman, Regimes of volatility: some observations on the variation of S&P implied volatilities, Goldman Sachs Quantitative Strategy Research Note, January 1999.

[41] E. Derman, I. Kani, Riding on a smile, *Risk* **7**, 1994, 32–9.

[42] D. Duffie, *Dynamic Asset Pricing Theory*, third edition, Princeton University Press, 2002.

[43] D. Duffie, J. Pan, K. Singleton, Transform analysis and asset pricing for affine jump-diffusions, *Econometrica* **68**, 2000, 1343–76.

[44] B. Dupire, Pricing with a smile, *Risk* **7**, 1994, 18–20.

[45] B. Dupire, *Monte Carlo: Methodologies and Applications for Pricing and Risk Management*, Risk Books, 1998.

[46] E. Fournie, J.-M. Lasry, J. Lebuchoux, P.-L. Lions, N. Touzi, Application of Malliavin calculus to Monte Carlo methods in finance, *Finance and Stochastics* **3**, 1999, 391–412.

[47] J.-P. Fouque, G. Papanicolaou, K.R. Sircar, *Derivatives in Financial Markets with Stochastic Volatility*, Cambridge University Press, 2000.

[48] M.C. Fu, S.B. Laprise, D.B. Madan, Y. Su, R. Wu, Pricing American options: a comparison of Monte Carlo simulation approaches, *Journal of Computational Finance* **4**(3), 2001, 39–88.

[49] J.K. Galbraith, *The Great Crash*, Houghton Mifflin, 1997.

[50] P. Glasserman, S.G. Kou, The term structure of simple forward rates with jump risk, working paper, Columbia University, 1999.

[51] P. Glasserman, N. Merener, Cap and swaption approximations in LIBOR market models with jumps, preprint, Columbia University, 2001.

[52] P. Glasserman, X. Zhao, Arbitrage-free discretization of lognormal interest rate models, *Finance and Stochastics* **4**, 2000, 35–69.

[53] G. Grimmett, D. Stirzaker, *Probability and Random Processes*, Second edition, Oxford University Press, 1992.

[54] H. Geman, M. Yor, Bessel processes, Asian options and perpetuities, *Mathematical Finance* **3**, 1993, 349–75.

[55] E. Haug, *The Complete Guide to Option Pricing Formulas*, Irwin, 1997.

[56] M.B. Haugh, L. Kogan, Pricing American Options: A Duality Approach, forthcoming in *Operations Research*.

[57] J.M. Harrison, D.M. Kreps, Martingales and arbitrage in multi-period securities markets, *Journal of Economic Theory* **20**, 1979, 381–408.

[58] J.M. Harrison, S.R. Pliska, Martingales and stochastic integration in the theory of continuous trading, *Stochastic Processes and Applications* **11**, 1981, 215–60.

[59] J.M. Harrison, S.R. Pliska, Martingales and stochastic integration in the theory of continuous trading, *Stochastic Processes and Applications* **13**, 1983, 313–16.

[60] D. Heath, R. Jarrow, A. Morton, Bond pricing and the term structure of interest rates: a new methodology for contingent claims valuation, *Econometrica* **60**, 1992, 77–105.

[61] S. Heston, A closed-form solution for options with stochastic volatility with applications to bond and currency options, *Review of Financial Studies* **6**(2), 1993, 327–43.

[62] S. Hodges, A. Neuberger, Rational bounds for exotic options, preprint, 1998.

[63] J. Hoogland, D. Neumann, Local scale invariance and contingent claim pricing, *International Journal of Theoretical and Applied Finance* **4**(1), 2001, 1–21.

[64] J. Hull, A. White, The pricing of options on assets with stochastic volatilities, *Journal of Finance* **42**(2), 1987, 281–300.

[65] C. Hunter, P. Jäckel, M. Joshi, Getting the drift, *Risk*, July 2001.

[66] I.E. Ingersoll, *Theory of Financial Decision Making*, Rowman and Littlefield, 1987.

[67] F. Jamshidian, LIBOR and swap market models and measures, *Finance and Stochastics* **1**, 1997, 293–330.

[68] P. Jäckel, *Monte Carlo Methods in Finance*, Wiley, 2002.

References

[69] P. Jäckel, Using a non-recombining tree to design a new pricing method for Bermudan swaptions, QUARC Royal Bank of Scotland working paper, 2000.

[70] P. Jäckel, R. Rebonato, Valuing American options in the presence of user-defined smiles and time-dependent volatility: scenario analysis, model stress and lower-bond pricing applications, *Journal of Risk* **4**(1), 2001, 35–61.

[71] P. Jäckel, R. Rebonato, Accurate and optimal calibration to co-terminal European swaptions in a FRA-based BGM framework, QUARC Royal Bank of Scotland working paper, 2000.

[72] T.C. Johnson, Volatility, momentum and time-varying skewness in foreign exchange returns, preprint 2001.

[73] C.S. Jones, The dynamics of stochastic volatility, preprint 2000.

[74] M. Joshi, Pricing path-dependent exotic options using replication methods, QUARC Royal Bank of Scotland working paper, 2001.

[75] M. Joshi, Log-type models, homogeneity of options prices and convexity, QUARC Royal Bank of Scotland working paper, 2001.

[76] M. Joshi, R. Rebonato, A stochastic-volatility displaced-diffusion extension of the LIBOR market model, QUARC Royal Bank of Scotland working paper, 2001.

[77] M. Joshi, J. Theis, Bounding Bermudan swaptions in a swap-rate market model, *Quantitative Finance* **2**, 2002, 370–7.

[78] I. Karatzas, E. Shreve, *Brownian Motion and Stochastic Calculus*, second edition, Springer Verlag, 1997.

[79] I. Karatzas, E. Shreve, *Methods of Mathematical Finance*, Springer Verlag, 1998.

[80] S.G. Kou, A jump-diffusion model for option pricing with three properties: leptokurtic feature, volatility smile and analytical tractability, contributed paper to the Econometric Society World Congress 2000.

[81] D. Lamberton, B. Lapeyre, *Introduction to Stochastic Calculus Applied to Finance*, CRC Press, 1996.

[82] H.E. Leland, Option pricing and replication with transaction costs, *Journal of Finance* **40**, 1985, 1283–301.

[83] A.L. Lewis, *Option Valuation under Stochastic Volatility*, Finance Press, 2000.

[84] A.L. Lewis, A simple option formula for general jump-diffusion and other exponential Levy processes, preprint www.optioncity.net, 2001.

[85] Y. Li, A new algorithm for constructing implied binomial trees: does the implied model fit any volatility smile? *Journal of Computational Finance* **4**(2), 2000, 69–95.

[86] F. Longstaff, E. Santa-Clara, E. Schwartz, Throwing away a billion dollars: the cost of suboptimal exercise in the swaptions market, UCLA working paper, 2000.

[87] R. Lowenstein, *Buffett: the Making of an American Capitalist*, Orion, 1995.

[88] D. Madan, E. Seneta, The Variance Gamma model for share market returns, *Journal of Business* **63**, 1990, 511–24.

[89] D. Madan, F. Milne, Option pricing with V.G. martingale components, *Mathematical Finance* **1**(4), 1991, 39–55.

[90] D. Madan, P. Carr, E.C. Chang, The Variance Gamma process and option pricing, *European Finance Review* **2** (1), 1998, 79–105.

[91] W. Margrabe, The value of an option to exchange one asset for another, *Journal of Finance* **7**, 1978, 77–91.

[92] D. Marris, Financial option pricing and skewed volatility, M. Phil. thesis, Statistical Laboratory, University of Cambridge, 1999.
[93] B. Moro, The full monte, *Risk* **8**(2), 1995, 53–7.
[94] R. Merton, *Continuous-Time Finance*, Blackwell, 1998.
[95] R. Merton, Option pricing when underlying stock returns are discontinuous, *Journal of Financial Economics* **3**, 1976, 125–44.
[96] M. Musiela, M. Rutkowski, *Martingale Methods in Financial Modelling*, Springer Verlag, 1997.
[97] M. Musiela, M. Rutkowski, Continuous-time term structure models: forward measure approach, *Finance and Stochastics* **1**, 1997, 261–91.
[98] L.A. McCarthy, N.J. Webber, Pricing in three-factor models using icosahedral lattices, *J. Computational Finance* **5**(2), 2001/2, 1–37.
[99] S. Nielsen, J. Overgaard Olesen, Regime-switching stock returns and mean reversion, Copenhagen Business School working paper 11–2000.
[100] B. Oksendal, *Stochastic Differential Equations: an Introduction with Applications*, Springer Verlag, 1998.
[101] F. Partnoy, *F.I.A.S.C.O.*, Profile Books, 1998.
[102] A. Pelsser, *Efficient Methods for Valuing Interest Rate Derivatives*, Springer Verlag, 2001.
[103] H. Pham, Optimal stopping, free boundary and American option in a jump diffusion model, *Applied Mathematics and Optimization* **35**, 1997, 145–64.
[104] W.H. Press, S.A. Teutolsky, W.T. Vetterling, B.P. Flannery, *Numerical Recipes in C++*, Cambridge University Press, 2002.
[105] R. Rebonato, *Interest Rate Option Models*, Wiley, 1998.
[106] R. Rebonato, *Volatility and Correlation in the Pricing of Equity, FX and Interest-Rate Options*, Wiley, 1999.
[107] R. Rebonato, *The Modern Pricing of Interest Rate Derivatives*, Princeton University Press, 2002.
[108] R. Rebonato, On the Pricing Implications of the joint lognormal assumption for the swaption and cap market, *Journal of Computational Finance* **3**, 1999, 57–76.
[109] E. Reiner, Understanding skew and smile behaviour in the context of jump processes and applying these results to the pricing and hedging of exotic options, Global Derivatives Conference 1998..
[110] L.C.G, Rogers, Monte Carlo valuation of American options, preprint, University of Bath, 2001.
[111] L.C.G. Rogers, Z. Shi, The value of an Asian option, *Journal of Applied Probability* **32**(4), 1995, 1077–88.
[112] W. Rudin, *Principles of Mathematical Analysis*, McGraw Hill, 1976.
[113] P.J. Schönbucher, A market model for stochastic implied volatility, *Philosophical Transactions of the Royal Society* **A 357**(1758), 1999, 2071–92.
[114] H.M. Soner, S.E. Shreve, J. Cvitanic, There is no non-trivial hedging portfolio for option pricing with transaction costs. *Annals of Applied Probability* **5**, 1995, 327–55.
[115] J.A. Tilley, Valuing American options in a path simulation model, *Transactions of the Society of Actuaries* **45**, 1993, 83–104.

[116] S.M. Turnbull, L.N. Wakeman, A quick algorithm for pricing European average options, *J. Financial and Quantitative Analysis* **26**, 377–389, 1991.
[117] J. Walmsley, *New Financial Instruments*, Wiley, 1998.
[118] P. Wilmott, *Derivatives: the Theory and Practice of Financial Engineering*, Wiley, 1999.
[119] P. Wilmott, S. Howison, J. Dewynne, *The Mathematics of Financial Derivatives*, Cambridge University Press, 1995.
[120] C. Zhou, Path-dependent option valuation when the underlying path is discontinuous, working paper, Federal Reserve Board, 1997.
[121] C. Zuhlsdorff, Extended Libor market models with affine and quadratic volatility, Department of Statistics, University of Bonn, 2000.
[122] P.L. Zweig, *Walter Wriston, Citibank, and the Rise and Fall of American Financial Supremacy*, Crown, 1995.

Index

σ-field 438
N 61
$N(\mu, \sigma^2)$ 90
$N(0, 1)$ 54
$\mathcal{O}$ 59

accreting notional 409
admissible exercise strategy *see* exercise strategy, admissible
almost 239
almost surely 91
American 9
American option *see* option, American
amortizing notional 409
annualized rates 281
annuity 287
anti-thetic sampling 178
arbitrage 18–19, 26–28, 409
  and bounding option prices 28–38
arbitrage-free price 42, 43
arbitrageur 12–13, 17
Arrow–Debreu security 142
at-the-forward 30
at-the-money 29, 62
auto cap 409

bank 12
barrier option *see* option, barrier
basis point 409
basket option 243
Bermudan option *see* option, Bermudan
Bermudan swaption *see* swaption, Bermudan
BGM 409
  implementation of 430–433
  J 409
  J model *see* BGM model
BGM model 301–334
  automatic calibration to co-terminal swaptions 321
  long steps 317
  running a simulation 316–321
bid-offer spread 20

Black formula 289–290
  approximate linearity 335
  approximation for swaption pricing under BGM model 321
Black–Scholes density 173
Black–Scholes equation 66, 150, 151
  for options on dividend-paying assets 112
  higher-dimensional 252–253
  informal derivation of 104–105
  rigorous derivation 105–108
  solution of 109–110
  with time-dependent parameters 154
Black–Scholes formula *see* option, call, Black–Scholes formula for
Black–Scholes model 71, 73, 103, 410
Black–Scholes price 18
bond 4–6, 410
  callable 280
  convertible 7, 410
  corporate 7
  government 1
  premium 2
  riskless 5, 7
  zero-coupon 5, 23–25, 27, 281, 413
Brownian bridge 214
Brownian motion 89–92, 97, 132, 242, 410
  correlated 245
  higher-dimensional 243–245
Buffett, Warren 2
bushy tree *see* tree, non-recombining

calibration, to vanilla options using jump-diffusion 357
call 280
call option *see* option, call
callable bond *see* bond, callable
cap 288, 410
caplet 288–290, 410
  strike of 288
caption 306, 410
cash bond 25, 410
Central Limit theorem 53, 57, 60, 259, 443

central method 222
CEV *see* constant elasticity of variance
chain rule 96
   for stochastic calculus 99
characteristic function 387
Cholesky decomposition 212
cliquet 404, 410
   call 404
   optional 405
   put 404
CMS *see* swap, constant maturity
co-initial 296, 320
commodities 112
complete market 142, 410
compound optionality 406
conditional probability 440
consol 410
constant elasticity of variance 103
constant elasticity of variance process 334
constant maturity swap *see* swap, constant maturity
contingent claim 142, 410
continuously compounding rate 24, 25
control variate
   and pricing of Bermudan swaptions 330
   on a tree 267
convenience yield 113
convexity 33–36, 78
   as a function of spot price in a log-type model 363
correlation 446
   between forward rates 300, 315
correlation matrix 249, 446
cost of carry 112
co-terminal 296, 320
coupon 4, 280, 410
covariance 446
covariance matrix 446
   and implementing BGM 322
crash 10, 82
credit default swap 22
credit rating 295, 410
cumulative distribution function 441
cumulative normal function 61, 415, 417

default 1
Delta 73, 77, 410
   and static replication 228, 230
   Black–Scholes formula for call option 77
   integral expression for 175
Delta hedging *see* hedging, Delta
dependent 441
derivative 10, 410
   credit 10
   weather 11
Derman–Kani implied tree 361
deterministic future smile 226, 405
digital 410
digital option *see* option, digital
dimensionality 208, 418
dimensionality reduction 213
discount curve 411
discretely compounding money market account 303

displaced diffusion model 334
distribution, log-normal *see* log-normal distribution
diversifiable risk 411
diversification 8
dividend 7, 411
   scrip 24
dividend rate 24
dividends, and the Black–Scholes equation 111–113
drift 57, 101
   of a forward rate under BGM 309
   real-world 61
Dupire model 361
dynamic replication *see* replication, dynamic

early exercise 65
equivalent martingale measure, for a tree with jumps 342
equivalent probability measures *see* probability measure, equivalence
European 9
European contingent claim 105
exercise 9
exercise boundary 267
exercise region 267
exercise strategy, admissible 265
expectation 411, 442
   conditional 145

fat tails 81, 411, 444
Feynman–Kac theorem 151
fickle 356
filtration 133, 144, 152
first variation *see* variation, first
fixed leg 285
fixed rate 411
floating 279
floating leg 285
floating rate 411
floating smile *see* smile, floating
floor 288, 411
floorlet 288, 411
floortion 306, 411
forward contract 9, 21, 166, 411
   and risk-neutrality 127
   value of 25
forward price 25, 30
forward-rate agreement 22, 283, 411
forward rates 282–284
Fourier transform 374, 387
FRA *see* forward-rate agreement
free boundary value problem 270

Gamma 74, 78, 411
   and static replication 228, 230
   Black–Scholes formula for a call option 78
   non-negativity of 363
Gamma distribution 381
Gamma function 381
   incomplete 384
Gaussian distribution 54, 93
Gaussian random variable, synthesis of 177

gearing 279
geometric Brownian motion 101, 104
gilt 293
Girsanov transformation 198
Girsanov's theorem 148, 156, 194–197, 347, 369, 411
   higher-dimensional 248–253
Greeks 74–81, 411
   and static replication 228, 230
   computation of, on a tree 171
   of multi-look options 220–222

heat equation 109–110
Heath, Jarrow & Morton 301
hedger 12–13, 17
hedging 4, 11, 64, 411, 421
   and martingale pricing 152–154
   Delta 17–18, 64, 70, 73, 105, 108, 152
   exotic option under jump-diffusion 354
   Gamma 74
   in a one-step tree 41–42
   in a three-state model 47
   in a two-step model 48
   of exotic options 403
   vanilla options in a jump-diffusion world 351
   Vega 77
hedging, discrete 73
hedging strategy 16–17, 41, 73
   stop-loss 134
HJM model 301
homogeneity 255, 262, 362

implied volatility *see* volatility, implied
importance sampling 178
in-the-money 29
incomplete 411
incomplete market 47, 340, 346–354, 368, 369
incomplete model 86
incremental path generation *see* path generation, incremental
independent 441
information 2, 4, 103, 130–135, 152, 380
   conditioning on 135
insider trading 3
insurance 11
inverse cumulative normal function 177, 415, 416
inverse floater 338
Ito 89
Ito calculus, higher-dimensional 243, 245–248
Ito process 96, 145
Ito's Lemma 96–100
   application of 101–104
   multi-dimensional 246

joint density function 444
joint law of minimum and terminal value of a Brownian motion, with drift 197
joint law of minimum and terminal value of a Brownian motion, without drift 192
jump-diffusion model 84, 343–360
   and deterministic future smiles 226
   and replication of American options 272
   price of vanilla options as a function of jump intensity 353
   pricing by risk-neutral evaluation 343–346
jump-diffusion process 340
jumps 82–84
jumps on a tree 341

Kappa 77
knock in 186
knock-in option *see* option, barrier
knock out 186
knock-out option *see* option, barrier
kurtosis 82, 412, 444

law of large numbers 65, 176, 442
law of the minimum of a Brownian motion drift 199, 200
law of the unconscious statistician 443
Leibniz rule 100
leveraging 279
LIBID 412
LIBOR 281, 294, 412
LIBOR-in-arrears 291–292
   caplet, pricing by BGM 305
   FRA, pricing by BGM 305
LIBOR market model 301
likelihood ratio 180, 221
liquidity 20
Lloyds 6
log-normal distribution 58
log-normal model 55
   approximation by a tree *see* tree, approximating a log-normal model
   for stock price movements 102
log-type model 362–364
long 20, 412
low-discrepancy numbers 179
   the pricing of exotic options 426–427
lucky paths 348

marginal distribution 445
Margrabe option *see* option, Margrabe
market efficiency 2–4
   weak 3, 4, 91
market maker 71
market model 412
market price of risk 86, 102
Markov property 3, 90, 91
   strong 194
martingale 119, 135, 412
   and no arbitrage 136
   continuous 144–150
   discrete 136
   higher-dimensional 248
martingale measure 138
   choice of 355
   uniqueness 140
martingale pricing
   and time-dependent parameters 154–155
   based on the forward 159–163
   continuous 147–150
   discrete 135–144

equivalence to PDE method 151–152
  with dividend-paying assets 159
martingale representation theorem 152
maturity 5
maximal foresight 275
mean-reverting process 369
measure change 347
model risk 226
moment 412
moment matching 178
  and pricing of Asian options 215–217
money-market account, 25, 104, 410
moneyness 364
monotonicity theorem 26
Monte Carlo simulation 65, 442
  and price of exotic options using a jump-diffusion model 358
  and pricing of European options 176
  computation of Greeks 179–181
  variance reduction 178
Moro 415
mortgage 280
multi-look option *see* option, multi-look

Name 6
natural payoff 309
NFLWVR 122, 125
no-arbitrage 42
no free lunch principle 18
no free lunch with vanishing risk *see* NFLWVR
non-recombining tree *see* tree, non-recombining
normal distribution 441 *see also* Gaussiandistribution
notional 283
numeraire 162, 289, 291, 293, 303
  change of 157
numerical integration, and pricing of European options 173–176

option 9–12
  American 65, 134, 263, 409
    boundary conditions for PDE 269
    lower bounds by Monte Carlo 272–274
    PDE pricing 267–270
    pricing on a tree 266–267
    replication of 270–272
    seller's price 276
    theoretical price of 266
    upper bounds by Monte Carlo 275–276
  American digital 203
  American put 204
  Asian 206, 409
    pricing by PDE or tree 217–218
    static replication of 231–233
  barrier 65, 409
    definition 186–189
    price of down-and-out call 201, 202
  basket 243, 409
  Bermudan 263, 409
  binary 409
  call 9, 166, 410
    American 31
    Black–Scholes formula for 61, 150

    down-and-in 186
    down-and-out 186
    formula for price in jump-diffusion model 345, 346
    pay-off 28
    perpetual American 278
    pricing under Black–Scholes 104
  chooser 273
  continuous barrier, expectation pricing of 191–192, 200–203
  continuous barrier, PDE pricing of 189–191
  continuous barrier, static replication of 226–229, 234–238
    down-and-out put 226–228
  continuous double barrier, static replication 228–229
  digital 80, 239
  digital call 166
    Black–Scholes formula for price of 168
  digital put 166
    Black–Scholes formula for price of 168
  discrete barrier 206
    static replication of 229–231
  double digital 120
  European 411
  exotic 10, 84
    Monte Carlo 424–426
    pricing under jump-diffusion 358–360
  knock-in 411
  knock-out 65, 412
  Margrabe 242, 254–256
  model-independent bounds on price 28–38
  multi-look 207
  Parisian 412
  path-dependent 207
    and risk-neutral pricing 207–209
    static replication of 231–233
  power call 167
  put 9, 166, 412
    Black–Scholes formula for 61
    pay-off 29
  quanto 242, 256–261
  static replication of up-and-in put with barrier at strike 233–234
  trigger 413
  vanilla 10
  with multiple exercise dates 263
out-of-the-money 29

path dependence, weak 209
path-dependent exotic option *see* option, path-dependent
path generation 210–215
  incremental 212
  using spectral theory 212
pathwise method 181, 220
PDE methods, and the pricing of European options 181
Poisson process 343
positive semi-definite 447
positivity 7, 26
predictable 152

predictor-corrector 319
present valuing 281
previsible 349
pricing, arbitrage-free 21
principal 5, 280
probability, risk-neutral *see* risk-neutral probability
probability density function 441
probability measure 438
   equivalence 138
product rule, for Ito processes 100
pseudo-square root 448
put-call parity 29, 61, 63
put-call symmetry 234–238
put option *see* option, put

quadratic variation *see* variation, quadratic
quanto call 258
quanto forward 258
quanto option *see* option, quanto
quasi Monte Carlo 179

Radon–Nikodym 198
Radon–Nikodym derivative 197
random time 85, 133
random variable 439
real-world drift *see* drift, real-world
recombining trees, implementing 423
reflection principle 192–194
replication 22, 106
   and dividends 111
   and the pricing of European options 181–183
   classification of methods 239
   dynamic 183, 239
   in a one-step tree 45–46
   in a three-state model 47
   semi-static, and jump-diffusion models 360
   static 183
      feeble 239
      mezzo 239
      strong 225, 239
      weak 225, 239
repo 294
restricted stochastic-volatility model *see* Dupire model
reverse option 299
reversing pair 298
rho 412
Rho 77
risk 1–2, 8, 9
   diversifiable 8
   purity of 8
riskless 1
riskless asset 27
risk neutral 18
risk-neutral distribution 60
risk-neutral density, as second derivative of call price 127
risk-neutral density, in Black–Scholes world 129
risk-neutral expectation 61

risk-neutral measure 138, 412
   completeness 156
   existence of 119
   uniqueness 156
risk-neutral pricing 61, 130
   higher-dimensional 248–253
risk-neutral probability 44, 49, 50, 56, 118
risk-neutral valuation 56
   in a one-step tree 42–45
   in a three-state model 47
   in jump models 83
   two-step model 49
risk premium 43, 57, 60, 102, 108, 412
Rogers method for upper bounds by Monte Carlo 275, 329

sample space 438
self-financing portfolio 27, 106–107, 118, 153, 348
   dynamic 27
share 6–7, 412
share split 54
short 412
short rate 24, 413
short selling 20
simplex method 274
skew 413, 444
smile 71–74
   displaced-diffusion 335, 399
   equity 400
   floating 84, 364, 386, 392–393
   foreign exchange 392
   FX 402
   interest-rate 334–336, 403
   jump-diffusion 357, 358, 394
   sticky 84, 392–393
   sticky-delta 392
   stochastic volatility 377, 395
   time dependence 393–394
   Variance Gamma 385, 396
smile dynamics
   Derman–Kani 400
   displaced-diffusion 399
   Dupire model 400
   equity 400
   FX 402
   interest-rate 403
   jump-diffusion 394
   market 392–394
   model 394–400
   stochastic volatility 395
   Variance Gamma 396
smoothing operator 110
spectral theory 212
speculator 12, 17
split, share *see* share split
spot price 29
square root, of a matrix 447
standard deviation 443
standard error 177
static replication *see* replication, static
stepping methods for Monte Carlo 419
stochastic 413

stochastic calculus 89
stochastic differential equation 95
   for square of Brownian motion 97
stochastic process 92–96, 131
stochastic volatility 84–85, 368
   and risk-neutral pricing 369–372
   implied 379
   pricing by Monte Carlo 370–373
   pricing by PDE and transform methods 373–376
stochastic volatility smiles *see* smile, stochastic volatility
stock 6–7, 413
stop loss hedging strategy 17
stopping time 133, 265, 325
straddle 167, 239
strike 9, 413
strong static replication *see* replication, static, strong
sub-replication 348–354
super-replication 348–354
swap 279, 284–288, 413
   constant maturity 307
   payer's 285, 412
   pricing by BGM 302
   receiver's 285, 412
   value of 287
swap rate 413
swap-rate market model 320
swaption 280, 288, 413
   Bermudan 280, 289, 321
      and factor reduction 331–334
      lower bound via local optimization 326
      lower bounds by BGM 324–328
      pricing by BGM 305
      upper bounds by BGM 328–331
   cash-settled 306
   European 289
      price of 292–293
   payer's 288, 412
   pricing by BGM 302
   receiver's 288, 412
swaptions, rapid approximation to price in a BGM model 320

Taylor's theorem 63, 77, 98
term structure of implied volatilities 313
terminal decorrelation 318, 332
Theta 77, 413
   and static replication 228, 230
time-dependent volatility, and pricing of multi-look options 219
time homogeneity 32, 312
time value of money 23–25
Tower Law 145
trading volatility *see* volatility, trading of
trading volume 380
transaction costs 20, 73, 86–87
trapezium method 173
tree
   and pricing of European options 168–173
   and time-dependent volatility 169
   approximating a log-normal model 57–64
   approximating a normal model 52–55
   higher-dimensional 258–261
   non-recombining 169
   one-step 41–47
   risk-neutral behaviour 58
   trinomial 169
   with interest rates 55–57
   with multiple time steps 47–52
trigger FRA 297
trigger swap 304
   pricing by BGM 304
trinomial tree *see* tree, trinomial

underlying 10
uniform distribution 441

valuation, risk-neutral *see* risk-neutral valuation
value at risk 413
Vanna 413
VAR 413
variance 413, 443
Variance Gamma density 388
Variance Gamma model 85, 383–386
   and deterministic future smiles 226
Variance Gamma
   mean rate 381
   process 380–382
   variance rate 382
variation 147, 346
   first 91, 346, 388
   quadratic 347
   second *see* variation, quadratic
Vega 77, 79, 413
   integral expression for 175
Vega hedging *see* hedging, Vega
volatility 57, 61–63, 70–71, 101
   Black–Scholes formula as linear function of 61
   forward 405
   implied 70, 182
   instantaneous curve 299, 312
   root-mean-square 299
   time-dependence and tree-pricing 273
   trading of 70
volatility surface 342

weak static replication *see* replication, static, weak
Wiener measure 132

yield 5, 23, 413
   annualized 24
yield curve 298, 413

zero-coupon bond *see* bond, zero-coupon